Sustainable Agriculture and Agronomy

Sustainable Agriculture and Agronomy

Editor: Farrell Waltz

www.callistoreference.com

Callisto Reference,
118-35 Queens Blvd., Suite 400,
Forest Hills, NY 11375, USA

Visit us on the World Wide Web at:
www.callistoreference.com

ISBN: 978-1-64116-119-0 (Hardback)

Cataloging-in-Publication Data

Sustainable agriculture and agronomy / edited by Farrell Waltz.
    p. cm.
Includes bibliographical references and index.
ISBN 978-1-64116-119-0
1. Sustainable agriculture. 2. Agronomy. 3. Alternative agriculture. I. Waltz, Farrell.
S494.5.S86 S87 2019
630--dc23

Table of Contents

Permissions

List of Contributors

Index

Preface

Agronomy and agricultural science deal with the production and management of crops and farm animals for food, fiber, fuel or farm labor. Agronomy builds on the fundamental basis of plant genetics, physiology and soil science for optimizing agricultural production. Sustainable agriculture is built on an understanding of the ecosystem as well as the relationship between organisms and their environment. Research in these domains explores the design and development of innovative agricultural practices of crop rotation, organic farming, integrated pest management, etc. in order to develop a sustainable approach to agriculture. This book is a compilation of chapters that discuss the most vital concepts and emerging trends in the field of sustainable agriculture. It strives to provide a fair idea about this discipline and to help develop a better understanding of the latest advances in this field. Scientists and students actively engaged in agriculture science, soil science and crop science will find this book full of crucial and unexplored concepts.

The researches compiled throughout the book are authentic and of high quality, combining several disciplines and from very diverse regions from around the world. Drawing on the contributions of many researchers from diverse countries, the book's objective is to provide the readers with the latest achievements in the area of research. This book will surely be a source of knowledge to all interested and researching the field.

In the end, I would like to express my deep sense of gratitude to all the authors for meeting the set deadlines in completing and submitting their research chapters. I would also like to thank the publisher for the support offered to us throughout the course of the book. Finally, I extend my sincere thanks to my family for being a constant source of inspiration and encouragement.

Editor

Development and Testing of a Device to Increase the Level of Automation of a Conventional Milking Parlor through Vocal Commands

Mauro Zaninelli

Department of Human Sciences and Quality of Life Promotion, Universita Telematica San Raffaele Roma, Via di Val Cannuta 247, Rome 00166, Italy; mauro.zaninelli@unisanraffaele.gov.it

Academic Editor: Ritaban Dutta

Abstract: A portable wireless device with a "vocal commands" feature for activating the mechanical milking phase in conventional milking parlors was developed and tested to increase the level of automation in the milking procedures. The device was tested in the laboratory and in a milking parlor. Four professional milkers participated in the experiment. Before the start of the tests, a set of acoustic models with speaker-dependent commands defined for the project was acquired for each milker using a dedicated "milker training procedure". Two experimental sessions were performed by each milker, with one session in the laboratory and a subsequent session in the milking parlor. The device performance was evaluated based on the accuracy demonstrated in the vocal command recognition task and rated using the word recognition rate (WRR). The data were expressed as %WRR and grouped based on the different cases evaluated. Mixed effects logistic regression modeling was used to evaluate the association between the %WRR and explanatory variables. The results indicated significant effects due to the location where the tests were performed. Higher values of the %WRR were found for tests performed in the laboratory, whereas lower values were found for tests performed in the milking parlor (due to the presence of background noise). Nevertheless, the general performance level achieved by the device was sufficient for increasing the automation level of conventional milking parlors.

Keywords: vocal commands; automation; milking parlor

1. Introduction

In the global dairy farming, musculoskeletal disorder (MSD) symptoms in the neck, shoulder, and upper extremities among farmers pose an ongoing problem throughout the world [1–3]. Using the work ability index (WAI) to study full-time dairy milkers [4], Finnish researchers reported an overall decline in work ability of 39%, which was mainly caused by MSD. In Germany, an evaluation of temporary disability data found a high rate of work absenteeism for milking parlor operators, with MSD being the most prevalent diagnosis [5]. Similarly, several Swedish studies found a higher frequency of work-related MSD among dairy farmers (principally among females) than in other occupations [6].

These results do not agree with the expectations of the global dairy farming, which aimed to reduce MSD through the adoption of modern parlors that reduce the physical load of workers. Other aspects, such as social, organizational, and socio-economic factors, must be considered to fully explain the increasing trend of MSD symptoms. The dairy farming industry has changed. The number of dairy farms has decreased, whereas herd sizes have increased to satisfy milk production and consumer demands [7]. These changes have led to increase: the working time; the number of cows milked per hour; the number of milking units per parlor; and the quantity of highly repetitive working routines, causing a general increase of manual labor, often performed in awkward working positions, that is a direct risk for back injuries and other MSD-related problems [8,9].

However, it is the general workload on the upper extremities, due by the fundamental tasks required by milking procedures, that represents the greatest contribution to the development of MSD [10,11]. Pinzke et al. [12] found that the task "attaching of a milking group" involves the highest load for the biceps and flexor muscles (during the holding of the milking group and the attaching of the teat-cups), and they reported that high muscle loads and extreme positions and movements of the hand and forearm might contribute to the development of injuries among milkers. Stal et al. [13] highlighted that high degrees of dorsiflexion and deviation of the wrist, combined with high values of muscle loads for the forearm, due to the holding of the milking group in one hand while attaching the four teat-cups to the cow udder, might contribute to injuries to the forearm, wrist, and hand. Cockburn et al. [14], in a study aimed to analyze and improve the posture of milkers during milking procedures, reported that working conditions of each milker can be improved considering: the parlor type, the udder base height, the floor level, and his/her body height. Nevertheless, the authors proved that no ideal milking parlor can be designed because in any case, the distance between the cow and the milker always requires the milker to reach out and lift the milking group, thereby loading weight on shoulders, elbows, and wrists, and thus increasing the risks for the development of MSD. In this scenario, an improvement of the level of automation in milking procedures could be a feasible way to maintain high milk production and, at the same time, reduce the physical risks for milkers.

During a milking session, the milker follows a well-defined routine consisting of three main phases: (1) fore-stripping, cleaning of the teats and udder, control for any possible evidence of mastitis, disinfection and, after 40–60 s, drying of the teats; (2) attachment of the milking cluster; (3) disinfection of the teats at the end of each milking. The main characteristic of the milking routine is that each step is performed in sequence for all of the animals in a rack. Therefore, the milker is forced to move inside the milking pit, from beginning to end, to complete each step of the routine. While the cows on one side of the milking parlor are being milked, the milker begins the same routine on the animals positioned on the opposite side to optimize the timing for the entire process.

Before attaching each milking cluster, the milker must start the mechanical milking phase of the milking post. This phase is generally controlled by certain milking control units that collaborate with the other devices present in the milking post (e.g., milk meters, flow meters, automatic cluster removers; [15]). The units monitor the phase of the milk ejection and automatically remove the milking clusters when the milk flow is found to be below a specific threshold (defined for the herd or for individual cows if the system allows for the identification of each cow in the herd). Furthermore, these units can simultaneously monitor the health of the cows [16–20] by the use of specific sensors. The mechanical milking phase is generally initiated by pushing a "start" button, which is located on each control unit or in a separate location in the milking post that is more convenient for the milker. This repetitive movement is performed by the milker many times during a milking session and is part of the general workload of the milking procedure. As an example, on a farm with a medium-sized herd of 200 lactation cows that are milked twice per day, the milker could be forced to push the "start" button up to 400 times per day.

The adoption of vocal commands could be a method of reducing the general workload of milkers. The mechanical milking phase could be initiated by a milker via vocal demands instead of a physically repetitive action. This increased level of automation could reduce the general workload imposed by the milking procedure and thus improve the health conditions of dairy workers.

The aim of this study was to develop and test a device to provide the feature of "vocal commands" to activate the mechanical milking phase for each milking post in a conventional milking parlor.

2. Materials and Methods

2.1. Hardware Layout

The layout of the electronic components used in the device is presented in Figures 1 and 2. The layout included the following: (1) a behind-the-head headset unidirectional microphone coupled with a wireless transmitter and receiver (TS-6310 VHF Wireless Microphone—Guangdong Takstar Electronic Co, Huizhou, China); (2) a vocal recognition board (EasyVR—ROBOTECH S.r.l., Sarzana, La Spezia, Italy); (3) a main board (Arduino UNO—Arduino S.r.l., Scarmagno, Turin, Italy); and (4) a connection board for each milking post connected that included an Arduino UNO (Arduino S.r.l., Scarmagno, Turin, Italy) and a relay module (Sunfounder, Shenzhen City , China).

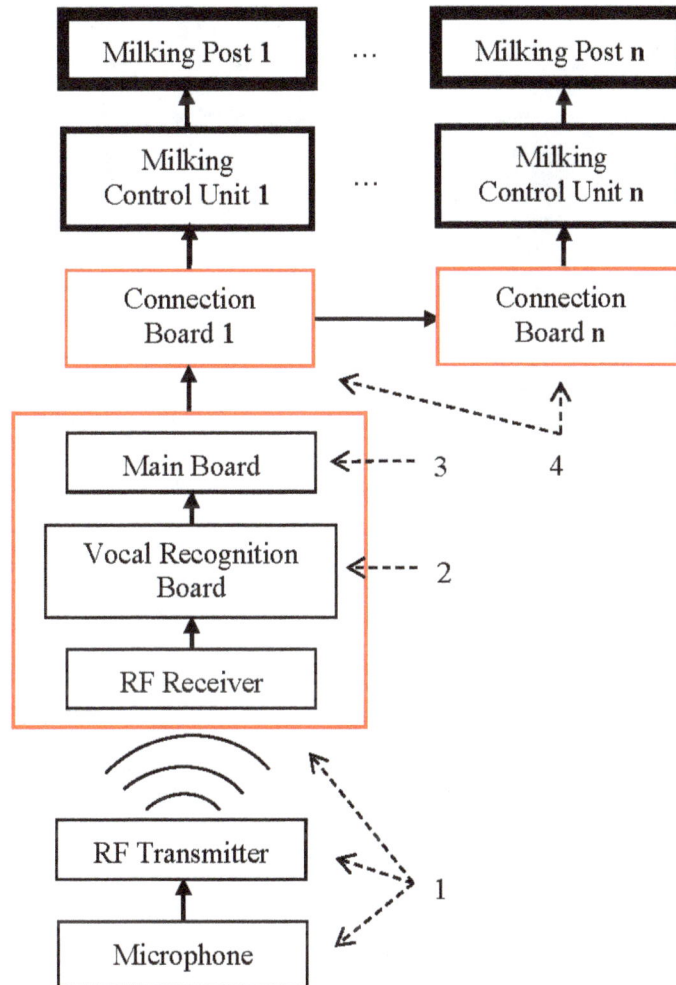

Figure 1. Hardware layout—connection scheme of the electronic components of the device. The device includes: (1) a microphone, a Radio Frequency (RF) transmitter, and receiver; (2) a vocal recognition board; (3) a main board; (4) a connection board for each milking post connected, made by Arduino UNO, and a specialized relay module. In red are shown the boxes where the electronic components of the device were placed in order to connect the system to the existing milking machine.

This hardware layout enabled the device to (a) acquire the vocal commands of the milker from any part of the milking pit through the wearable microphone and wireless connection provided by the Radio Frequency (RF) communication system; (b) transform the vocal commands of the milker into audio signals; (c) interpret the vocal commands of the milker through the recognition engine embedded in the vocal recognition board; (d) code the recognized command as a standard command message through the main board; (e) send this command to the milking posts through the main board; (f) decode the command messages received from the main board through the connection boards located near each milking post (and connected to the corresponding milking control unit by its relay module and the input provided by the unit for the external push button); and (g) execute the command received from the milker by generating the appropriate electrical spike through the relay of the specific connection board to which the command was addressed.

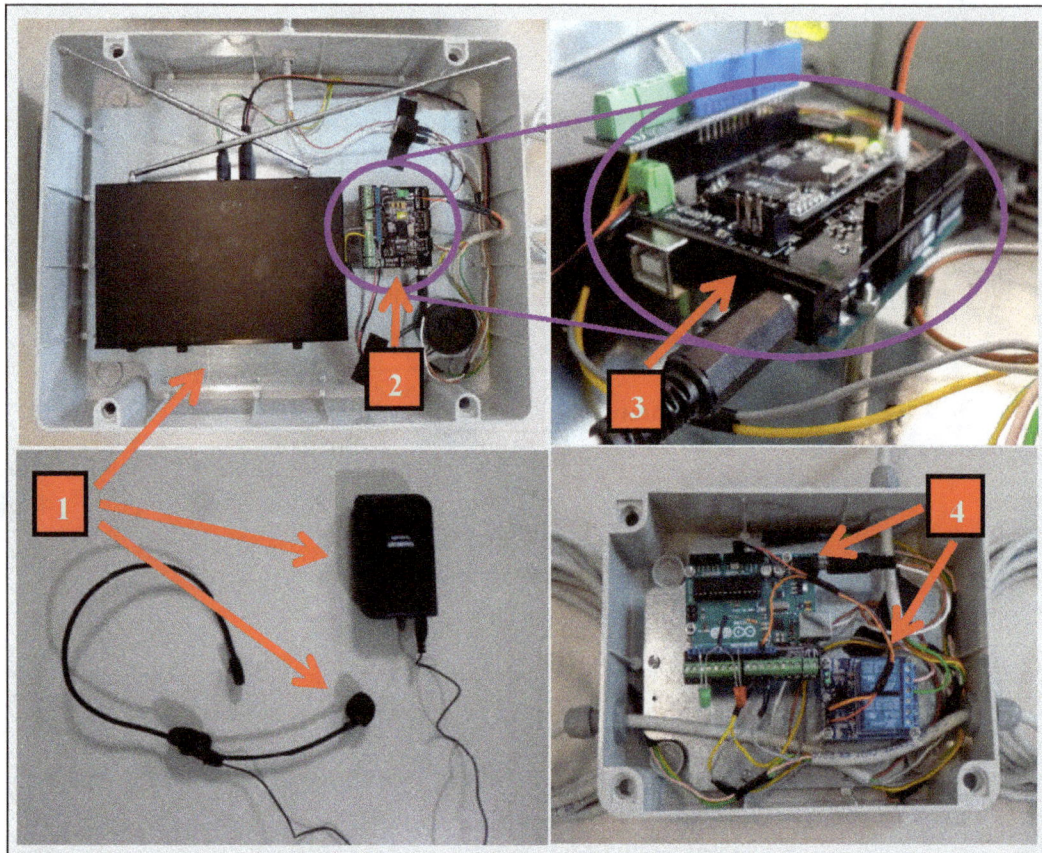

Figure 2. Details of the electronic components used for the tested device: (1) a microphone, transmitter, and receiver; (2) a vocal recognition board; (3) a main board; and (4) a connection board for each milking post connected, made by Arduino UNO, and a specialized relay module.

2.2. Software Development

Two software applications were developed according to a master/slave design model using the programming language of Arduino and its Integrated Development Environment (IDE—Arduino 1.5.7, Arduino S.r.l., Scarmagno, Turin, Italy). After a compilation procedure was performed by the IDE, these software applications were used as firmware in the Arduino boards. The first application (the master) was uploaded on the main board, and the second application (the slave) was used as firmware in each connection board. A specific set-up was used in each connection board, which enabled the software application to store the number of the milking post to which the specific connection board was connected. Flow diagrams of these software applications are presented in Figure 3.

The first software application enabled the acquisition and decoding of the vocal commands of the milker. This application was constantly in a "listening mode"—making a continuous loop of a defined number of seconds [21]. If a command was recognized by the vocal recognition board during this time window, the corresponding command was built and sent as a specific sequence of characters to all of the connection boards through a specialized software serial bus (built with the pins 10 and 11 of all of the Arduino boards).

"MASTER" SOFTWARE APPLICATION   "SLAVE" SOFTWARE APPLICATION

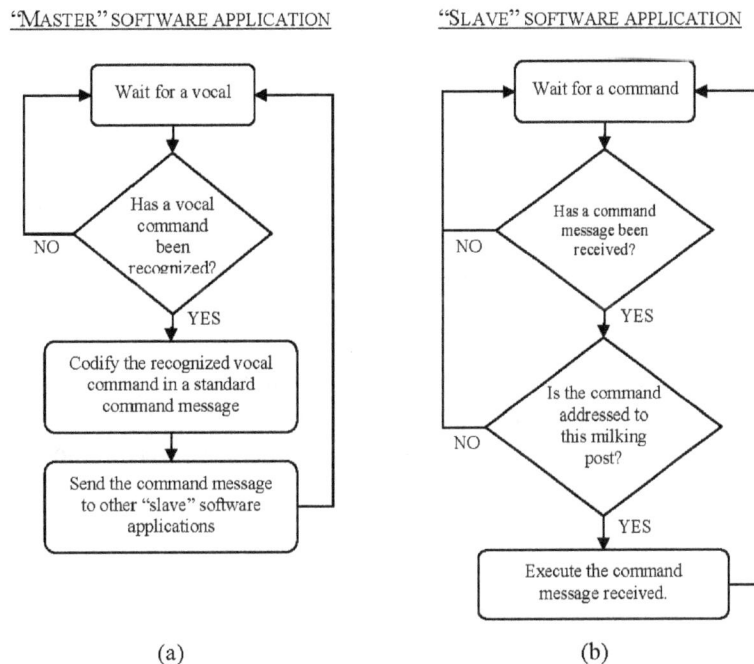

(a)   (b)

Figure 3. Flow diagrams of the developed software applications. In (**a**) is presented the "master" software application diagram. This software application is used as firmware for the main board. In (**b**) is shown the "slave" software application diagram. This software application was used as firmware for the connection boards with a specific setup for each board (i.e., the reference of the number of the milking post to which the connection board was connected).

The second software application performed a different task. This application was always in a "waiting" loop for messages received from the main board. When a message was received, this application (a) decoded the message; (b) checked whether the message was addressed to that specific milking post; and (c) in the case of correct correspondence, executed the command using the control pin in the relay module to generate an electric spike for the milking control unit (simulating in this way the pushing of the corresponding manual button).

Furthermore, to inform the milker of the current state of software tasks that were running, feedback was provided by the two software applications through colored LED and acoustic signals (i.e., "beeps"). These "beeps" were generated by specific software functions provided by a proprietary library of the voice recognition board and were physically realized by an 8 Ω acoustic speaker, which was directly controlled by the voice recognition board through a dedicated output. This feedback informed the milker about (a) the start of the "listening window," in which the device was waiting for a vocal command; (b) the positive recognition of a vocal command made by the vocal recognition board; and (c) the execution of a vocal command made by a specific connection board.

2.3. Types of Vocal Commands

The vocal recognition board used in the device was able to manage two different types of customized vocal commands: speaker-dependent (SD) commands and speaker-independent (SI) commands.

The SD commands are vocal commands provided by each user. A single user must follow a procedure to train the board to recognize the user's specific voice when a customized vocal command is given. In this procedure, the user must pronounce each word/utterance of the set of customized vocal commands (i.e., the word vocabulary necessary for the recognition task) at least two times. At the end of this training procedure, which was performed through a proprietary software application (EasyVR Commander^{TM} 3.8.0.0—ROBOTECH S.r.l., Sarzana, La Spezia, Italy), the resulting set of acoustic models is uploaded to the board. Consequently, the board is able to recognize the vocal commands of the user that has followed the procedure.

The SI commands are vocal commands not provided by a specific user. Through a proprietary software application (QuickT2SI™ 3.1.14—ROBOTECH S.r.l., Sarzana, La Spezia, Italy), each word/utterance necessary to the recognition task is written and included in the word vocabulary. In a following step, each word/utterance of the vocabulary can be edited for pronunciation, and different settings can be selected (for example, the language model, acoustic model, and command phrase settings). At the end of this task, a customized word vocabulary is built and ready to be imported into the software application cited above (EasyVR Commander™ 3.8.0.0) for the necessary uploading to the board. This final step allows the board to recognize each word/utterance of the set of vocal commands from any possible user. The resulting recognition accuracy is generally less than in the case of SD commands. Therefore, in this study, only SD commands were used to achieve the best possible recognition accuracy considering the noisy environment in which the recognition task is performed.

2.4. Vocal Command Structure

The functional command of the milking system controlled by the developed device was only the "start" of the milking procedure. As a result, (a) the "start" action was considered to be the action to perform at default; (b) the word "start" was omitted from the recognition task syntax; and (c) only the reference to the position of the milking post was used in the vocal command structure. Furthermore, the word "post" was added to each vocal command (e.g., "post one", "post two", "post three", etc.; [22]).

However, the acoustic models of the SD commands were built as singular utterances. The recognition task semantics were specified to receive one valid utterance during a single listening window for comparison with those included in the set of vocal commands. Italian was used as a language model, and Italian terms were used to build each utterance: "postazione" for the word "post" and the numbers "uno, due, tre, etc." for the corresponding terms "one, two, three, etc."

2.5. Experimental Design

The experimental tests involved four professional milkers (i.e., professional workers that have milked cows for at least three years). Before the start of the tests, a set of acoustic models of the SD commands was acquired for each milker through a specialized software application (EasyVR Commander™ 3.8.0.0). At the end of this procedure (performed in a mildly quiet room), each milker was trained to pronounce each utterance included in the set of vocal commands using the same software application. This milker training procedure required the milker to pronounce each vocal command in sequence. A delay of a few seconds was used between utterances. The milker received feedback on the correct recognition of the last vocal command pronounced directly from the personal computer. At the end of the vocal commands included in the set, the milker restarted from the beginning, for a total of 30 repetitions.

Experimental tests were performed in the laboratory and in the milking parlor of a dairy farm. The milking parlor was a 6 + 6 herringbone milking system. Only one milking line (i.e., six milking posts) was connected to the developed device to allow in future studies the comparing of a fully automated milking line to a conventional milking line, in the same milking parlor. The same layout was used in the laboratory to yield comparable data. Six connection boards were connected, and the following vocal commands were used in the tests: "post one", "post two", "post three", "post four", "post five", and "post six".

In the laboratory, the experimental procedure consisted of a sequence of activities similar to those generally performed in the milking routine. Starting at the first milking post, the milker pronounced the first vocal command (i.e., "post one") into the microphone after the acoustic "beep" provided by the device at the beginning of the listening window. If the utterance was correctly recognized by the device, the milker stayed in front of the first "milking post" for few seconds as a simulation of the attaching phase of the milking group. After that pause, the milker moved to the following simulated "milking post". If the vocal command provided was not recognized, the milker moved to the following "milking post" assuming to have started the milking procedure of the previous "milking post" by a manual pushing of the "start" button. At the end of the "milking line", the milker paused for a few minutes. After that pause, the milker restarted from the beginning of the "milking line". This sequence of activities was repeated ten times. This number of repetitions was selected because it was equal to the number of milking "cycles" expected for the tests conducted in the milking parlor, which was estimated based on the number of available milking posts and milk cows of the dairy farm. The entire experimental procedure was repeated for each milker involved in the tests.

At the end of the laboratory tests, the same experimental procedure was performed by each milker in a milking parlor. The times spent to start the milking procedure, through the device and by the manual pushing of the "start" button, were recorded and compared. Each procedure was performed during an entire milking session. For all tests performed, a unique researcher followed the entire experimental session and recorded data.

2.6. Statistical Analysis

The performance of the device was evaluated based on the accuracy demonstrated in the vocal command recognition task and rated using the word error rate (WER—[22–24]) according to the following formula:

$$WER = \frac{S + D + I}{N} \tag{1}$$

where:

S is the number of substitutions,
D is the number of the deletions,
I is the number of the insertions, and
N is the number of words in the reference.

The vocal commands used in this study were acquired as singular utterances. Therefore, the classification of the result of the recognition task was as follows: (a) "correct" if the "vocal recognition board" correctly identified the vocal command pronounced by the milker; (b) "substitution" if the "vocal recognition board" mismatched the vocal command pronounced by the milker and the command selected from the set of vocal commands; and (c) "deletion" if the "vocal recognition board" did not match the vocal command pronounced by the milker with any command included in the set of vocal commands. As a consequence, for each result of the recognition task (a) no "insertion" was possible; (b) if a "substitution" occurred, then a "deletion" was automatically excluded (and vice versa); and (c) the number of words in the reference was always equal to 1. Therefore, the WER index was simplified according to the following formula:

$$WER = S + D \tag{2}$$

Next, the obtained results were converted into a word recognition rate (WRR—[12–14]) according to the following formula:

$$WRR = 1 - WER \tag{3}$$

In the final step, all of the results were grouped based on the different cases evaluated in the study and calculated as percentages.

A statistical analysis of the results was performed. A mixed effects logistic regression model was used to evaluate the association between the %WRR and the explanatory variables considered. The statistical analysis was performed using R software (version 3.2.3, 2015, The R Foundation for Statistical Computing, Vienna, Austria), the package *lme4* (version 1.1-10—[25]), and the *glmer* procedure. The following linear model was fitted:

$$Y_{ijkr} = \beta_o + \beta_1 P_i + \beta_2 M_j + \beta_3 C_{k(j)} + \beta_4 VC_{r(kj)} + e_{ijkr} \quad (4)$$

where: Y_{ijkr} is the %WRR; β_n are coefficients of the linear model; P_i is the effect of the location where the tests were conducted (i = 1–2; 1 = laboratory; 2 = milking parlor); M_j is the effect of the milker (j = 1–4; 1 = milker one; 2 = milker two, etc.); $C_{k(j)}$ is the effect of the acquisition cycle performed (k = 1–10; 1 = cycle one, 2 = cycle two, etc.) nested within milkers; $VC_{r(kj)}$ is the effect of the vocal command (j = 1–6; 1 = milking post one, 2 = milking post two, etc.) nested within cycles and milkers; and e_{ijkr} is the residual error. The variables M_j, $C_{k(j)}$, and $VC_{r(kj)}$ were included in the statistical model as random effects.

Furthermore, the mean values of the time required to start the milking procedure using the device or the manual pushing of the "start" button were calculated. In order to check for significant differences between the mean values obtained, the *aov* procedure of the package *stats* (version 3.2.3—[26]) was used.

3. Results

The results for %WRR are reported in Table 1. The locations where tests were conducted were found to have a significant effect on the %WRR achieved. Higher %WRR values were found for tests performed in the laboratory (the overall %WRR value was 92.9%), whereas lower values were found for tests performed during a milking in the milking parlor (with the overall %WRR value of 80.8%). However, the overall %WRR value for all of the tests performed was 86.9%.

Table 1. Results obtained in the tests reported in terms of %WRR (word recognition rate).

		Commands	Locations											
			Laboratory				Parlor				All Locations			
			%S	%O	%WRR	*n*	%S	%O	%WRR	*n*	%S	%O	%WRR	*n*
		1	0.0	0.0	100.0		0.0	20.0	80.0		0.0	10.0	90.0	
		2	0.0	0.0	100.0		10.0	20.0	70.0		5.0	10.0	85.0	
		3	0.0	10.0	90.0		0.0	20.0	80.0		0.0	15.0	85.0	
	Milker 1	4	0.0	10.0	90.0	10	0.0	10.0	90.0	10	0.0	10.0	90.0	20
		5	0.0	0.0	100.0		0.0	0.0	100.0		0.0	0.0	100.0	
		6	0.0	0.0	100.0		0.0	30.0	70.0		0.0	15.0	85.0	
		All	0.0	3.3	96.7	60	1.7	16.7	81.7	60	0.8	10.0	89.2	120
		1	0.0	10.0	90.0		10.0	30.0	60.0		5.0	20.0	75.0	
		2	10.0	10.0	80.0		0.0	0.0	100.0		5.0	5.0	90.0	
		3	0.0	10.0	90.0		0.0	30.0	70.0		0.0	20.0	80.0	
	Milker 2	4	0.0	0.0	100.0	10	10.0	10.0	80.0	10	5.0	5.0	90.0	20
		5	10.0	0.0	90.0		0.0	30.0	70.0		5.0	15.0	80.0	
		6	0.0	0.0	100.0		0.0	10.0	90.0		0.0	5.0	95.0	
		All	3.3	5.0	91.7	60	3.3	18.3	78.3	60	3.3	11.7	85.0	120
		1	0.0	10.0	90.0		0.0	20.0	80.0		0.0	20.0	80.0	
		2	0.0	20.0	80.0		0.0	10.0	90.0		0.0	10.0	90.0	
Milkers		3	0.0	10.0	90.0		0.0	20.0	80.0		0.0	20.0	80.0	
	Milker 3	4	10.0	10.0	80.0	10	10.0	20.0	70.0	10	10.0	20.0	70.0	20
		5	0.0	0.0	100.0		10.0	20.0	70.0		10.0	20.0	70.0	
		6	0.0	10.0	90.0		0.0	30.0	70.0		0.0	30.0	70.0	
		All	1.7	10,0	88.3	60	3.3	20.0	76.7	60	3.3	20.0	76.7	120
		1	0.0	0.0	100.0		0.0	10.0	90.0		0.0	5.0	95.0	
		2	10.0	10.0	80.0		0.0	20.0	80.0		5.0	15.0	80.0	
		3	0.0	0.0	100.0		0.0	10.0	90.0		0.0	5.0	95.0	
	Milker 4	4	0.0	10.0	90.0	10	0.0	0.0	100.0	10	0.0	5.0	95.0	20
		5	0.0	0.0	100.0		0.0	20.0	80.0		0.0	10.0	90.0	
		6	0.0	0.0	100.0		0.0	20.0	80.0		0.0	10.0	90.0	
		All	1.7	3.3	95.0	60	0.0	13.3	86.7	60	0.8	8.3	90.8	120
		1	0.0	5.0	95.0		2.5	20.0	77.5		1.2	12.5	86.2	
		2	5.0	10.0	85.0		2.5	12.5	85.0		3.7	11.2	85.0	
		3	0.0	7.5	92.5		0.0	20.0	80.0		0.0	13.7	86.2	
	All Milkers	4	2.5	7.5	90.0	10	5.0	10.0	85.0	10	3.7	8.7	87.5	20
		5	2.5	0.0	97.5		2,5	17.5	80.0		2.5	8.7	88.7	
		6	0.0	2.5	97.5		0.0	22.5	77.5		0.0	12.5	87.5	
		All	1.7	5.4	92.9^{a}	240	2.1	17.1	80.8^{b}	240	1.9	11.2	86.9	480

^{a,b} values in the same row with different uppercase superscripts differ significantly ($p < 0.01$).

The times necessary to start the milking procedure were recorded during the tests performed in milking parlor. A significantly higher mean value was found when vocal commands were used to activate the milking system (2.56 s vs. 2.03 s—$p < 0.01$).

4. Discussion

The device performed better on the vocal command recognition tasks in the laboratory than in the milking parlor. This finding was expected [27–29]. Milking parlors often have significant background noises. These noises can be caused by the milking machine operations, the presence of animals, the activities of milkers, or other tasks performed on the farm close to the milking parlor that involve the use of noisy machines, such as tractors. The presence of these background noises can affect the performance of a speech recognition system [24]. However, some strategies can be used to improve the performance of this type of system. For example, the recording of the acoustic models by the milkers could be performed directly in the milking parlor [17,19,20] when the milking machine is working. Otherwise, many mathematical and/or statistical models could be used to improve the robustness of the recognition system in the presence of background noise [22,28,30–35]. Nevertheless, all these techniques improve the recognition accuracy in the case of stationary noise [30] and they require more computational resources, which can affect the speed of the recognition task [22]. Therefore, considering the field results obtained in this study and the characteristics of the noise of the milking parlor, one can conclude that no changes to the current layout of the device should be necessary.

The overall value of %WRR achieved for all of the tests performed (86.9%) was in agreement with the results obtained in similar published studies. For example Be et al. [23], in a study carried out to develop an interface to control a robot using vocal commands, reported values of %WRR that varied between 86% and 93% on the basis of the different Spanish words considered to control the robot. These results were considered satisfactory by the authors because the recognition rates were maintained by the system in the range of 80% to 100% for all the cases investigated. Hirsch and Pearce [30], in a study that evaluated the performances of a speech recognition system in different real noisy conditions, such as: a car, a street, a train station, an airport, a restaurant, and an exhibition hall reported values of %WRR that ranged between 54.94% and 87.77%. The lowest performances were achieved by the system when: the acoustic models were recorded in a "clean condition" (i.e., without a background noise), the level of noise was high, and the experimental conditions were characterized by non-stationary noises as was the case for the tests performed in this study. However, it is interesting to note that the overall performance achieved in this study was proportionally limited by milkers 2 and 3. This result could be explained considering that these milkers were of Indian mother tongue even though they were able to speak coherent Italian. Considering this fact, two possible improvements could be suggested in order to reach better performance: (a) setup the vocal recognition engine using the speaker's mother tongue as a language model, even though the words included in the vocabulary are selected from another language and (b) use a set of vocal commands and language models equal to the speaker's mother tongue, and if necessary, different options for milkers in the same milking parlor that are fluent in more than one language. Unfortunately, in our study, these potential improvements could not be evaluated because the hardware used for the recognition task did not support Indian language as a model language.

The time necessary to start a milking procedure was, on average, higher when vocal commands were used to interact with the milking system. However, in this study, each vocal command started with the word "postazione" (i.e., "milking post"). This word, theoretically, is not strictly necessary and could be deleted from each vocal command. In this way, less time would be necessary to start the milking procedure. Furthermore, during the tests, it was observed that sometimes milkers started to pronounce a vocal command before they had finished to prepare the milking group for the attachment phase. During the tests carried out, these actions were corrected by the researcher that was in the milking parlor. However, in a real use of the device, these actions could be allowed. Since the milker could interact with the milking unit without the need to have a hand free from a specific task, a reduction of the time necessary to start a milking procedure should be reached.

The use of the device enabled the milkers to start the mechanical milking phase through the use of vocal commands instead of a physically repetitive action. However, the device was used to control only one function of the milking system. Theoretically, using a different set of vocal commands, or building vocal commands with a more complex structure, could eventually result in other functions of the milking system being controlled by the device. For example, the milking control units of a milking system monitor the phase of milk ejection and enable the automatic removal of milking clusters when the milk flow is below a specific threshold defined for each cow. Nevertheless, in old milking parlors, the identification of each animal is not possible, and the use of a singular threshold for the entire herd could result in incorrect data for all of the animals. In these instances, the milker must oversee the milking procedure and start the removal procedure by pushing the "stop" button in the corresponding milking control unit. If this command is not provided by the milker within a short timeframe, a dangerous "over milking" can occur. An incorrect removal threshold for milk flow can also affect the initial phase of the milking procedure. If this value is too high and the milk flow of a specific cow at the starting of the milking is lower (for a specific time), then the milking group could be removed too early and a new attachment of the group could be required. To avoid such a situation, the milker can initiate the milking by selecting the "manual" mode on the milking group, but is then forced to come back to the same milking post to manually stop the milking (and remove the milking group) or select the "automatic" mode on the milking unit. In all of these unfavorable events, the milker is forced to perform additional work. Nevertheless, these cases could be managed via the use of vocal commands. Both the actions of "start" and "stop" and the setup of the milking units ("manual" or "automatic") could be controlled by the device, changing the structure of the vocal commands, such as "Post one" + "Stop" (to stop the milking and remove the post one milking group) and "Post one" + "Manual" (to select the manual mode for post one). This modification could further increase the level of automation already reached by the tested device, allowing for a further reduction in the general workload of the milking procedure.

However, future studies will be necessary to identify and quantify the ergonomic improvements that the vocal commands feature can bring to milking procedures. In these studies, the reduction of the physical workload of milkers and the risk of MSD will be assessed.

5. Conclusions

The device that was developed and tested was able to provide the feature of "vocal commands" to activate the mechanical milking phase, for each milking post, in a conventional milking parlor. The adoption of this device could increase the automation level of conventional milking parlors and it should reduce the general workload of milkers. Further investigations will be useful in order to identify the ergonomic improvements that the device could bring to milking procedures.

Acknowledgments: I thank Maurizio Ruggeri (Total Dairy Management S.r.l.) and Giancarlo Teti (ROBOTECH S.r.l.) for all the support provided in the device development and testing.

Conflicts of Interest: The author declares no conflict of interest.

References

1.   Fathallah, F.A. Musculoskeletal disorders in labor-intensive agriculture. *Appl. Ergon.* **2010**, *41*, 738–743. [CrossRef] [PubMed]
2.   Niu, S. Ergonomics and occupational safety and health: An ILO perspective. *Appl. Ergon.* **2010**, *41*, 744–753. [CrossRef] [PubMed]
3.   Arborelius, U.P.; Ekholm, J.; Nisell, R.; Nemeth, G.; Svensson, O. Shoulder load during machine milking. An electromyographic and biochemical study. *Ergonomics* **1986**, *29*, 1591–1607. [CrossRef] [PubMed]
4.   Karttunen, J.P.; Rautiainen, R.H. Risk factors and prevalence of declined work ability among dairy farmers. *J. Agric. Saf. Health* **2011**, *17*, 243–257. [CrossRef] [PubMed]

5. Liebers, F.; Caffier, G. *Job Specific Sickness Absence due to Musculoskeletal Disorders in Germany*; Dortmund: Berlin/Dresden, Germany, 2009.

6. Stal, M.; Moritz, U.; Gustafsson, B.; Johnsson, B. Milking is a high-risk job for young females. *Scand. J. Rehabil. Med.* **1996**, *28*, 95–104. [PubMed]

7. Douphrate, D.I.; Fethke, N.B.; Nonnenmann, M.W.; Rosecrance, J.C.; Reynolds, S.J. Full shift arm inclinometry among dairy parlor workers: A feasibility study in a challenging work environment. *Appl. Ergon.* **2012**, *43*, 604–613. [CrossRef] [PubMed]

8. Bernard, B. *Musculoskeletal Disorders and Workplace Factors: A Critical Review of Epidemiological Evidence for Work-Related Musculoskeletal Disorders of the Neck, Upper Extremity, and LowBack*; DHHS (NIOSH) U.S. Department of Health and Human Service; Centers for Desease Control and Prevention; National Institute for Occupational Safety and Health: Cincinnatti, OH, USA, 1997; Volume 97–141.

9. Da Costa, B.R.; Vieira, E.R. Risk factors for work-related musculoskeletal disorders: A systematic review of recent longitudinal studies. *Am. J. Ind. Med.* **2010**, *53*, 285–323. [CrossRef] [PubMed]

10. Douphrate, D.I.; Lunner Kolstrup, C.; Nonnenmann, M.W.; Jakob, M.; Pinzke, S. Ergonomics in modern dairy practice: A review of current issues and research needs. *J. Agromed.* **2013**, *18*, 198–209. [CrossRef] [PubMed]

11. Pinzke, S. Comparison of working conditions and health among dairy farmers in southern Sweden in over a 25-year period. *Front. Public Health* **2015**, *2002*, 2002–2003.

12. Pinzke, S.; Stal, M.; Hansson, G.-A. Physical workload on upper extremities in various operations during machine milking. *Ann. Agric. Environ. Med.* **2001**, *8*, 63–70. [PubMed]

13. Stal, M.V.; Pinzke, S.; Hansson, G.A. The effect on workload by using a support arm in parlour milking. *Int. J. Ind. Ergon.* **2003**, *32*, 121–132. [CrossRef]

14. Cockburn, M.; Savary, P.; Kauke, M.; Schick, M.; Hoehne-Huckstadt, U.; Hermanns, I.; Ellegast, R. Improving ergonomics in milking parlors: Empirical findings for optimal working heights in five milking parlor types. *J. Dairy Sci.* **2015**, *98*, 966–974. [CrossRef] [PubMed]

15. Zaninelli, M.; Tangorra, F.M. Development and testing of a "free-flow" conductimetric milk meter. *Comput. Electron. Agric.* **2007**, *57*, 166–176. [CrossRef]

16. Zaninelli, M.; Agazzi, A.; Costa, A.; Tangorra, F.M.; Rossi, L.; Savoini, G. Evaluation of the Fourier Frequency Spectrum Peaks of Milk Electrical Conductivity Signals as Indexes to Monitor the Dairy Goats' Health Status by On-Line Sensors. *Sensors* **2015**, *15*, 20698–20716. [CrossRef] [PubMed]

17. Zaninelli, M.; Rossi, L.; Costa, A.; Tangorra, F.M.; Agazzi, A.; Savoini, G. Monitoraggio dello stato di salute delle capre attraverso l'analisi on-line della conducibilita elettrica del latte. *Large Anim. Rev.* **2015**, *21*, 81–86.

18. Zaninelli, M.; Rossi, L.; Costa, A.; Tangorra, F.M.; Agazzi, A.; Savoini, G. Signal Spectral Analysis to Characterize Gland Milk Electrical Conductivity in Dairy Goats. *Ital. J. Anim. Sci.* **2015**, *14*, 362–367. [CrossRef]

19. Zaninelli, M.; Rossi, L.; Tangorra, F.M.; Costa, A.; Agazzi, A.; Savoini, G. On-line monitoring of milk electrical conductivity by fuzzy logic technology to characterise health status in dairy goats. *Ital. J. Anim. Sci.* **2014**, *13*, 340–347. [CrossRef]

20. Zaninelli, M.; Tangorra, F.; Costa, A.; Rossi, L.; Dell'Orto, V.; Savoini, G. Improved Fuzzy Logic System to Evaluate Milk Electrical Conductivity Signals from On-Line Sensors to Monitor Dairy Goat Mastitis. *Sensors* **2016**, *16*, E1079. [CrossRef] [PubMed]

21. Zaninelli, M.; Costa, A.; Tangorra, F.M.; Rossi, L.; Agazzi, A.; Savoini, G. Preliminary evaluation of a nest usage sensor to detect double nest occupations of laying hens. *Sensors* **2015**, *15*, 2680–2693. [CrossRef] [PubMed]

22. Anusuya, M.A.; Katti, S.K. Speech Recognition by Machine, A Review. *Int. J. Comput. Sci. Inf. Secur.* **2009**, *6*, 181–205.

23. Be, D.; Escalante, M.; Gonzalez, C.; Miranda, C.; Escalante, M.; Gonzalez, S. Wireless Control LEGO NXT robot using voice commands. *Int. J. Comput. Sci. Eng.* **2011**, *3*, 2926–2934.

24. Brown, K.; Rabiner, L.R.; Juang, B.-H. Speech Recognition: Statistical Methods. In *Encyclopedia of Language & Linguistics*; Elsevier: Boston, MA, USA, 2006; pp. 1–18.

25. Bates, D.; Machler, M.; Bolker, B.; Walker, S. Fitting Linear Mixed-Effects Models Using lme4. *J. Stat. Softw.* **2015**, *67*, 1–48. [CrossRef]

26. R Development Core Team. *R: A Language and Environment for Statistical Computing*; R Foundation for Statistical Computing: Vienna, Austria, 2015.

27. Shahamiri, S.R.; Binti Salim, S.S. Real-time frequency-based noise-robust Automatic Speech Recognition using Multi-Nets Artificial Neural Networks: A multi-views multi-learners approach. *Neurocomputing* **2014**, *129*, 199–207. [CrossRef]

28. Quintin, E.C.; Halan, S.K. Experiments in the application of isolated-word recognition to secondary driving controls for the disabled. *J. Rehabil. Res. Dev.* **1991**, *28*, 59–66. [CrossRef] [PubMed]

29. Nadas, A.; Nahamoo, D.; Picheny, M. Speech recognition using noise-adaptive prototypes. *IEEE Trans. Acoust. Speech Signal Process.* **1989**, *37*, 1495–1503. [CrossRef]

30. Hirsch, H.-G.; Pearce, D. The Aurora Experimental Framework for the Performance Evaluation of Speech Recognition Systems under Noisy Conditions. In *ISCA ITRW ASR2000 "Automatic Speech Recognition: Challenges for the Next Millennium"*; LIMSI-CNRS: Paris, France, 18–20 September 2000; pp. 181–188.

31. Rabiner, L.; Levinson, S. Isolated and connected word recognition-theory and applications. *IEEE Trans. Commun.* **1981**, *29*, 621–659. [CrossRef]

32. Cui, X.; Gong, Y. A Study of Variable-Parameter Gaussian Mixture Hidden Markov Modeling for Noisy Speech Recognition. *IEEE Trans. Audio Speech Lang. Process.* **2007**, *15*, 1366–1376. [CrossRef]

33. Hernando, J.; Nadeu, C. Linear prediction of the one-sided autocorrelation sequence for noisy speech recognition. *IEEE Trans. Speech Audio Process.* **1997**, *5*, 80–84. [CrossRef]

34. Ahmed, M. Comparison of noisy speech enhancement algorithms in terms of LPC perturbation inches. *IEEE Trans. Acoust. Speech Signal Process.* **1989**, *37*, 121–125. [CrossRef]

35. Lim, J.; Oppenheim, A. Enhancement and bandwidth compression of noisy speech. *Proc. IEEE* **1979**, *7*, 1586–1604. [CrossRef]

2

Modelling Nutrient Load Changes from Fertilizer Application Scenarios in Six Catchments around the Baltic Sea

Hans Thodsen [1,*], Csilla Farkas [2], Jaroslaw Chormanski [3], Dennis Trolle [1], Gitte Blicher-Mathiesen [1], Ruth Grant [1], Alexander Engebretsen [2], Ignacy Kardel [3] and Hans Estrup Andersen [1]

[1] Department of BioScience, Aarhus University, Vejlsøvej 25, 8600 Silkeborg, Denmark; Trolle@bios.au.dk (D.T.); gbm@bios.au.dk (G.B.-M.); rg@bios.au.dk (R.G.); hea@bios.au.dk (H.E.A.)

[2] Bioforsk, Division for Soil, Water and Environment, Frederik A. Dahlsvei 20, 1430 Ås, Norway; Csilla.Farkas@bioforsk.no (C.F.); alexander.engebretsen@bioforsk.no (A.E.)

[3] Department of Hydraulic Engineering, Faculty of Civil and Environmental Engineering, Warsaw University of Life Sciences-SGGW, ul. Nowoursynowska 166, 02-787 Warszawa, Poland; J.Chormanski@levis.sggw.pl (J.C.); i.kardel@levis.sggw.pl (I.K.)

* Correspondence: hath@bios.au.dk

Academic Editor: Paul Davidson

Abstract: The main environmental stressor of the Baltic Sea is elevated riverine nutrient loads, mainly originating from diffuse agricultural sources. Agricultural practices, intensities, and nutrient losses vary across the Baltic Sea drainage basin (1.75×10^6 km^2, 14 countries and 85 million inhabitants). Six "Soil and Water Assessment Tool" (SWAT) models were set up for catchments representing the major agricultural systems, and covering the different climate gradients in the Baltic Sea drainage basin. Four fertilizer application scenarios were run for each catchment to evaluate the sensitivity of changed fertilizer applications. Increasing sensitivity was found for catchments with an increasing proportion of agricultural land use and increased amounts of applied fertilizers. A change in chemical fertilizer use of ±20% was found to affect watershed NO$_3$-N loads between zero effect and ±13%, while a change in manure application of ±20% affected watershed NO$_3$-N loads between zero effect and −6% to +7%.

Keywords: agricultural management scenarios; Baltic Sea; environmental modelling; SWAT

---

1. Introduction

The ecological state of the Baltic Sea has been under pressure over the past few decades due to elevated nutrient inputs [1–3]. Hypoxia, which has serious negative ecological implications, occurs frequently. Hypoxia in sheltered marine areas is often associated with increased nutrient levels, and nutrient loads are recognized as the main driver of hypoxia in the Baltic Sea [4–9].

The elevated nutrient inputs originate predominantly from increased river nutrient loads [10]. Nitrogen loads have primarily increased due to augmented use of agricultural fertilizers, compared with preindustrial levels [11]. A source apportionment for the total riverine nutrient loading of the Baltic Sea in 2000 revealed that for total nitrogen (TN), natural background losses accounted for 28%, diffuse losses for 64%, and point source discharges for 8% of the riverine TN input to the Baltic Sea [12]. Agriculture accounted for 70–90% of the diffuse TN load [13]. For phosphorous, Mörth et al. [14] estimated that 15% to 70% of the total phosphorous (TP) load originated from point sources in sparsely populated boreal catchments and densely populated temperate catchments respectively between 1996–2000. Atmospheric deposition of nitrogen from land-based combustion of fossil fuels amounts to between 25% and 33% of the total nitrogen input to the Baltic Sea [15]. Ship traffic contributes 4–5% of the total atmospheric nitrogen deposition through fuel combustion [16].

The rising incidence of eutrophication has led policy makers and researchers to focus on the development of tools to reduce the nutrient input to the Baltic Sea and thus prevent further deterioration of ecological quality [17]. The most complex strategy developed is the Baltic Sea Action Plan [15,18,19], which includes an eutrophication segment aiming "to have a Baltic Sea unaffected by eutrophication" and in "good ecological status" [18]. However, knowledge-based assessment tools and management solutions are needed to achieve the objectives of the Baltic Sea Action Plan. With respect to agricultural nutrient losses, the potential impact of various management strategies needs to be assessed. According to the Helsinki Commission (HELCOM) pollution load compilation [20], the required reduction in TN load (total TN load) is 118,000 tons N, compared to the 1997–2003 reference period. By 2010, approximately 69% of this reduction was archived (90,000 tons N), but with large variation between the seven marine basins of the Baltic Sea. The remaining requirements after the 2008–2010 period are varying between zero and approximately 62,000 tons N (26% of the 2008–2010 total load) to the Baltic Proper (the main central part of the Baltic Sea).

Development of strategies aiming at reducing the nutrient loads to the Baltic Sea must be multi-faceted and comprehensive as the success of the applied strategies depends on multiple processes. Surface runoff, soil erosion, and nutrient retention/transport between the root zone and the catchment outlet are complex processes that are influenced by several environmental and anthropogenic factors. Process-based models describing both the hydrology and the sediment and nutrient transport within catchments are useful tools for describing such complex processes, and for evaluating the impact of measures and management strategies. Such models are capable of handling large amounts of data, combined with theoretical and expert knowledge, to integrate all available information about the system of interest.

Several studies involving modelling of river water and nutrient loads to marine areas with different purposes, and using different types of models, have been published [21–28]. Specifically addressing the Baltic Sea, the hydrological model HBV was set up for the entire Baltic Sea drainage basin to estimate runoff [29], with a preliminary study on scale effects [30]. The HBV model applied a rather crude spatial distribution approach, based on variability parameters, to model soil moisture dynamics and runoff in each individual sub-basin. The variability parameters were found to be relatively stable over a wide range of scales. Therefore, the HBV model could be applied at macro scale [30]. The "Hydrological Predictions for the Environment" HYPE model was set up and tested for both Sweden and the entire Baltic Sea catchment, reporting a water balance error of <10% and <25%, respectively [31,32].

Concerning the modelling of nutrient transport, a large-scale model "Catchment Simulation software" (CSIM) was set up for the entire Baltic Sea basin, aiming to describe the substantial differences in nutrient loads between the various catchments relative to geographical conditions, land use, population density, climate, etc. [14]. This was the first modelling work of its type conducted for the Baltic Sea catchment, and the main challenge was to capture the huge scale differences in nutrient transport [14]. Therefore, the model was kept relatively simple, focusing on key processes. The CSIM model describes inter-annual and seasonal variability of water and nutrient fluxes for 105 catchments. However, large-scale lumped models, such as CSIM, set up for entire regions are not capable of describing the impact of various mitigation measures applied at small/semi-distributed scale. Thus, to develop cost-effective adaptation strategies, models that can handle processes at small/semi-distributed scale, are needed.

In this study, the semi distributed (SWAT) was set up for six type catchments within the Baltic Sea watershed. Models like SWAT permit working on a fine spatial scale and addressing catchment and agricultural management in a detailed way [33]. Several individual SWAT models have been set up for areas within the Baltic Sea watershed with a focus on different objectives [17,34–43]. Ekstrand et al. [34] calibrated the SWAT model to five river basins in Sweden (tributaries to Lake Mälaren) to improve the modelling of phosphorus losses relative to the rainfall-runoff coefficient-based Watershed Management System (WATSHMAN) model. Francos et al. [36] applied the SWAT model to the Kerava watershed (South Finland) and found a good agreement between the measured and predicted values of runoff, total N, and total P concentrations at the outlet, especially when using precipitation data with fine spatial resolution. Lam et al. [38] performed an assessment of point and diffuse source pollution of nitrate in the Kielstau catchment (North German lowlands). Marcinkowski et al. [42] modelled combined climate, land use change, and fertilizer application scenarios for the Reda catchment in northern Poland. Abbaspour et al. [44] constructed a continental scale model covering Europe and thereby also the Baltic Sea catchment. All these studies focused, however, on certain geographical areas and were not representative of the entire Baltic Sea watershed. Also, these studies dealt only with hydrological processes, or had little to no emphasis on agricultural management.

This study focuses on modelling the effect of changes in agricultural fertilization practices on nutrient loads in different river catchments around the Baltic Sea. Similar SWAT studies have previously been performed, for example, Santhi et al. [45] evaluated the long-term effects of Water Quality Management Plans on non-point source pollution in a Texas catchment. Schilling and Wolter [46] examined a suite of measures, among these fertilizer application reductions, to meet regulatory limits for public water supplies. White et al. [47] modelled nutrient loads from six Oklahoma catchments, also identifying critical source areas for sediment and phosphorous. Thodsen et al. [41] used SWAT to identify high risk and low risk areas for diffuse nutrient losses in the Odense Fjord catchment, Denmark.

We ran four agricultural fertilization scenarios. These were ±20% chemical fertilizer application and ±20% manure fertilization. In the southern part of the Baltic Sea catchment, in Poland, the Baltic countries, Russia, and Belarus, it is likely that agriculture will intensify with higher fertilizer and manure application rates as a consequence of an expansion in meat and dairy production, leading to increased diffuse nutrient losses [48,49]. The scenarios increasing chemical and manure fertilization were based on this prediction, and on the possibility of water quality, not being the top priority of decisions made in all countries around the Baltic Sea. Along the northern and western rim of the Baltic Sea, in Germany, Denmark, Sweden, and Finland, application rates were likely to be stable or decline following political regulations. There is a continuous need to reduce nutrient loads to meet reduction needs in marine areas and lakes. The 20% reduction scenarios were based on this prediction.

We hypothesised that the effect of altered fertilizer/manure application would vary according to differences in the fraction of agricultural land use and agricultural fertilization practices in the Baltic catchments. The findings of our study will be of use for decision makers in targeting mitigation measures.

The aim of this study was to evaluate the effect of changed agricultural chemical fertilizer and manure application rates on nutrient loads in six type catchments within the Baltic Sea drainage basin.

2. Materials and Methods

2.1. Study Areas

The Baltic Sea drainage basin covers an area of 1.75×10^6 km^2, and the climate ranges from sub-arctic conditions in the north to temperate conditions in the south (Figure 1). The south-western part has a relatively humid Atlantic climate, while the eastern and northern parts have a dryer and more continental climate.

The six type catchments included in this study were: River Kalix, representing the boreal forested northern catchments: River Pärnu, Estonia, and River Nevezis, Lithuania, representing the eastern catchments with both forest and medium intensity agriculture; the Norrström catchment (only including the land area and thus not Lake Mälaren), Sweden, representing the eastern Swedish catchments with both agriculture (medium-high intensity), forest and urban areas; River Plonia, Poland, representing the high intensity agricultural southern catchments; and River Odense, Denmark, representing the high intensity agricultural catchments in the south-western part of the Baltic Sea basin. The combined area of the six pilot catchments was about 55,000 km^2, i.e., approximately 3% of the total Baltic Sea catchment area. The six type catchments represented ranges in climate, land use, and agricultural practices (Table 1).

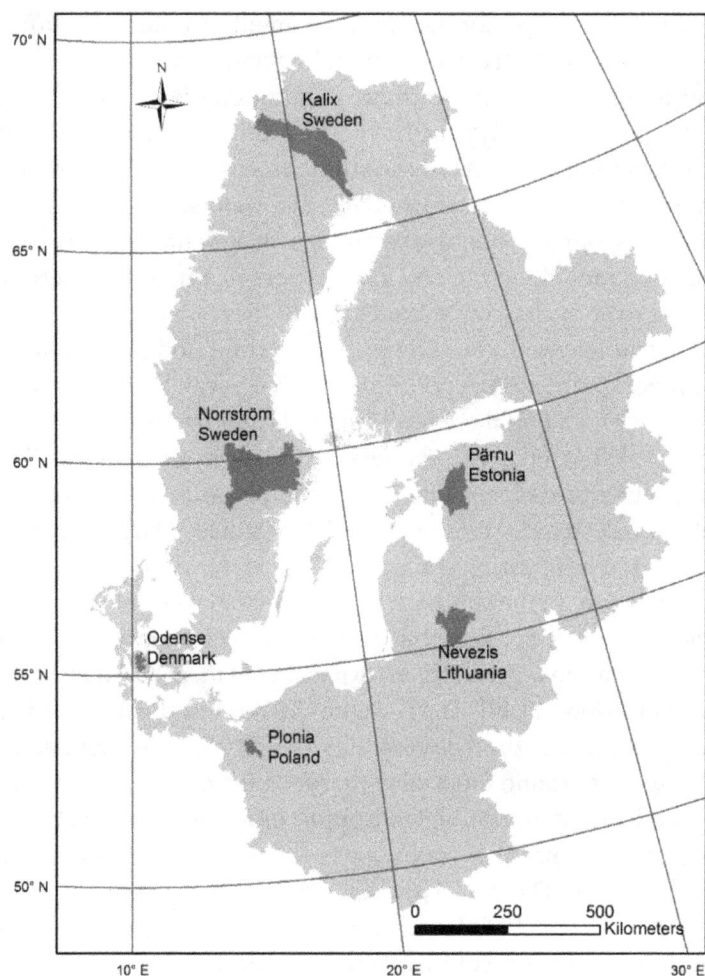

Figure 1. Map of the six type catchments around the Baltic Sea.

Table 1. Catchment characteristics. P is the mean annual precipitation for the period 1995–2006. T is the mean temperature for the period 1995–2006.

Catchment/Country	Area (km^2)	Agriculture (Area %)	P (mm year^{-1})	T (°C)	Catchment Type
Odense/Denmark	1059	71	704	8.7	Intensive agriculture—west
Pärnu/Estonia	6721	33	650	6.5	Agricultural-forested
Nevezis/Lithuania	6142	63	597	6.9	Agricultural-forested
Plonia/Poland	1034	78	550	8.4	Intensive agriculture—south
Kalix/Sweden	18,108	0.5	712	−0.5	Boreal forest
Norrström/Sweden	21,872	26	680	6.4	Forest/intensive agriculture

2.2. Data

Harmonised methods and data for the individual models were used. In some cases, national or local data of better quality than the common data source were chosen in order to optimise model performance. The common data sets on soil (see below) and land use (see below) were aggregates of national maps and differed between countries [50,51].

Climate data on all catchments derived from the data set belonging to the European Joint Research Centers (JRC) MARS50 data set were used, in some cases supplemented with better quality national or local data [52]. The MARS50 data set is a 50 km grid data set that includes all climate data necessary for setting up the SWAT model. MARS50 is available online. For the River Odense catchment, national 10 km gridded daily precipitation was used [53]. For the Pärnu catchment, precipitation and air temperature (minimum and maximum) data for six stations (Koodu, Kuusiku, Massumõisa, Parnu Türi and Viljandi), provided by the Estonian Environment Agency, were used. SWAT uses five different climate variables with daily resolution: temperature (minimum and maximum), precipitation, wind speed, relative humidity and short wave solar radiation, all derived from the MARS50 data set [54].

The Kalix, Nevezis, Plonia and Pärnu catchments were set up using the "Advanced Spaceborne Thermal Emission and Reflection Radiometer" ASTER Digital Elevation Model (DEM) [55]. The Odense catchment was set up using a 32 m resampled version of a 1.6 m LIDAR DEM [56]. The Norrström catchment was set up with a local DEM, with a resolution of 25 m.

The Corine land use map was applied to all catchments [51]. The soil map that was applied to all models except the Odense catchment model, was obtained from the Harmonized World Soil Database (HWSD). This soil map is an aggregation of nationally developed Food and Agriculture Organization of the United Nations (FAO) soil maps [50]. SWAT soil parameters were applied either directly from the HWSD, or estimated using the Hypress soil physics model [57]. For the Odense catchment, a national scale three-layer soil map was used because its quality was deemed superior to the HWSD [58].

Inputs on agricultural management included in the six different SWAT models originated from both national and European statistical sources on crop distribution, crop yields and fertilizer application rates and from local knowledge of agricultural practices [41]. The agricultural management data primarily represented the year 2005. Since the data were collected by different institutions in the various countries, some represented other years/periods. The agricultural area of each catchment was split into a number of rotations reflecting the complexity of the agricultural management in each catchment (Table 2). Increasing complexity, a rising percentage of agricultural area and enhanced knowledge about local agricultural conditions resulted in a higher number of agricultural rotations. The rotations aimed to be realistic with respect to the succession of crops and the application of chemical fertilizers and manure to individual crops, and to ensure that the annual application of chemical fertilizers and manure application was identical between years. Each rotation included dates for tillage/soil treatment, sowing, fertilizer applications and harvesting. Fertilizers were applied as either chemical fertilizers or manure (in many cases both) in accordance with the type of farming represented, for example cereals, pigs and cattle (beef or dairy). The amount of each fertilizer type (e.g., chemical, manure, pig slurry, cattle slurry) applied corresponded with the amounts typically used in the geographical region for the type and the intensity of the agricultural type in focus.

Table 2. Soil and Water Assessment Tool (SWAT) model setup information.

Catchment	Sub-Basins	HRUs *	Soil Types	Slope Classes	Calibration Period	Validation Period	Warm-Up Years	Agricultural Rotations/Years
Odense	31	2734	11	3	1997–2001	2002–2006	2	14/5
Pärnu	130	5271	12	4	1999–2002	2003–2006	7	2/7
Nevezis	102	1936	16	3	2000–2003	2004–2005	5	4/5
Plonia	66	2479	17	3	1997–2002	2005–2006	2	8/3
Kalix	23	280	2	3	1998–2000	2001–2006	2	2/1
Norrström	39	867	8	3	1997–2000	2001–2006	2	4/5

* Hydrological Response Unit (HRU).

In each SWAT model, atmospheric nitrogen deposition was provided as values of both dry and wet deposition based on gridded data from "the European Monitoring and Evaluation Programme" EMEP [59,60].

2.3. Model Setup

The complexity of the individual SWAT models (number of sub-basins and HRUs) reflected the aim of creating comparable models, and was influenced by catchment size, level of detail of basic input data (land use classes, soil classes, slope classes), number of monitoring stations and larger lakes/reservoirs, and the homogeneity of the catchment, as well as the number of rotations into which the agricultural area was split. The number of sub-basins in each SWAT model was thus primarily determined by the number of "points of interest" and not by the threshold of stream initialisation in the delineation of the models, as many of the auto-generated river confluence sub-basins were deleted in the delineation process. Jha et al. [61] evaluated the effect of average relative catchment sizes in SWAT (earlier version) on NO_3 and MinP loads for similar sized catchments in Wyoming, US. They found that loads had no effect on NO_3 loads when the sub-basin size was <2% of total catchment and <5% for MinP. In all models except the River Kalix model, the number of HRUs was limited by applying thresholds to the percentage area of a sub-basin that the soil type/land use/slope class should occupy. The thresholds depended on the number of soil types and land uses present in the maps covering the watershed, thereby determining the initial number of HRUs in each model. However, the number of HRUs depended greatly on the number of splits that the agricultural HRUs were divided into (Table 2). For example, in the River Odense SWAT model the agricultural HRUs were split into 14 different kinds of agriculture, meaning that the number of HRUs with agricultural land uses was multiplied with 14. The number of splits was determined by the range of farming intensity, management practices, and knowledge about local farming practices. Basic SWAT statistics for comparing model setups are shown in Table 2. For example, the spatial resolution of the HWSD soil map was very coarse for Sweden, and only provided two soil types for the Kalix catchment, while there were 11 soil types for the much smaller River Odense catchment. This difference naturally resulted in more HRUs for the Odense catchment than for the Kalix catchment. Agricultural catchments had more HRUs than other catchment types since special emphasis was put on agriculture by splitting this land use type into specific rotations for different kinds of agriculture and agricultural management practices [41]. The upper limit of complexity was set by the computation time for running the SWAT models, and thus for performing the calibration procedure.

2.4. Calibration Data

Recording of river runoff (Q), nitrate (NO_3), and reactive soluble phosphorus/mineral phosphorous (MinP) (using SWAT nomenclature) loads for calibration and validation were collected for at least one location in each catchment (Table 3). All data were collected from national monitoring sources that converted individual nutrient concentration sample values into daily, monthly and bi-weekly load values using approaches complying with the HELCOM requirements of using either "Daily flow and daily concentration regression" or "Daily flow and daily concentration interpolation" methods [62–64].

Table 3. Overview of available calibration data (Q = River runoff, NO_3 = Nitrate river loads, MinP = Mineral phosphorous river loads). + = Crop yield data available.

Catchment	Q (#Stations)	NO_3 (#Stations)	MinP (#Stations)	Crop Yield
Odense	Daily (4)	Daily (4)	Daily (4)	+
Pärnu	Daily (7)	Bi-weekly (6)	Bi-weekly (6)	+
Nevezis	Monthly (3)	Monthly (3)	Monthly (3)	+
Plonia	Daily (2)	Monthly (2)	Monthly (2)	+
Kalix	Daily (3)	Monthly (1)	Monthly (1)	+
Norrström	Daily (2)	Daily/Monthly (2)	Monthly (2)	+

Observations of organic forms of N and P were not available for all catchments; hence, in order to allow cross-catchment comparisons, we chose to include only inorganic N and P. Organic N comprises only a minor fraction of the total N loads in the southern agriculturally dominated part of the Baltic Sea drainage basin [14,65]. In the northern part of the Baltic Sea drainage basin, organic forms of nutrients are dominant [66]. This is, however, mainly due to the land use (forests, extensive peatlands) which was not changed in the applied model scenarios.

2.5. Calibration Procedure

A common calibration procedure was applied to all six catchments. The models were calibrated sequentially in two steps, as previously done [67], with a possible pre-calibration adjustment of crop base temperature (T_BASE) to ensure crop yields and growth rates were at realistic levels for all major crops and vegetation types [44]. This was to ensure that vegetation was growing properly and thereby ensuring that evaporation and nutrient uptakes were not biased by unrealistic vegetation growth [68]. In Step One, river runoff was calibrated against observations on a daily or monthly scale, according to the time step of the observational data. Not all calibration parameter ranges were closed during Step One, as some parameters were sensitive to calibration of both hydrology and nutrient dynamics. Therefore, some parameters were left with a small range to be exploited during the NO_3 and MinP calibration. In Step Two, river NO_3 and MinP river loads were calibrated against observations (Table 3). NO_3 and MinP were calibrated sequentially with NO_3 as primary nutrient and thus calibrated first in each calibration iteration. Where a calibration parameter was sensitive to both NO_3 and MinP, its calibration range was primarily narrowed regarding NO_3 (Table 4).

Table 4. Parameters used during each of the two calibration steps.

Runoff	NO$_3$ and MinP
ALPHA_BF.gw	ANION_EXCL.sol
ALPHA_BNK.rte	CDN.bsn
CH_k2.rte	CMN.bsn
CN2.mgt	HLIFE_NGW.gw
DEP_IMP.hru	N_UPDIS.bsn
EPCO.hru	NPERCO.bsn
ESCO.hru	SDNCO.bsn
GW_DELAY.gw	
GW_REVAP.gw	
GWQMN.gw	
LAT_TTIME.hru	CH_OPCO.rte
OV_N.hru	GWSOLP.gw
RCHRG_DP.gw	P_UPDIS.bsn
REVAPMN.gw	PHOSKD.bsn
SFTMP.bsn	PPERCO.bsn
SMFMN.bsn	PRF.bsn
SMFMX.bsn	PSP.bsn
SMTMP.bsn	USLE_P.mgt
SOL_AWC.sol	
SOL_BD.sol	
SOL_K.sol	
SURLAG.bsn	
TDRAIN.mgt	
DDRAIN.mgt	
GDRAIN.mgt	

Model calibration of river runoff and nutrient loads was performed based on the Sequential Uncertainty Fitting Algorithm (SUFI2) that uses a global search procedure through Latin Hypercube Sampling [69–71]. All parts of the calibration were optimised by running at least 1000 simulations through SWAT calibration and uncertainty program (SWAT-CUP) software using multiple iterations where parameter ranges were gradually narrowed. Only sensitive parameters were considered after the first few rounds of calibration. The Nash-Sutcliffe model efficiency value was chosen as the primary objective function [72,73].

2.6. Scenarios

Four scenarios of agricultural management relating to application rates of fertilizers or manure were run. Regulation of nutrient inputs in agriculture is the most important measure adopted for nutrient loss mitigation, and is applied in all the EU member states in the Baltic Sea drainage basin (e.g., maximum allowable inputs of fertilizer and maximum allowable livestock densities [74]). The model scenarios included changes in chemical fertilizer use by ±20% and changes in livestock number by ±20%, respectively. The changes in livestock numbers were implemented in SWAT as changes in the manure application amount. The scenarios were introduced to SWAT by changing the amount of fertilizer (chemical or manure) applied to a given agricultural crop at any given time. The scenarios thus had the highest effect in sub-basins with a large fraction of agricultural land. Livestock scenarios were implemented without counterbalancing changes in manure application with chemical fertilizer application. No combinations of altered chemical fertilizer and manure applications were used. A 20% increase in fertilization was also used as a scenario for catchments in central Germany [75].

3. Results

3.1. Model Validation

All models were validated with respect to river runoff (Q), NO_3, and MinP against observations covering at least one-third of the period with observations. Model validation statistics for the six SWAT models are shown in Table 5.

Table 5. Model calibration/validation statistics for the six SWAT models on river runoff, river NO_3 loads and river MinP loads. (calibration/validation) Nash-Sutcliffe [72]. BR^2 is described in [70]. Daily runoff values are used, except for Nevezis where runoff is given as monthly values. All NO_3 and MinP values are derived from monthly data, except for Odense where NO_3 and MinP is given as daily values.

	Statistical Parameter	Odense Denmark	Norrström Sweden	Kalix Sweden	Pärnu Estonia	Nevezis Lithuania	Plonia Poland
Run off	Water balance error (%)	14/4	−13/−11	−1/4	−10/−2	4/−2	0/−8
	Nash-Sutcliffe	0.85/0.82	0.61/0.57	0.87/0.87	0.61/0.65	0.84/0.82	0.73/0.57
	R^2	0.89/0.84	0.77/0.70	0.88/0.89	0.68/0.71	0.88/0.84	0.74/0.58
	BR^2	0.87/0.80	0.69/0.66	0.91/0.89	0.33/0.46	0.76/0.59	0.57/0.34
	Error 25 percentile (%)	−21	−71	−31	104	41	21
	Error 75 percentile (%)	34	−2	7	−8	19	20
NO_3	Mass balance error (%)	18/37	−4.3/−8.2	12/−10	−25/−25	−36/−48	3.0/−8.0
	Nash-Sutcliffe	0.52/0.48	−0.53/−0.67	0.33/0.31	0.76/0.72	0.69/0.67	0.61/0.34
	R^2	0.78/0.69	0.36/0.32	0.49/0.45	0.68/0.84	0.82/0.80	0.62/0.34
MinP	Mass balance error (%)	7.0/8.1	5.7/−0.40	8.6/1.6	−23/−26	21/24	13/16
	Nash-Sutcliffe	0.33/0.29	0.11/−0.20	−1.4/−1.7	0.52/0.45	0.32/0.34	0.38/0.39
	R^2	0.57/0.61	0.62/0.56	0.23/0.23	0.53/0.48	0.32/0.35	0.38/0.39

The River Pärnu model was initially run with MARS50 precipitation, which resulted in a Nash-Sutcliffe value of −0.24, compared with a value of 0.61 when using local station data. It was obvious that flow peaks did not correspond well with high MARS50 precipitation events.

The time series of observed and simulated river water discharge for both the calibration and the validation period are shown for the River Plonia in Figure 2 by way of example.

As displayed in the figure, the simulated curve followed the observed curve quite well during most of the period. The simulated curve did not systematically over- or underestimate the annual spring peak flow events or the summer low-flow conditions.

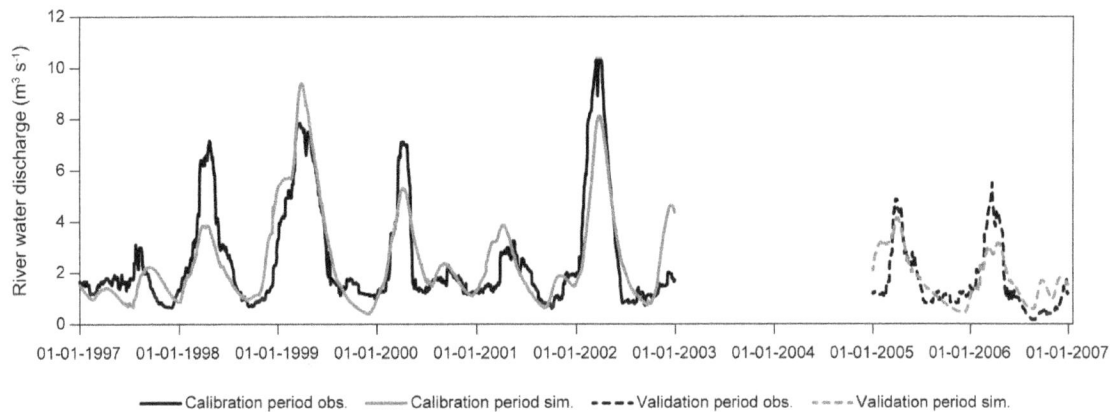

Figure 2. Time series of observed and simulated river water discharge for the calibration period (1997–2002) and the validation period (2005–2006) in the Plonia catchment, Poland.

3.2. Fertilizer Application

Both the total amount of fertilizers applied in a catchment and the amount applied in agricultural areas were important in order to elucidate the nutrient dynamics of catchments (Table 6).

Table 6. Applied annual fertilizer amounts (kg N ha^{-1}) in the SWAT model for each of the six catchments (N/P). The manured area of the Pärnu catchment is very small and therefore not included.

		Odense Denmark	Norrström Sweden	Kalix Sweden	Pärnu Estonia	Nevezis Lithuania	Plonia Poland
Chemical fertilizer	kg N ha^{-1} per catchment area	67/6	14/1.6	0.15/0.03	33/8	35/0.5	115/15
Manure	kg N ha^{-1} per catchment area	62/13	5/1.0	0.3/0.06	-	10/3	5/2
Chemical fertilizer	kg N ha^{-1} per agricultural area	94/8	55/6	30/5	98/23	64/1	147/19
Manure	kg N ha^{-1} per agricultural area	87/18	20/4	60/11	-	19/6	6/2.5

Large differences existed in the amounts of fertilizers used in the different catchments, and were most pronounced between the amounts used per ha^{-1} in the total catchment, primarily reflecting the fraction of the catchment used for agricultural purposes (Table 1). Additionally, there were relatively marked variations between the amounts of fertilizers used per ha^{-1} in the agricultural areas, reflecting the differences in agricultural intensity (Table 6). For the northern boreal catchment, Kalix, the agricultural nutrient input to the catchment of 0.45 kg N ha^{-1} was much smaller than the atmospheric N deposition of about 7 kg N ha^{-1}.

3.3. Scenario Results

Changes in modelled mean annual NO$_3$ and MinP loads resulting from the scenario runs are presented in Table 7.

The baseline nutrient losses from the River Plonia catchment were low, considering the size of the catchment, the relatively high fraction of agricultural land use, and the amount of fertilizer used. The low load can be explained by the presence of the 36 km^2 large Lake Miedwie a short distance upstream of the monitoring station used for calibration and validation. Therefore, the Plonia catchment nutrient loads were not directly comparable with those from the other catchments.

Table 7. Changes in $NO_3/MinP$ (%) load from catchments in chemical fertilizer and manure application scenarios.

Scenario		Odense Denmark	Norrström Sweden	Kalix Sweden	Pärnu Estonia	Nevezis Lithuania	Plonia Poland
Baseline	Ton year^{-1}	1926/30.3	5704/181	436/30	384/16	3910/228	28.5/2.0
ChemicalFertilizer	−20%	−7.5/−0.2	−1.1/−1.1	−0.02/−0.03	−0.3/−0.6	0.3/−0.1	−13/2
	+20%	7.8/0.2	1.1/1.1	0.02/0.03	0.0/0.0	2.1/−0.1	13/−1
Manure	−20%	−6.3/−0.2	−0.1/−0.6	−0.06/−0.05	−0.1/0.0	0.8/0.1	−0.5/0.0
	+20%	6.8/0.2	0.1/0.0	0.06/0.05	0.2/0.0	0.8/0.1	0.5/0.0

The differences in scenario responses to NO_3 between the catchments were large. The effect of reducing fertilizer application was negligible for the boreal River Kalix, which has a very low fraction of agricultural land use. The largest effects were found for the River Plonia and the River Odense catchments, the latter having the largest fraction of agricultural land and relatively large fertilizer application rates. The River Nevezis catchment also has a comparatively large percentage of agricultural land, but a weaker response in river loads to the scenarios was found, reflecting the lower fertilizer application in this catchment than in the Odense and Plonia catchments. The Pärnu and Norrström catchments both have an agricultural land use around 30%, and showed small responses to the scenarios.

3.4. Model Sensitivity

The most sensitive model parameters for each modelling objective in each of the six SWAT models are given in Table 8. The most sensitive parameters were chosen based on the SWAT-CUP global sensitivity analysis, utilising the Nash-Sutcliffe objective function, at an early stage of the calibration procedure [69,70,72].

Table 8. Most sensitive model parameters.

	Odense Denmark	Norrström Sweden	Kalix Sweden	Pärnu Estonia	Nevezis Lithuania	Plonia Poland
Water discharge	GW_DELAY EPCO	SMTMP	SMTMP	GW_DELAY ALPHA_BF CH_N2	SMTMP CN2	GW_DELAY CANMX
NO$_3$ load	SDNCO HLIFE_NGR	NPERCO HLIFE_NGR	NPERCO	NPERCO CDN N_UPDIS	NPERCO	SDNCO HLIFE_NGW
MinP load	GWSOLP PPERCO	GWSOLP PPERCO	PSP	PSP PPERCO	GWSOLP	GWSOLP

GW_DELAY is groundwater delay time (days), ALPHA_BF is Baseflow alpha factor (1 day^{-1}), ESCO is soil evaporation compensation factor, CANMX is maximum canopy storage, CN2 is curve number, SMTMP is snow melt temperature, CH_N2 is Manning channel roughness "n" for the main channel, HLIFE_NGR is half-life of nitrate in shallow aquifer (days), SDNCO is denitrification threshold water content, NPERCO is nitrate percolation coefficient, N_UPDIS is nitrogen uptake distribution parameter, CDN is denitrification exponential rate coefficient, GWSOLP is concentration of soluble phosphorous in groundwater contribution to stream flow from sub-basin (mg L^{-1}), PPERCO is phosphorous percolation coefficient and PSP is phosphorous availability index [54].

The parameters shown in Table 8 mostly reflected the overall conditions of the catchment. The northern and eastern catchments with a spring flood flow regime are very sensitive to the snowmelt temperature, and the base flow-dominated southern and western catchments are sensitive to groundwater parameters such as the groundwater delay time. For base flow-dominated catchments, the half-life of nitrate in the shallow aquifer tended to be sensitive.

4. Discussion

In model comparison studies, it is important that the inputs to the different models are as comparable as possible. However, some inputs to the six SWAT models in this study differed, and for instance the soil data were derived from different sources because no uniform single source exists. The HWSD is a collection of national maps that vary in terms of spatial resolution; thus, the HWSD is detailed for Estonia and Lithuania but very crude for Sweden. Therefore, we decided that the best available soil map would be used for each catchment. MARS50 climate data were used in all catchments for all parameters except for precipitation (Odense and Pärnu) and temperature (Pärnu), which were in these cases available at a better spatial scale from other sources. For the Pärnu catchment, the replacement of the MAR50 precipitation with data of better spatial resolution resulted in noticeable improvements to the model; the Nash-Sutcliffe value was improved from -0.24 to 0.61 for runoff during the calibration period. The importance of good precipitation data for calibrating hydrological models is also emphasised by Chaibou et al. [76] and by De Almeida Bressiani [77], who tested a range of precipitation data. The Corine land use map is another aggregation of national maps, but as these are made from a common standard, harmonisation is greater than that of the HWSD, and the Corine map was therefore used in all catchments. The agricultural management and crop yield data used in setting up the models were obtained from national and EU statistics, and were therefore comparable. The expertise of the modellers with agricultural statistics differed as knowledge about local agricultural practices varied from extensive to sparse. Availability of calibration data differed in terms of the amount of data (spatial and temporal), quality of the data, and parameters of the data. All stages of the modelling process are subject to uncertainties and errors, and uncertainty is even stronger in a study comparing conditions in different geographical areas and different countries. As described above, some of the basic input data to the SWAT model, such as the soil maps, differed, increasing the uncertainty related to the soil maps, and implying higher uncertainty in our study than in a study dealing with just one homogeneously produced soil map. The same increase in uncertainty associated with a "total study" was added from all other input data. However, the six catchments were primarily chosen based on the extensive available data, and we therefore believe that the quality was the best possible for the aim of this study. Furthermore, the catchments represented the variations in geographical and agricultural management conditions in the Baltic Sea watershed. For modelling of river runoff, the quality of the climate forcing data, particularly precipitation, is important [69]. The spatial scale of the precipitation data is highly significant. SWAT uses a time series of daily precipitation from the nearest station (in this case the centre point of a grid) and applies this precipitation to a sub-basin, implying that a single value is used for lumped areas of varying size. Potentially, this could induce certain scale problems. For example, a single-station value is applied to a large catchment, the problem being that, for example, a single thunder storm shower (>100 mm day^{-1}) affecting 1% of a large catchment, including the precipitation gauge location, would simulate an extreme high-flow event not occurring in reality, and would not be representative of the entire catchment when applied to the hydrological model. A similar problem may arise if gridded precipitation data is averaged over too few stations. On the other hand, the spatial scale of the gridded observations could be too large, and the variation in water discharge caused by differences in precipitation would therefore not be reflected in the model. In this study, a 50 km grid resolution (2500 km^2) was used for four of the six catchments, while a finer resolution was available for the latter two. The average sub-basin size in the four catchments ranged from 16 km^2 for the River Plonia to 787 km^2 for the River Kalix. Thus, it is obvious that the grid scale of the precipitation data fitted the average sub-basin size of the River Kalix better than that of the River Plonia. That is, if the River Plonia catchment had a substantial geographical precipitation gradient, the 50 km gridded data would presumably be too coarse to capture this. Potentially, the entire River Plonia catchment could be covered by only one grid cell.

By choosing SWAT to model all six catchments, the problem of comparing results from different models was avoided and a consistent modelling approach was applied, lending credibility to the comparison of calibration/validation and scenario results between the six watersheds.

The most sensitive parameters for each SWAT model at an early stage of each calibration step are shown in Table 8. We chose not to quantify parameter sensitivity, as the sensibility of single parameters depends on a number of circumstances—for instance, choice of objective function, temporal resolution of calibration data, calibration procedure, spatial resolution of input data, conceptual model uncertainty, modeller's knowledge of the modelled catchment, chosen initial calibration parameter range, climate of the chosen calibration, and validation period—at the stage of the calibration procedure during which the sensitivity analysis is performed as well as on the model output in focus [69]. For example, the modeller's knowledge about the catchment and experience in calibrating models for a particular area may strongly influence parameter sensitivity. In this study, the modelling team had extensive experience in modelling the River Odense catchment, leading to knowledge about parameter setting for the modelling of NO_3, where the denitrification threshold water content (SDNCO) should be between 0.75 and 0.99 and the denitrification exponential rate coefficient below 1 to produce realistic denitrification rates, thereby avoiding non-uniqueness problems [69]. Extensive knowledge reduced the sensitivity of these parameters compared with a situation where calibration was initiated with a full range. Similar knowledge was not available for the River Kalix, Norrström, Nevezis and Pärnu, and the range of these parameters therefore had to be larger, producing a potentially greater sensitivity. Where calibration data were available with daily resolution, parameters addressing fast-responding parameters were more sensitive than where calibration data were available with monthly resolution.

Model uncertainties relative to reproduction of observational values of Q, NO_3 and MinP are evaluated in Table 5. Moriasi et al. [78] states that Nash-Sutcliffe efficiency values > 0.75 for monthly time steps of runoff are "very good" (Rivers Odense, Kalix, Nevezis), values between 0.75 and 0.65 are "good" and values between 0.65 and 0.50 are "satisfactory" (Rivers Norrström, Pärnu, Plonia) (Table 5). Nash-Sutcliffe values calculated from daily values are usually lower than values calculated from monthly values. Mass balance "percent BIAS" (PBIAS) (%) for N and P estimates of ±25% were considered "very good" (River Odense NO_3 & MinP, River Kalix NO_3 & MinP, River Pärnu NO_3, River Nevezis MinP and River Plonia NO_3 & MinP), ±25% to ±40% as "good" (River Pärnu MinP) and the interval ±40% to ±70% as "satisfactory" (River Nevezis NO_3). Overall, the validation statistics were considered adequate.

The total water balance error of the six SWAT model validations was relatively good, with a maximum error of 11%, but when evaluated according to the 0.25 percentile (25% of all values are smaller than this) and the 0.75 percentile, the models had larger biases (Table 5).

The scenario simulations showed that the effect of altered fertilizer application was strongest in catchments with a large fraction of agricultural land use and with intense agriculture like the River Odense and River Plonia catchments (Tables 1 and 6). River Norrström and River Nevezis exhibited a stronger response to the scenarios than River Pärnu, although the combined agricultural pressure in the three catchments was about the same. River Kalix showed a negligible response to the scenarios, which was expected, as agriculture only occupies 0.5% of its catchment area.

For the Kalix, Norrström, Pärnu and Nevezis catchments, the modelled effects of the scenarios on MinP were of the same magnitude as the effect on NO_3, and were an order of magnitude smaller on MinP than on NO_3 for the River Odense and Plonia catchments. This reflected the higher NO_3 input to these intensely farmed catchments. Besides this, the effects on MinP were rather linked to soil erosion processes than to leaching processes. Therefore, catchments with a relatively continental climate and, consequently, relatively large erosive spring snow melt events, showed a stronger response than catchments with a more Atlantic climate having milder winters and a smaller build-up of snow. Additionally, SWAT did not simulate leaching of dissolved phosphorous from the soil to the groundwater and into the river [54].

The results presented in this paper for the Baltic Sea catchments suggest that the effect of changed agricultural nutrient applications on riverine nutrient loads will be strongest in areas with relatively intensive fertilizer application. Therefore, decision makers should focus on mitigation methods in these areas if they are aiming at maximum impacts per hectare of agricultural land. Large differences in nutrient loads in the high intensity agricultural areas remain, though, due to differences in, for instance, fertilizer application procedures and crop yield and notable differences in retention found primarily for nitrogen [79]. A study running a +20% fertilization scenario for catchments in central Germany found NO_3 river load increases from 2% to 6% [75].

Along the southern rim of the Baltic Sea, in Poland (despite the fact that the Polish Plonia catchment has the highest chemical fertilizer application rates among the six catchments included in this study), the Baltic countries, Russia and Belarus, it is very likely that agriculture will be intensified, with higher fertilizer and manure application rates, as a consequence of an expansion of meat and dairy production, leading to increased diffuse nutrient losses [48,49]. Along the northern and western rim of the Baltic Sea, in Germany, Denmark, Sweden and Finland, application rates are likely to be stable, or will decline following political regulations.

The loads of organic nitrogen and phosphorous make up a substantial part of the total nutrient input to the Baltic Sea, and a primary part of total nutrient inputs in northern boreal forest areas [14]. The organic fractions were not considered in this study for two reasons: (1) data were not available for all six catchments; (2) the scenarios were not thought to substantially influence the load of organic N and P.

5. Conclusions

Four fertilizer application scenarios were run for each of the six watersheds to evaluate the sensitivity of changed fertilizer application rates. Increasing sensibility was found for catchments with an increasing proportion of agricultural land use, and enhanced amounts of fertilizer application. A change in chemical fertilizer use of $\pm20\%$ affected watershed NO_3-N loads between zero effect and $\pm13\%$. A change in manure application of $\pm20\%$ affected watershed NO_3-N loads between zero effect and -6% to $+7\%$.

Acknowledgments: The study was carried out as part of the RECOCA project under the EU BONUS program, the Go4Baltic project under the EU BONUS program and within the Baltic Nest Institute framework. The authors would like to thank the following people for providing important data: Erik Smedberg and Miguel Alberto Rodriguez Medina, Baltic Nest Institute/University of Stockholm, Ekaterina Sokolova, Swedish University of Agriculture, Arvo Iital, Tallinn University of Technology, Ausra Smitiene, Environmental Project Management Agency, Lithuania, and Antanas Sigitas Sileika, Water Management Institute, Lithuanian University of Agriculture and Anne Mette Poulsen for revising the manuscript language.

Author Contributions: Csilla Farkas and Alexander Engebretsen collected data and set up and ran the River Pärnu SWAT model. Jaroslaw Chormanski and Ignacy Kardel collected data and set up and ran the River Plonia SWAT model. Hans Estrup Andersen collected data and set up and ran the River Nevezis SWAT model. Hans Thodsen and Dennis Trolle collected data and set up and ran the River Odense, Norrström and Kalix SWAT models. Gitte Blicher-Mathiesen and Ruth Grant collected agricultural data from the countries bordering the Baltic Sea and provided the agricultural input to SWAT. Hans Thodsen wrote the main part of the manuscript with contributions from Csilla Farkas, Hans Estrup Andersen, Jaroslaw Chormanski, Ruth Grant and Dennis Trolle.

Conflicts of Interest: The authors declare no conflict of interest.

References

1. Conley, D.J.; Carstensen, J.; Ærtebjerg, G.; Christensen, P.B.; Dalsgaard, T.; Hansen, J.L.S.; Josefson, A.B. Long-term changes and impacts of hypoxia in Danish coastal waters. *Ecol. Appl.* **2007**, *17*, S165–S184. [CrossRef]

2. Conley, D.J.; Carstensen, J.; Aigars, J.; Axe, P.; Bonsdorff, E.; Eremina, T.; Haahti, B.M.; Humborg, C.; Jonsson, P.; Kotta, J.; et al. Hypoxia is increasing in the coastal zone of the Baltic Sea. *Environ. Sci. Technol.* **2011**, *45*, 6777–6783. [CrossRef] [PubMed]

3. Jansson, B.O.; Dahlberg, K. The environmental status of the baltic sea in the 1940s, today, and in the future. *Ambio* **1999**, *28*, 312–319.

4. Conley, D.J.; Bjorck, S.; Bonsdorff, E.; Carstensen, J.; Destouni, G.; Gustafsson, B.G.; Hietanen, S.; Kortekaas, M.; Kuosa, H.; Meier, H.E.M.; et al. Hypoxia-related processes in the Baltic Sea. *Environ. Sci. Technol.* **2009**, *43*, 3412–3420. [CrossRef] [PubMed]

5. Kemp, W.M.; Boynton, W.R.; Adolf, J.E.; Boesch, D.F.; Boicourt, W.C.; Brush, G.; Cornwell, J.C.; Fisher, T.R.; Glibert, P.M.; Hagy, J.D.; et al. Eutrophication of chesapeake bay: Historical trends and ecological interactions. *Mar. Ecol. Prog. Ser.* **2005**, *303*, 1–29. [CrossRef]

6. Rabalais, N.N.; Turner, R.E.; Wiseman, W.J. Gulf of mexico hypoxia, aka "the dead zone". *Ann. Rev. Ecol. Syst.* **2002**, *33*, 235–263. [CrossRef]

7. Savchuk, O.P.; Wulff, F. Long-term modeling of large-scale nutrient cycles in the entire Baltic Sea. *Hydrobiologia* **2009**, *629*, 209–224. [CrossRef]

8. Savchuk, O.P.; Wulff, F.; Hille, S.; Humborg, C.; Pollehne, F. The baltic sea a century ago—A reconstruction from model simulations, verified by observations. *J. Mar. Syst.* **2008**, *74*, 485–494. [CrossRef]

9. Carstensen, J.; Andersen, J.H.; Gustafsson, B.G.; Conley, D.J. Deoxygenation of the Baltic Sea during the last century. *Proc. Natl. Acad. Sci. USA* **2014**, *111*, 5628–5633. [CrossRef] [PubMed]

10. Stalnacke, P.; Grimvall, A.; Sundblad, K.; Tonderski, A. Estimation of riverine loads of nitrogen and phosphorus to the Baltic Sea, 1970–1993. *Environ. Monit. Assess.* **1999**, *58*, 173–200. [CrossRef]

11. Eriksson, H.; Pastuszak, M.; Lofgren, S.; Morth, C.M.; Humborg, C. Nitrogen budgets of the Polish agriculture 1960–2000: Implications for riverine nitrogen loads to the Baltic Sea from transitional countries. *Biogeochemistry* **2007**, *85*, 153–168. [CrossRef]

12. Helsinki Commission. The Fifth Baltic Sea Pollution Load Compilation (PLC-5). In *Baltic Sea Environment Proceedings*; No. 128; Helsinki Commission: Helsinki, Finland, 2001.

13. Helsinki Commission. The fourth Baltic Sea Pollution Load Compilation (PLC-4). In *Baltic Sea Environment Proceedings*; Helsinki Commission: Helsinki, Finland, 2004; Volume 93, p. 189.

14. Morth, C.M.; Humborg, C.; Eriksson, H.; Danielsson, A.; Medina, M.R.; Lofgren, S.; Swaney, D.P.; Rahm, L. Modeling riverine nutrient transport to the Baltic Sea: A large-scale approach. *Ambio* **2007**, *36*, 124–133. [CrossRef]

15. Helsinki Commission. *Airborne Nitrogen Loads to the Baltic Sea*; HELCOM Environmental Focal Point Information; Baltic Marine Environment Commission: Helsinki, Finland, 2005; pp. 1–7.

16. Bartnicki, J.; Semeena, V.S.; Fagerli, H. Atmospheric deposition of nitrogen to the Baltic Sea in the period 1995–2006. *Atmos. Chem. Phys.* **2011**, *11*, 10057–10069. [CrossRef]

17. Piniewski, M.; Kardel, I.; Giełczewski, M.; Marcinkowski, P.; Okruszko, T. Climate change and agricultural development: Adapting Polish agriculture to reduce future nutrient loads in a coastal watershed. *Ambio* **2014**, *43*, 644–660. [CrossRef] [PubMed]

18. Helsinki Commission. Ecosystem Health of the Baltic Sea 2003–2007: HELCOM Initial Holistic Assessment. In *Baltic Sea Environment Proceedings*; No. 122; Helsinki Commission: Helsinki, Finland, 2010; p. 63.

19. Wulff, F.; Humborg, C.; Andersen, H.; Blicher-Mathiesen, G.; Czajkowski, M.; Elofsson, K.; Fonnesbech-Wulff, A.; Hasler, B.; Hong, B.; Jansons, V.; et al. Reduction of Baltic Sea nutrient inputs and allocation of abatement costs within the Baltic Sea catchment. *Ambio* **2014**, *43*, 11–25. [CrossRef] [PubMed]

20. Helsinki Commission. Review of the Fifth Baltic Sea Pollution Load Compilation for the 2013 HELCOM Ministerial Meeting. In *Baltic Sea Environment Proceedings*; No. 141; Helsinki Commission: Helsinki, Finland, 2013; p. 54.

21. Andersen, H.E.; Kronvang, B.; Larsen, S.E. Development, validation and application of Danish empirical phosphorus models. *J. Hydrol.* **2005**, *304*, 355–365. [CrossRef]

22. Bouraoui, F.; Grizzetti, B. An integrated modelling framework to estimate the fate of nutrients: Application to the loire (France). *Ecol. Model.* **2008**, *212*, 450–459. [CrossRef]

23. Jordan, T.E.; Correll, D.L.; Weller, D.E. Effects of agriculture on discharges of nutrients from coastal plain watersheds of chesapeake bay. *J. Environ. Q.* **1997**, *26*, 836–848. [CrossRef]

24. Kronvang, B.; Behrendt, H.; Andersen, H.E.; Arheimer, B.; Barr, A.; Borgvang, S.A.; Bouraoui, F.; Granlund, K.; Grizzetti, B.; Groenendijk, P.; et al. Ensemble modelling of nutrient loads and nutrient load partitioning in 17 European catchments. *J. Environ. Monit.* **2009**, *11*, 572–583. [CrossRef] [PubMed]

25. Kronvang, B.; Grant, R.; Larsen, S.E.; Svendsen, L.M.; Kristensen, P. Non-point-source nutrient losses to the aquatic environment in Denmark—Impact of agriculture. *Mar. Freshw. Res.* **1995**, *46*, 167–177.

26. Sferratore, A.; Billen, G.; Garnier, J.; Thery, S. Modeling nutrient (N, P, Si) budget in the Seine watershed: Application of the riverstrahler model using data from local to global scale resolution. *Glob. Biogeochem. Cycles* **2005**, *19*, 1–14. [CrossRef]

27. Thodsen, H.; Kronvang, B.; Andersen, H.E.; Larsen, S.E.; Windolf, J.; Jørgensen, T.B.; Troldborg, L. Modelling diffuse nitrogen loadings of ungauged and unmonitored lakes in Denmark: Application of an integrated modelling framework. *Int. J. River Basin Manag.* **2009**, *7*, 245–257. [CrossRef]

28. Windolf, J.; Thodsen, H.; Troldborg, L.; Larsen, S.E.; Bogestrand, J.; Ovesen, N.B.; Kronvang, B. A distributed modelling system for simulation of monthly runoff and nitrogen sources, loads and sinks for ungauged catchments in Denmark. *J. Environ. Monit.* **2011**, *13*, 2645–2658. [CrossRef] [PubMed]

29. Graham, L.P.; Bergström, S. Water balance modelling in the baltic sea drainage basin—Analysis of meteorological and hydrological approaches. *Meteorol. Atmos. Phys.* **2001**, *77*, 45–60. [CrossRef]

30. Bergström, S.; Graham, L.P. On the scale problem in hydrological modelling. *J. Hydrol.* **1998**, *211*, 253–265. [CrossRef]

31. Arheimer, B.; Dahné, J.; Donnelly, C.; Lindström, G.; Strömqvist, J. Water and nutrient simulations using the hype model for Sweden vs. the Baltic Sea Basin—Influence of input-data quality and scale. *Hydrol. Res.* **2012**, *43*, 315–329. [CrossRef]

32. Lindström, G.; Pers, C.; Rosberg, J.; Strömqvist, J.; Arheimer, B. Development and testing of the hype (hydrological predictions for the environment) water quality model for different spatial scales. *Hydrol. Res.* **2010**, *41*, 295–319. [CrossRef]

33. Arnold, J.G.; Srinivasan, R.; Muttiah, R.S.; Williams, J.R. Large area hydrologic modeling and assessment—Part 1: Model development. *J. Am. Water Resour. Assoc.* **1998**, *34*, 73–89. [CrossRef]

34. Ekstrand, S.; Wallenberg, P.; Djodjic, F. Process based modelling of phosphorus losses from Arable Land. *Ambio* **2010**, *39*, 100–115. [CrossRef] [PubMed]

35. Fohrer, N.; Schmalz, B.; Tavares, F.; Golon, J. Modelling the landscape water balance of mesoscale lowland catchments considering agricultural drainage systems. *Hydrol. Wasserbewirtsch.* **2007**, *51*, 164–169.

36. Francos, A.; Bidoglio, G.; Galbiati, L.; Bouraoui, F.; Elorza, F.J.; Rekolainen, S.; Manni, K.; Granlund, K. Hydrological and water quality modelling in a medium-sized Coastal Basin. *Phys. Chem. Earth Part B Hydrol. Oceans Atmos.* **2001**, *26*, 47–52. [CrossRef]

37. Onuşluel Gül, G.; Rosbjerg, D. Modelling of hydrologic processes and potential response to climate change through the use of a multisite swat. *Water Environ. J.* **2010**, *24*, 21–31. [CrossRef]

38. Lam, Q.D.; Schmalz, B.; Fohrer, N. Modelling point and diffuse source pollution of nitrate in a rural lowland catchment using the swat model. *Agric. Water Manag.* **2010**, *97*, 317–325. [CrossRef]

39. Lu, S.; Kayastha, N.; Thodsen, H.; Van Griensven, A.; Andersen, H.E. Multiobjective calibration for comparing channel sediment routing models in the soil and water assessment tool. *J. Environ. Q.* **2014**, *43*, 110–120. [CrossRef] [PubMed]

40. Lu, S.; Kronvang, B.; Audet, J.; Trolle, D.; Andersen, H.E.; Thodsen, H.; van Griensven, A. Modelling sediment and total phosphorus export from a lowland catchment: Comparing sediment routing methods. *Hydrol. Process.* **2015**, *29*, 280–294. [CrossRef]

41. Thodsen, H.; Andersen, H.E.; Blicher-Mathiesen, G.; Trolle, D. The combined effects of fertilizer reduction on high risk areas and increased fertilization on low risk areas, investigated using the swat model for a Danish catchment. *Acta Agric. Scand. Sect. B Soil Plant Sci.* **2015**, *65*, 217–227. [CrossRef]

42. Marcinkowski, P.; Piniewski, M.; Kardel, I.; Giełczewski, M.; Okruszko, T. Modelling of discharge, nitrate and phosphate loads from the reda catchment to the puck lagoon using swat. *Ann. Wars. Univ. Life Sci. SGGW Land Reclam.* **2013**, *45*, 125. [CrossRef]

43. Santhi, C.; Srinivasan, R.; Arnold, J.G.; Williams, J.R. A modeling approach to evaluate the impacts of water quality management plans implemented in a watershed in Texas. *Environ. Model. Softw.* **2006**, *21*, 1141–1157. [CrossRef]

44. Abbaspour, K.C.; Rouholahnejad, E.; Vaghefi, S.; Srinivasan, R.; Yang, H.; Kløve, B. A continental-scale hydrology and water quality model for europe: Calibration and uncertainty of a high-resolution large-scale swat model. *J. Hydrol.* **2015**, *524*, 733–752. [CrossRef]

45. Trolle, D.; Nielsen, A.; Rolighed, J.; Thodsen, H.; Andersen, H.E.; Karlsson, I.B.; Refsgaard, J.C.; Olesen, J.E.; Bolding, K.; Kronvang, B.; et al. Projecting the future ecological state of lakes in Denmark in a 6 degree warming scenario. *Clim. Res.* **2015**, *64*, 55–72. [CrossRef]

46. Schilling, K.E.; Wolter, C.F. Modeling nitrate-nitrogen load reduction strategies for the des moines river, iowa using swat. *Environ. Manag.* **2009**, *44*, 671–682. [CrossRef] [PubMed]

47. White, M.J.; Storm, D.E.; Busteed, P.R.; Stoodley, S.H.; Phillips, S.J. Evaluating nonpoint source critical source area contributions at the watershed scale. *J. Environ. Q.* **2009**, *38*, 1654–1663. [CrossRef] [PubMed]

48. Hägg, H.; Lyon, S.; Wällstedt, T.; Mörth, C.-M.; Claremar, B.; Humborg, C. Future nutrient load scenarios for the baltic sea due to climate and lifestyle changes. *Ambio* **2014**, *43*, 337–351. [CrossRef] [PubMed]

49. Hagg, H.E.; Humborg, C.; Morth, C.M.; Medina, M.R.; Wulff, F. Scenario analysis on protein consumption and climate change effects on riverine n export to the Baltic Sea. *Environ. Sci. Technol.* **2010**, *44*, 2379–2385. [CrossRef] [PubMed]

50. Nachtergaele, F.H.; van Velthuizen, L.; Verelst, N.; Batjes, K.; Dijkshoorn, V.; van Engelen, G.; Fischer, A.; Jones, L.; Montanarella, M.; Petri, S.; et al. *Harmonized Worlds soil Database v. 1.1*; Food and Agriculture Organization of the united Nations (FAO): Rome, Italy, 2009; p. 38.

51. European Environment Agency. *CLC2006 Technical Guidelines*; EEA Technical report No. 17/2007; European Environment Agency: Copenhagen, Denmark, 2007; p. 70.

52. JRC, MARS50. Available online: http://marswiki.jrc.ec.europa.eu/agri4castwiki/index.php/Meteorological _data_from_ground_stations (accessed on 27 April 2017).

53. Scharling, M. *Klimagrid Danmark*; Technical Report 01-18; Danish Meteorological Institute: Copenhagen, Denmark, 2001; p. 12. (In Danish)

54. Winchell, M.; Srinivasan, R.; Di Luzio, M.; Arnold, J. *ArcSWAT Interface for SWAT2009, Users Guide*; Blackland Research and Extension Center, Texas AgriLife Research: Temple, TX, USA, 2010; p. 495.

55. Abrams, M.; Hook, S.; Ramachandran, B. *ASTER User Handbook, Version 2*; Jet Propulsion Laboratory, California Institute of Technology, 2013. Available online: http://asterweb.jpl.nasa.gov/content/03_data/ 04_Documents/aster_user_guide_v2.pdf (accessed on 27 March 2017).

56. KMS. *Danmarks Højdemodel—DHM Terræn* ; National Survey and Cadastre: Copenhagen, Denmark, 2015; p. 10. (In Danish)

57. Wosten, J.H.M. The hypress database of hydraulic properties of European soils. *Adv. Geoecol.* **2000**, *32*, 135–143.

58. Greve, M.H.; Greve, M.B.; Bocher, P.K.; Balstrom, T.; Breuning-Madsen, H.; Krogh, L. Generating a danish raster-based topsoil property map combining choropleth maps and point information. *Geogr. Tidsskr. Dan. J. Geogr.* **2007**, *107*, 1–12. [CrossRef]

59. Benedictow, A.H.; Berge, H.; Fagerli, M.; Gauss, J.E.; Jonson, Á.; Nyiri, D.; Simpson, S.; Tsyro, Á.; Valdebenito, S.; Valiyaveetil, P.; et al. *Transboundary Acidification, Eutrophication and Ground Level Ozone in Europe in 2008*; EMEP Status Report 2010; Norwegian Meteorological Institute: Oslo, Norway, 2010; p. 126.

60. Jonson, J.E.; Bartnicki, J.; Olendrzynski, K.; Jakobsen, H.A.; Berge, E. Emep eulerian model for atmospheric transport and deposition of nitrogen species over Europe. *Environ. Pollut.* **1998**, *102*, 289–298. [CrossRef]

61. Jha, M.; Gassman, P.W.; Secchi, S.; Gu, R.; Arnold, J. Effect of watershed subdivision on swat flow, sediment, and nutrient predictions. *J. Am. Water Resour. Assoc.* **2004**, *40*, 811–825. [CrossRef]

62. Danish Center for Environment and energy (DCE). Technical Instruction, Mass Transport. Available online: http://bios.au.dk/fileadmin/bioscience/Fagdatacentre/Ferskvand/DB01_stoftransport.pdf (accessed on 27 April 2017).

63. Brandt, M.; Ejhed, H. Transport—Retention—Källfördelning, Belastning på Havet. Naturvårdsverket. Rapport 5247. 2002. Available online: http://www.naturvardsverket.se/Documents/publikationer/620-5247s1_44.pdf?pid=2901 (accessed on 27 April 2017).

64. Helsinki Commission. PLC Water Guidelines. Part of HELCOM Monitoring Guidelines. Available online: http://www.helcom.fi/Documents/Action%20areas/Monitoring%20and%20assessment/Manuals% 20and%20Guidelines/PLC_2loadorientatedapproach.pdf (accessed on 27 April 2017).

65. Blicher-Mathiesen, G.; Rasmussen, A.; Andersen, H.E.; Timmermann, A.; Jensen, P.G.; Wienke, J.; Hansen, B.; Thorling, L. *Landovervågningsoplande 2013: NOVANA*; Scientific Report No. 120; Aarhus Universitet, DCE—Nationalt Center for Miljø og Energi: Silkeborg, Denmark, 2015; p. 125.

66. Voss, M.; Dippner, J.W.; Humborg, C.; Hürdler, J.; Korth, F.; Neumann, T.; Schernewski, G.; Venohr, M. History and scenarios of future development of Baltic Sea eutrophication. *Estuar. Coast. Shelf Sci.* **2011**, *92*, 307–322. [CrossRef]

67. Engel, B.; Storm, D.; White, M.; Arnold, J.; Arabi, M. A hydrologic/water quality model application protocol. *J. Am. Water Resour. Assoc.* **2007**, *43*, 1223–1226. [CrossRef]

68. Arnold, J.G.; Moriasi, D.N.; Gassman, P.W.; Abbaspour, K.C.; White, M.J.; Srinivasan, R.; Santhi, C.; Harmel, R.D.; Van Griensven, A.; Van Liew, M.W.; et al. Swat: Model use, calibration, and validation. *Trans. ASABE* **2012**, *55*, 1491–1508. [CrossRef]

69. Abbaspour, K. *SWAT Calibration and Uncertainty Programs—A User Manual. Department of Systems Analysis*; Integrated Assessment and Modelling (SIAM); Eawag, Swiss Federal Institute of Aquatic Science and Technology: Duebendorf, Switzerland, 2008; p. 95.

70. Abbaspour, K.C.; Vejdani, M.; Haghighat, S. Swat-cup calibration and uncertainty programs for swat. In Proceedings of the Modsim 2007: International Congress on Modelling and Simulation, Christchurch, New Zealand, 2007; pp. 1603–1609.

71. Abbaspour, K.C.; Yang, J.; Maximov, I.; Siber, R.; Bogner, K.; Mieleitner, J.; Zobrist, J.; Srinivasan, R. Modelling hydrology and water quality in the pre-ailpine/alpine thur watershed using swat. *J. Hydrol.* **2007**, *333*, 413–430. [CrossRef]

72. Nash, J.E.; Sutcliffe, J.V. River flow forecasting through conceptual models part I—A discussion of principles. *J. Hydrol.* **1970**, *10*, 282–290. [CrossRef]

73. Molina-Navarro, E.; Andersen, H.E.; Nielsen, A.; Thodsen, H.; Trolle, D. The impact of the objective function in multi-site and multi-variable calibration of the swat model. *Environ. Model. Softw.* **2017**, *93*, 255–267. [CrossRef]

74. Andersen, H.E.; Blicher-Mathiesen, G.; Bechmann, M.; Povilaitis, A.; Iital, A.; Lagzdins, A.; Kyllmar, K. Mitigating diffuse nitrogen losses in the Nordic-Baltic countries. *Agric. Ecosyst. Environ.* **2014**, *195*, 53–60. [CrossRef]

75. Jomaa, S.; Jiang, S.; Thraen, D.; Rode, M. Modelling the effect of different agricultural practices on stream nitrogen load in central Germany. *Energy Sustain. Soc.* **2016**. [CrossRef]

76. Chaibou Begou, J.; Jomaa, S.; Benabdallah, S.; Bazie, P.; Afouda, A.; Rode, M. Multi-site validation of the swat model on the bani catchment: Model performance and predictive uncertainty. *Water* **2016**, *8*, 178. [CrossRef]

77. De Almeida Bressiani, D.; Srinivasan, R.; Jones, C.A.; Mendiondo, E.M. Effects of different spatial and temporal weather data resolutions on the stream flow modeling of a Semi-Arid Basin, Northeast Brazil. *Int. J. Agric. Biol. Eng.* **2015**, *8*, 125.

78. Moriasi, D.N.; Arnold, J.G.; Van Liew, M.W.; Bingner, R.L.; Harmel, R.D.; Veith, T.L. Model evaluation guidelines for systematic quantification of accuracy in watershed simulations. *Trans. ASABE* **2007**, *50*, 885–900. [CrossRef]

79. Grizzetti, B.; Bouraoui, F.; de Marsily, G.; Bidoglio, G. A statistical method for source apportionment of riverine nitrogen loads. *J. Hydrol.* **2005**, *304*, 302–315. [CrossRef]

3

Overview of Organic Cover Crop-Based No-Tillage Technique in Europe: Farmers' Practices and Research Challenges

Laura Vincent-Caboud [1],*, Joséphine Peigné [1], Marion Casagrande [1,2] and Erin M. Silva [3]

[1]   Department of Agroecology and Environment, ISARA-Lyon (member of the University of Lyon), 23 rue Jean Baldassini, F-69364 Lyon Cedex 07, France; jpeigne@isara.fr (J.P.); marion.casagrande@itab.asso.fr (M.C.)

[2]   ITAB, Quartier Marcellas, F-26800 Etoile sur Rhône, France

[3]   Department of Agronomy, University of Wisconsin-Madison, Madison, 53706 WI, USA; emsilva@wisc.edu

*   Correspondence: lavincent-caboud@isara.fr

Academic Editors: Patrick Carr and Les Copeland

Abstract: Cover crop mulch–based no-tillage (MBNT) production is emerging as an innovative alternative production practice in organic farming (OF) to reduce intensive soil tillage. Although European organic farmers are motivated to implement MBNT to improve soil fertility and achieve further management benefits (e.g., labor and costs savings), low MBNT practice is reported in Europe. Thus, this paper aims to understand the challenges of both farmers and researchers limiting the further adoption of MBNT in organic farming in temperate climates. The primary no-tillage (NT) practices of organic European farmers and findings of organic MBNT studies conducted in Europe are reviewed, focusing on living or mulch cover crop-based NT (LBNT or MBNT) for arable crop production. Major conclusions drawn from this review indicate consistent weed control and an establishment of best practices for cover crop management as the two main overarching challenges limiting adoption. In view of substantial gaps of knowledge on these issues, additional research should focus on cover crop selection and management (species, date of sowing) to increase cover crop biomass, particularly in warmer climates. Lastly, further research is needed to optimize cover crop termination to prevent competition for water and nutrients with cash crops, particularly in wetter northern conditions which promote vigorous cover crop growth.

Keywords: no tillage; cover crops; organic farming; roller crimper; conservation tillage

1. Introduction

Conservation agriculture (CA) has been developed in conventional farming to minimize and prevent soil erosion, reduce labor and energy inputs, and preserve soil fertility [1–4]. The Food and Agriculture Organization (FAO) has defined conservation agriculture as a range of practices based on three main principles: (1) minimum soil disturbance obtained using reduced-tillage (RT) or no-tillage (NT); (2) a permanent soil cover via living or dead (mulch) cover crops; and (3) diversified crop rotations [2]. NT involves production practices with which a cash crop is sown into soil that has not been tilled since the previous harvest [5] (Figure 1). As highlighted by several authors, NT may also be referred to as direct seeding or direct drilling, terms that are often employed in European publications [3,6], although some North American researchers have stressed differences in soil disturbance between NT and direct drilling [7]. Typically, with the use of NT practices, the soil surface is only disturbed down to three to five centimeters in depth in the seeding row, with significant amounts of crops or cover crop residues is maintained on the soil surface to preserve soil quality and suppress weeds [8].

```
                          ┌─────────────────────────────┐
                          │       No tillage (NT)       │
                          │                             │
                          │  Cash crop sowing without   │
                          │  soil tillage since the harvest│
                          │  of the previous crop       │
                          └─────────────────────────────┘
```

No tillage (NT)

Cash crop sowing without soil tillage since the harvest of the previous crop

No tillage into crops residues

Cash crops sowing without soil tillage into residues of the previous crop after harvesting

Cover crop–based no tillage (CCNT)

NT and cover crop termination : first cover crop sowing after the harvest of the previous crop, then cash crop sowing into the cover crop (living or dead)

Living cover crop–based no tillage (LBNT)

Cash crops are sown into living cover crop

Cover crop mulch–based no tillage (MBNT)

Cash crops are sown into cover crop previously terminated mechanically or with synthetic herbicide or both in combination

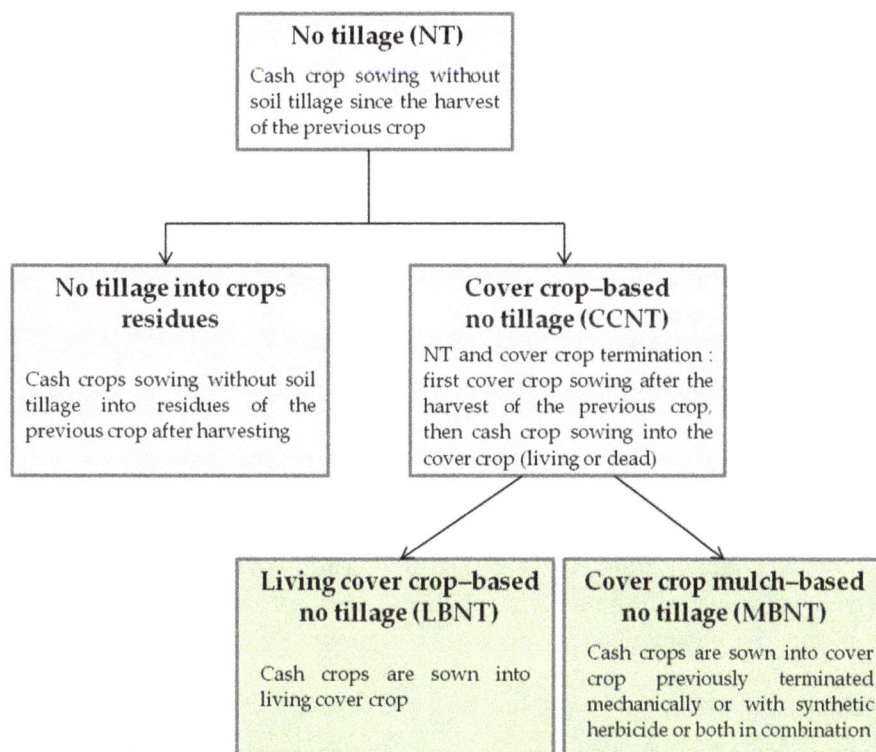

Figure 1. Diagram of different techniques of no-tillage and cover crop management.

The specific integration of cover crops into NT practices is an innovative alternative recently developed to mitigate environmental pollution, such as nitrate leaching, reduce inputs (fertilization, herbicides), and suppress weeds via cover crop competition for nutrient and water resources [9,10] (Figure 1). Cover crop mulch–based NT and living cover crop-based NT have gained increasing attention as methods to further enhance the agroecosystem services of NT systems, particularly in organic agriculture. Depending on the cover crop species used in the system, weed control is achieved through physical suppression preventing access to light, and reducing soil temperatures, competition with nutrient resources, and allelopathic interactions [11]. Cover crop-based NT (CCNT) techniques, i.e., MBNT and LBNT, consist of sowing the main crop into a living cover crop, frost-killed cover crop, or cover crop terminated either mechanically (roller crimper, mower) or with a synthetic herbicide, or both in combination (Figure 1). These strategies, in addition to limiting soil disturbance and suppressing weeds, also provides an opportunity to address another principle of conservation agriculture by improving living soil cover. Although CCNT is employed in conventional agriculture to promote soil fertility, control weeds, increase water infiltration, and preserve soil moisture [12], these systems often integrate the use of herbicides to manage weeds and/or terminate the cover crop, thereby raising environmental sustainability issues [8,13].

Organic farming (OF) has also emerged as an alternative for maintaining soil fertility and reducing the harmful environmental impacts of agriculture. The European standard No. 834/2007 bans the use of chemical synthetics (pesticides and fertilizers) and limits off-farm and synthetic inputs in OF. Current organic practices use crop rotation as a foundational practice to provide nutrient recycling and to control weeds, pests, and diseases [14]. Theoretically, OF offers many advantages in the shift toward sustainable agricultural systems via the increase of soil organic matter and biodiversity. In reality, however, without use of synthetic herbicides, farmers often depend on intensive tillage to manage weeds [15,16]. This can result in soil degradation: decreases in biological activity and biodiversity, increases in soil erosion, losses in organic matter content, and destruction of soil physical structure due to frequent machine traffic and tillage activities on the field [17,18]. Organic NT represents an opportunity to mitigate these drawbacks, improve soil fertility components, and provide further benefits regarding European farmers' working conditions (labor savings, energy consumption reduction, etc.) [13].

Despite the potential benefits of integrating NT practices into organic cropping systems (including, but not limited to, MBNT and LBNT) which have been reported by researchers, and the interest shown by European farmers, NT has not been widely implemented in Europe and few research studies have been conducted investigating this technique [19,20]. The trials that have been carried out in Europe and interviews with farmers using organic conservation tillage including MBNT bring to light the typical practices, motivations, and challenges related to conservation techniques [19,20]. This paper aims to synthesize knowledge and conclusions drawn from previous European research projects to identify the overarching challenges facing research in Europe in the promotion of organic MBNT in arable crops.

Based on a literature review and interviews of European organic farmers, we first explain the methodology selected to address the goal of this paper. Then, we present a broad overview of organic farmers' NT and cover crop practices in Europe and identify their main challenges. Next, we examine, in detail, the knowledge drawn from European research to address the organic farmers' difficulties. Lastly, we discuss the primary future research challenges to improve our understanding of organic MBNT under a temperate climate.

2. Methodology

Several papers have been published in North America on organic MBNT, summarizing the results of research trials based on cover crop species [21–26], cover crop termination methods and timing [24,27–33], weed suppression [34–41], and cash crop sowing (fertilization strategies, row spacing, seeding rate) [24,42,43]. Nevertheless, in Europe, very few studies exist that focus on organic MBNT, raising questions of the effectiveness and appropriateness of this technique to enhance the sustainability of European organic farms [44]. Concerns related to the implementation of organic MBNT in Europe must be articulated to identify future research specifically addressing the implementation of MBNT on European farms. Indeed, despite the robust discussion of the challenges related to MBNT by American researchers [16,35,36,39], these challenges may differ in Europe due to its unique agricultural context (e.g., climate, information or equipment access, organic seed availability, etc.).

In Europe, although a range of studies and review articles have been carried out on NT in conventional farming [8,45,46], these papers did not specifically address MBNT, leaving a dearth of information on MBNT, particularly regarding its unique integration into organic production. In Europe, published papers dealing with tillage issues in OF are focused on RT, and it is difficult to discern from this knowledge about NT [13,15,17,20,44,47–49]. While there are some references referring to organic MBNT in arable crops, data analysis on this technique is often presented in the context of a comparison with traditional ploughing and RT [15,20,50–52]. This research strategy limits the generation of data which could contribute to the refinement of the MBNT technique and the development of best management practices, while diluting access of MBNT information.

In order to address this resource gap, this review paper focuses solely on organic MBNT and LBNT, identifying challenges specific to these NT techniques on organic European farms. The review is organized into four sections: the first two sections summarize MBNT and LBNT practices and challenges on European organic farms, considering separately two distinct climatic regions in Europe: the northern region (>45° N) and the southern region (≤45° N). These sections are informed by two surveys, one conducted across Europe and the second focused on France, which aimed to identify conservation practices of organic farmers focusing on NT, RT, and green manure techniques [19,20,53]. Green manure, also called cover crop, is referred to as any crops implemented with the aim of soil fertility preservation and improvement. While certain aspects of these surveys' data have been published elsewhere, we will employ a unique approach focusing solely on the farmer-reported practices and perceptions regarding NT and cover crop management, and place the survey information against the backdrop of the European literature on conservation agricultural practices in both organic and conventional systems, creating a framework upon which to overview knowledge and challenges specific to the organic MBNT in Europe [13,17,18,54,55]. In particular, review papers focused on the implementation of NT practices within the European biological, physical, and chemical contexts are used to describe the pedo-climatic effects on MBNT implementation in Europe [3,8,46,56]. In the third section, we summarize research focused on organic CCNT (i.e., MBNT and LBNT) in Europe, which further informs the fourth section, in order to identifying remaining gaps in the understanding of MBNT in Europe needed to promote farmer adoption and success with techniques combining NT with intensive cover cropping.

3. European Organic Farmers' No-Tillage and Cover Crop Practices

Existing published literature describing the results of controlled NT research studies, do not necessarily report on the specific attitudes and practices employed by European farmers' using MBNT or LBNT practices. Within the literature that does describe farmer implementation of MBNT practices, farmers report low adoption of organic MBNT and LBNT techniques, in part due to challenges with consistent termination of cover crops without the use of synthetic herbicides [13,47]. A recent European survey carried out in 2012 with 159 organic farmers who are applying conservation principles (i.e., at least one of the following practices: reduced tillage, cover crop, no tillage) (see [20]) revealed that even within this progressive cohort of farmers utilizing sustainable RT and cover crop practices, only 27% of interviewed farmers used NT techniques [20]. This mirrored the results of an earlier 2010 French survey interviewing 24 producers, which also confirmed that very few farmers had experimented MBNT [53].

3.1. Organic No-Tillage and Cover Crop Practices in Europe

Organic NT approaches vary across Europe, largely due to differences in climate, national cultures, and/or information access [20]. Much of Europe experiences a temperate climate, with consistent humid and cool conditions, as compared to other regions of the globe with higher seasonal rainfall and/or drought periods, as encountered in North America [8]. However, warmer and drier climates can be found in Southern Europe, with these types of environments expected to expand with global warming, increasing the risk of soil degradation [8]. With these changes in climatic conditions, interest in organic NT is expected to increase as additional production strategies are sought to reduce soil erosion and preserve soil moisture in the face of increasing extreme weather events. In this section, we will discuss the main practices and challenges in two European areas: the northern region (>45° N), where 109 organic farmers were interviewed in 2012, and the southern region (≤45° N, with 50 interviewed farmers interviewed in 2012). Conservation practices implemented by these farmers may cover a range of techniques [20], but we focus in this paper on NT (including MBNT and LBNT), in tandem with the cover crop information drawn from these surveys.

As reported in the 2012 aforementioned survey of organic farmers using progressive, sustainable practices, in the northern part of Europe, only 23% of the 109 interviewed farmers used NT, while 93% of farmers implemented a cover crop in their crop rotation to increase soil N supply to the subsequent crop [20]. Maëder and Berner [13] have shown that the humid and cool conditions experienced in northern climates increases the difficulty of implementing NT. Researchers have recognized the role of tillage in managing weeds and accelerating soil warming to foster crop emergence, which otherwise could be delayed by large quantities of mulch left on the soil surface (Table 1). Furthermore, humid conditions enable cover crops to be established rapidly, thus influencing northern farmers' focus on cover crop practices rather than NT [20].

Conversely, in the southern part of Europe, 34% of the 50 interviewed farmers use NT, with only 48% implementing a cover crop [20]. NT is more widespread among southern farmers, likely due to soil degradation and soil water loss caused by intensive tillage under warm and dry conditions where soil is more exposed to water and wind erosion [8]. Moreover, higher temperatures accelerate soil warming, promoting crop emergence and contributing to the success of NT (Table 1). However, while implementation NT practices were greater among the southern farmers, this group demonstrated a decreased use of cover crops as compared with northern European farmers. This illustrates the trade-offs regarding sustainable practices which exist in the south; while cover crops are difficult to establish in the water-limited southern climate, this same condition promotes alternative sustainable practices, the adoption of intensive non-inversion tillage strategies to manage weeds, maintain yields, and incorporate fertilizers to limit ammoniac volatilization [20]. Due to these trade-offs, the implementation of multiple practices integrating NT principles, i.e., including LBNT and MBNT, represents a real challenge.

3.2. Cover Crop Management

The majority of the 159 interviewed farmers implemented cover crops in their crop rotation strategies, but only 19% of them combined NT with a cover crop [20], likely due to the limited termination methods for cover crops in OF, which relies exclusively either on a mechanical solution or frost [18]. Cover crop regrowth due to inconsistent or incomplete termination may compete with the cash crop for nutrients and water resources. Detailed management descriptions of the winter cropping practices of 68 farmers applying cover crops showed that most terminated the cover crop by undercutting or mowing, but only 2% of them by rolling [20]. Thus, the cover crop often remained a living cover when combined with MBNT. Moreover, organic farmers recognized the critical trade-offs between the management of the cover and cash crops, including the impacts on (i) managing weeds with the use of a false seed-bed or (ii) incorporating a cover crop to provide nitrogen (N) to the subsequent crops [57]. Such practices do not appear to be aligning cover crop management with MBNT principles. Thus, alternatives are needed to help organic farmers manage weeds and allow for the incorporation of N-rich cover crops without tilling.

3.3. Application in Crop Rotations

NT primarily integrates the use of winter cereal crops, such as wheat (*Triticum vulgare* L.), barley (*Hordeum vulgare* L.), or triticale (*x Triticosecale* spp. L.), seeded into a living cover crop or by direct sowing following the harvest of the previous main crop. Only 2.5% of the 40 interviewed farmers using NT in Europe sowed spring crops (e.g., soybean (*Glycine max* L.) or corn (*Zea mays* L.)) using NT practices [20], with the practice only implemented where conditions are ideal (e.g., low weed pressure, good climate conditions, etc.).

The length of crop rotations in which NT crops are included averaged seven years [20]. Diversified crop rotations are typically used, including cereals, legumes, and other species, such as *Brassicacae*. Among the 40 farmers using NT, crop rotations are mainly based on alternating spring and winter crops to manage weeds, and include an average of 25% spring crops, 35% winter crops, 10% other legume crops, and 30% cover crops used to increase soil N supply to the subsequent crop. Tillage is generally not completely absent from crop rotations, and occasional ploughing is performed at some stages (e.g., before spring sowing, or to incorporate cover crops). Farmers limit the application of NT on specific crops in crop rotation (e.g., wheat, barley, etc.) with occasional ploughing at a mean frequency of 0.6 times over 10 years among the interviewed farmers [20].

3.4. Farmers' Experience in No Tillage Production

The European survey showed that organic farmers lack experience in NT, with only five years of practice compared with an average of 16 years with the use of cover crops [20]. Results indicated that only 10% of 40 organic farmers using NT applied this technique without adopting RT practices, with farmers using NT averaging 12 years of experience in RT and 13 years in cover crops. This latter finding suggests that farmers using NT are not the most experienced in CA, confirming what has also been observed in the French survey [53]. Organic farmers' progress in NT techniques occurs through exploring solutions on a farm-by-farm basis, integrating the unique objectives and constraints of their farm organization.

The situation across Northern Europe contrasts with southern farmers who have slightly greater experience in NT, with an average of seven years of practice, as opposed to five years in northern areas among the 40 interviewed farmers applying NT [20]. These differences in farmers' practices and experiences using NT highlight the need for better understanding of the main challenges unique to their regions and farming systems if relevant solutions are to be provided in a broader European context.

This overview of NT and cover crop practices in Europe reveals the obstacles facing organic farmers to combine both techniques in accordance with the pedo-climatic context (Table 1).

Table 1. Opportunities and challenges for organic MBNT adoption in the European context.

European Regions	Opportunities	Challenges	References
Northern areas (>45° N)	High cover crop biomass potential Large number of farmers with experience of cover crop practices Soil preservation to cope with climate change	Slow soil warming Moisture conditions foster rapid weed development and degradation of cover crop residues Cover crop management/termination difficulties Low NT adoption Fewer experienced farmers in CA	[8,13,19,20]
Southern areas (≤45° N)	Fast soil warming reduces soil erosion and maintains soil moisture in the face of climate change More experienced farmers in NT	Water and nutrient deficiencies for the cash crop (more irrigation required) Less application of cover crops in the crop rotation Cover crop establishment difficulties (under warm and dry conditions) Fewer farmers experienced in cover crop practices	

NT = no-tillage; CA = conservation agriculture.

4. The Challenges of Living or Killed (Mulch) Cover Crop-Based No-Tillage in European Organic Farming

4.1. Weed Management

European researchers indicated insufficient weed suppression and resulting high weed pressures as the main reason for the low adoption of organic NT practices in Europe [48,49,58,59]. For some farmers, CCNT practices (MBNT and LBNT) offer a means of suppressing weeds without mechanical methods; however, maintaining weed suppression throughout the entire cash crop production season from crop sowing through to harvest was cited as a major difficulty raised by farmers (Figure 2) [19]. Lefèvre et al. [53] also stressed weed suppression as a primary concern of interviewed French farmers. Indeed, the possible need to intensify mechanical weed control during a crop rotation including NT due to inadequate weed management in the NT phase may lead to additional issues at the rotational scale (e.g., soil erosion, increase in labor and investment, etc.). This problem is illustrated by current farmer practices where tillage is not entirely removed from the crop rotation [20]; weed management concerns appear more important for northern farmers who are slower to adopt NT practices (Figure 2). In northern environments, humid conditions promote weed development throughout the cash crop production season, including in CCNT if the residue biomass is inadequate to provide season-long weed suppression; with significant cover crop residue on the soil surface, CCNT limits the ability of farmers to employ mechanical methods of weed management.

4.2. Inadequacy of Available NT Equipment and Low Technical Skills and Information

European farmers reported an inability to access information related to the organic NT technique to guide their management decision [19]. More specifically, Peigné et al. [20] found that southern farmers have less access to NT information resources than northern farmers, despite the higher number of NT practitioners observed in South Europe. According to Lefèvre et al. [53], the shortage of technical guidelines leads farmers to explore solutions independently on their own farms. For example, lack of knowledge as to the best practices for seeding cash crops into a flattened cover crop is an overarching challenge which is consistent with the low level of NT experience observed in Europe [20].

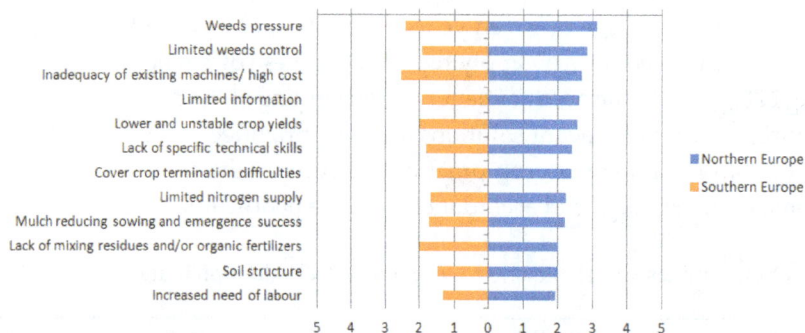

Figure 2.   The challenges of no-tillage for 159 European organic farmers (according to Casagrande et al. [19]). The means of Likert scale values of the challenges facing organic farmers in Southern Europe (**left**) and Northern Europe (**right**) among the forty interviewed farmers practicing NT in Europe in 2012 are shown. The Likert scale is based on five values: (1) not important, (2) low importance, (3) moderate importance, (4) very important, (5) highly important (the number of interviewed farmers is, respectively, 25 and 15 in northern and southern areas).

Additionally, the high cost and the low availability of NT equipment is a limiting factor in the farmers' ability to invest in effective machinery, with the resulting lack of adequate equipment reducing NT success [8]. Indeed, Peigné et al. [18] have demonstrated the importance of specific machines needed for MBNT production (e.g., roller crimper, NT seeder), which remain very scarce on European organic farms with only a few interviewed farmers rolling the cover crop for termination. Equipment must be designed, manufactured, and/or modified to be more adapted to MBNT [15]. European researchers in conventional farming also presented the small quantity of NT equipment manufactured in Europe as a main challenge [8,46]. Above all, according to Derpsch et al. [46] and Freidrich et al. [56], European farmers who want to use specific and efficient machinery in CA must import the equipment from North America. The authors further explain that while efficient equipment exists in Europe, the lack of technical support for farmers could explain some of the failures observed in NT on European farms. Thus, supporting farmers to build and/or modify NT equipment, or invest in the purchase of NT equipment with other farmers, could generate more appropriate machines to specific soil and climate conditions and reduce costs.

4.3. Unstable Crop Yields

Lower yields in NT compared with traditional tillage illustrates the effect of weeds and cover crop competition for water and nutrient resources, which may lead to a lower net return, as evidenced by French farmers [53]. Furthermore, inconsistent crop yields may be exacerbated by climate change, which may be even more profound using NT practices [8]. Under humid and cool conditions, cash crop yields experience greater instability if the cover crop is not successfully managed, leading to insufficient weed control and competition from the cover crop. In southern parts of Europe, the occurrence of summer drought in relation to cash crop planting and establishment may lead to significant yield losses if irrigation is not available, as illustrated by Delate et al. [21] in Iowa. As such, farmers who do implement NT in southern regions tend to use, more frequently, the technique only on winter crops without a cover crop to limit competition with the main crop.

4.4. Cover Crop Termination

MBNT and LBNT are challenging for farmers with respect to issues that arise with cover crop termination without the use of synthetic pesticides, relying instead on mechanical methods or termination of the cover crop at frost [18,19]. Although farmers tend to use frost-sensitive cover crop species (e.g., certain clover species, common vetch (*Vicia sativa*), etc.), a winter-killed cover crop may not allow for adequate maintenance of residue on the soil surface after the winter and through the cash crop growing season, which is essential to ensure weed suppression until crop harvest, with biomass limited by either fall growth of the cover crop or decomposition of residues. Cover crop termination remains an even greater concern for northern farmers, as wet and cool spring conditions can create an environment fostering strong cover crop growth while soil conditions limit field operations, creating difficulties with management (Figure 2). Additionally, the French survey also indicated additional issues with cover crop management, including the high cost of seed purchasing, greater labor needs, and difficulties related to cover crop establishment (unevenly spread, lack of N availability) [53].

4.5. Crop Emergence

In the case of MBNT, technical and mechanical issues related to direct seeding into a flattened cover crop may limit the seed-to-soil contact and contribute to farmers' concerns regarding crop emergence (Figure 2). This is of particular concern for northern farmers, who must grapple with the challenge of planting through thick cover crop biomass [15]. Moreover, Morris et al. [60] and Soane et al. [8] have highlighted that, in conventional farming systems, soil temperatures are lower under large amounts of residue on the soil surface, delaying crop emergence. Additionally, concerns exist related to limited N supply to the subsequent cash crops, with slower mineralization rates resulting from cooler soil temperatures and significant soil carbon additions. The challenge regarding N availability, also noted by Cooper et al. [44], is of even greater importance for northern farmers experiencing lower spring temperatures, thus aggravating the delay in N supply.

4.6. Soil Structure

Improving soil properties is often a primary motivator for farmers interested in adopting CA [8,19,61]. Maintaining soil structure, however, may represent a challenge in NT due to the heavy required equipment (e.g., direct seeder, roller crimper) [19]. In particular, topsoil compaction may begin as quickly as five years after the implementation of NT practices in conventional agriculture and seems strongly dependent on soil type and climate [6,8,62]. Unstable soil with low organic matter content and greater soil moisture may increase soil compaction, possibly leading to the greater emphasis on soil structure issues cited by northern organic farmers [8]. In Finland, Alakukku et al. [63], cited in Soane et al. [8], observed similar yields in NT soil as a ploughing treatment on silty clay, where lower yields were obtained in NT on clay soil under similar conditions. Soane et al. [8] highlighted that soil compaction is also associated with the previous crop harvest and the lack of tillage prior to sowing subsequent crops. Consequently, the short timeframe between cash crop harvest and cover or cash crop planting in NT does not allow for improved soil structural formation through biological activity.

4.7. Increased Labor Requirements

Potential savings in labor is a primary farmer motivation when deciding to adopt NT as documented by European researchers, as significant labor is required for soil tillage [2,53,64]. However, farmers' concerns about increased labor requirements related to the implementation of a cover crop phase may explain why relatively few famers combine NT and cover crops, as perceptions exist that the time required for cover crop management will negatively impact the expected labor-saving benefits of NT.

5. Overview of Living or Killed (Mulch) Cover Crop-Based No-Tillage Organic Research in Europe

A recent meta-analysis on shallow non-inversion tillage in OF showed that impacts of NT management on farming system characteristics are poorly documented in Europe [44]. While a significant body of North American research is concentrated on organic NT, European studies are focused on RT and are mainly carried out in Germany, France, Switzerland, Greece, Italy, Slovakia, and the Netherlands [13,47,54,59,65,66]. Therefore, relatively little knowledge has been provided on organic MBNT or LBNT, despite the recognition by international researchers that this innovation may help to manage weeds and decrease soil disturbance while maintaining biodiversity [10,17]. The meta-analysis presented by Cooper et al. [44] indicates that while promising prospects exist related to the improvement of soil quality, substantial yield losses due to perennial weed pressure and the delay in N supply remain concerns, as evidenced by European organic farmers' challenges outlined in the preceding sections [19].

In Europe, trials including organic NT have been established in Greece, Switzerland, and France (Table 2). These studies have been carried out on NT without the integration of cover crops [50,67], or on CCNT [15,68,69] (Figure 1). In this latter case, most trials have concentrated on using a living cover crop with winter wheat, flax (*Linum usitatissimum* L.), or quinoa (*Chenopodium quinoa* L.) crops [52,68,69]. Only two studies have been conducted on MBNT in France, in 2005 (corn crop) and 2008 (soybean crop) [15]. Mäder and Berner [13] highlighted that additional NT trials have been led in Europe, such as in Germany, but little feedback on scientific knowledge is easily available within the scientific community because of the different national languages in which these trials were reported.

5.1. Living Cover Crop-Based No-Tillage (LBNT) Production

LBNT assessed in Europe is often compared to other tillage techniques mostly relying on traditional tillage (20–25 cm) and RT (5–10 cm) [51]. These studies have documented soil fertility improvement in LBNT with higher macroporosity, more organic matter, and higher soil N content due to benefits provided by the cover crop (biological activity, soil structure, etc.) [50–52,70]. Researchers have also reported the major challenge of identifying the best cover crop species which ensure weed suppression without affecting cash crop yield and quality. For instance, studies conducted in Switzerland on the direct seeding of wheat into a living cover crop demonstrated competition of the cover crop with the main crop [69]. According to Den Hollander et al. [71], clovers, while providing soil fertility benefits, had limited utility for weed control and high competition with the main crop. However, Hiltbrunner et al. [68,69] observed improved weed control with white clover (*Trifolium repens* L.) and submediterranean clovers (*Trifolium lappaceum* L.) as compared to alfalfa (*Medicago sativa* L.), with increased wheat seeding rates potentially improving yield. Across living cover crops trials, overall, researchers have identified white clover as the legume species providing the best compromise between competing with weeds while limiting the competition for light with the wheat crop [69,71,72]. The use of a living cover crop is still debated regarding its potential competition with cash crops and related cash crop yield losses.

Table 2. Published experimental trials, including organic no-tillage techniques in Europe.

Country	Soil Type	Year	Cash Crop	Cover Crop	References
France	Cambisol (silty) Fluvisol (sandy loam)	2003 2005 2008	corn corn soybean	clover (living) alfalfa (mulch *) rye (mulch *)	[15,54,73]
Greece	Tagnic Eutric Cambisol (clay loam)	2007 2008 2010 2011	corn winter wheat flax quinoa	Common vetch (*Vicia sativa* L.) (living), fababean (*Vicia faba* L.) (living)	[50–52,70]
Switzerland	Partial gleyic Cambisol Orthic Luvisol	2002 2004 2005	winter wheat winter wheat winter wheat	birdsfoot trefoil (*Lotus corniculatus* L.), white clover (*Trifolium repens* L.), subclover (*Trifolium subterraneum* ssp.), strong-spined medick (*Medicago truncatula Gaertner*) (living)	[67–69]

* Direct seeding of cash crop into mulch cover crop terminated with roller crimper.

5.2. Cover Crop Mulch-Based No-Tillage (MBNT) Techniques

In France, where MBNT was tested (Figure 1), four tillage treatments were compared: (1) traditional ploughing (30 cm depth); (2) shallow ploughing (18 cm depth); (3) RT (12 cm–15 cm depth); and (4) MBNT [15]. For this latter treatment, cash crops were sown into a cover crop terminated with a 1.7 m wide roller crimper with steel blades welded onto a cylinder. This equipment was built following the design model of Brazilian rollers. In 2005, corn was sown into alfalfa mulch and soybean was planted under rye (*Secale cereale* L.) mulch in 2008. Results showed significantly more earthworms in MBNT with the cover crop biomass left on the soil surface providing food resources which foster biological activity [54]. This finding has been confirmed by several authors in conventional farming who underscore the role of high biological activity in generating resilient organic systems [6,58]. Vian et al. [73] showed higher microbial carbon content, organic carbon activity and N mineralization in MBNT from a 0 to 30 cm soil depth. These findings are consistent with conventional studies where more soil C and N content are observed after longer-term trials, although additional fertilization is often required during the first three years of NT adoption [8,9].

Yield losses were also observed in the two French trials, with 25% soybean yield reduction in MBNT treatments in 2008 compared to ploughing, and as much as 75% corn yield reduction under MBNT in 2005 [15]. Crop failures have been attributed to two main factors: (1) weed infestation and (2) cover crop regrowth competition with main crops.

In 2005, alfalfa mulch was unevenly distributed on the soil surface and did not ensure sufficient weed control. More than 100 weeds/m^2 infested the MBNT plots on average, while only one weed/m^2 was observed after ploughing at the corn emergence stage [15]. In North America, studies also indicate contrasting results in maintaining corn yield due to competition by both weeds and the cover crop [21,74]. Indeed, although the Rodale Institute (Pennsylvania, USA) achieved as much as 10.25 t/ha corn yield in 2007 with a hairy vetch (*Vicia villosa* L.) cover crop [24], substantial yield losses have also been reported in Iowa due to drought conditions [21]. While legume cover crops increase N supply to the subsequent corn crop, legume species are also more difficult to terminate through rolling and experience less effective weed control due to their faster residue degradation rate (low C:N rate) and lower biomass production as compared to cereal species. This lack of cover crop biomass on the soil surface throughout corn production season could explain the failure observed in the French trial [15] with alfalfa, as well as the interest of some North American researchers to use cover crop mixtures composed of both legume and cereal species prior to corn MBNT [75]. Cereal species could reduce the amount of N supplied to corn; however, this hypothesis must be further tested to obtain a better understanding of N dynamics in MBNT production.

In France in 2008, cereal rye demonstrated a greater ability to suppress weeds on soybean MBNT, but did not provide enough biomass to ensure weed control throughout the season [15]. Weed density was more abundant at the end of the cash crop season due to unusually wet conditions. Cereal rye has been the most researched cover crop in North America prior to soybean MBNT and has led to the most promising results regarding the range of benefits conferred (e.g., allelopathic effect, high biomass, early flowering, flexibility of sowing date, kill by rolling, etc.) [26,39,42,76]. Researchers also extolled the benefits of other cereal cover crops (e.g., triticale, barley, etc.) related to their longer persistence on the soil surface (high C:N rate) and greater biomass production than legume species. Above all other factors related to cover crop management, according to Mirsky et al. [38], the cover crop must reach more than 8000 kg/ha to lead to satisfactory weed control [77]. This biomass level was likely not obtained in the European studies.

In France, despite weed infestation, Peigné et al. [15] reported greater soybean populations in MBNT compared with ploughing. The researchers explained that seed predation by pigeons occurring on the trial affected more plants on the ploughed plots because of the lack of mulch cover, which protected the soybean seed and seedlings in the MBNT plots. In North America, Delate et al. [21] and Lefèbvre et al. [78] also tended to observe higher soybean populations in MBNT in the United States and in Canada due to efficient weed control at the early vegetative stage of soybean. In addition, the authors indicated that rotary hoeing on ploughed plots likely decreased soybean stands. Based on the French results, we could expect additional benefits from MBNT to ensure crops and soil protection regarding both the risks of pests and climate (storage, drought periods etc.). Substantial work is still required to obtain good weed suppression until crop harvest and maintain similar yields to those obtained in traditional tillage to fully address farmers' concerns [19].

As underlined by a large number of authors, effective cover crop termination is essential to MBNT success [38]. Nevertheless, French researchers have highlighted the difficulty of killing a cover crop with a roller crimper [15]. Indeed, some plots needed to be rolled more than once and cover crop regrowth occurred throughout the crop season, likely impacting cash crop yield [15]. According to North American researchers comparing various stages of cover crop growth and the related efficacy of termination, cover crops must reach at least the flowering growth stage to be successfully controlled [24,31,38,40]. However, the French trial experienced difficulties in delaying the sowing of the cash crop due to climate conditions which led to the cover crop being terminated prior to flowering, contributing to the failure to control the cover crop [15]. In light of these problems, North American researchers have screened different cover crop species and cultivars to allow for earlier cover crop flowering the following spring.

6. European Research Challenges for the Future

As evidenced by Freidrich et al. [56] in conventional farming, this review of European organic NT trials indicates that projects often compare MBNT or LBNT with other tillage techniques, an experimental framework that does not focus on optimizing MBNT success. Thus, considerable work should be carried out to compare several management systems under various pedo-climatic conditions to find the best strategies leading to the suppression of weeds until crop harvest, yield improvement, and cover crop control.

This paper also indicates the potential soil fertility benefits which could be provided by the further implementation of organic MBNT in Europe [44,54], although there is still a poor understanding of the longer-term effects of MBNT production on soil quality. Furthermore, despite farmers' recognized interest in MBNT, European research has provided very little knowledge to address their concerns (effects on soil structure, NT viability, skills, etc.), thus inhibiting the adoption of MBNT in organic farming. Two primary factors remain poorly researched in Europe, challenging the ability of farmers to maintain crop yields in organic MBNT production: (1) cover crop management (species, cultivars, termination), and (2) reliable, consistent, and effective weed suppression. The following key agronomical issues should be addressed under temperate conditions to eliminate these knowledge gaps:

- **Choice of relevant cover crop species:** In Europe, few cover crop species have been assessed for MBNT production, even though the appropriate choice of cover crop is essential for efficient weed control [38]. Further screening of species and cultivars under temperate climate conditions is needed to identify optimal cover crop/cash crop combinations and support the decision-making process of farmers. Other questions must also be addressed: Can cover crop species that have given promising results in North America (e.g., rye and triticale) lead to similar results on soybean MBNT in Europe? Which cover crop(s) lead to the best trade-off between weed suppression, crop yield and cover crop control?

- **Increasing cover crop biomass:** As indicated by many researchers, adequate cover crop biomass is a key factor to ensure weed control in MBNT; however, doubts have been raised regarding the ability to achieve high levels of cover crop biomass (8000–9000 kg/ha) determined to be optimal by North American researchers in Europe, particularly in southern areas. Thus, the following questions must be addressed: What level of cover crop biomass must be reached to ensure satisfactory weed control? Which management strategies could optimize cover crop biomass and achieve a comparable degree of weed suppression as ploughing or chemical methods? How does cover crop biomass impact weed development and crop yield? What are the weed control implications of earlier cover crop sowing?

- **Improving cover crop termination:** Improving cover crop termination by mechanical methods (e.g., rolling-crimping and mowing) leads to greater success in organic MBNT; however, consistent termination of the cover crop remains challenging using the current practices of European farmers. Cover crop termination remains particularly challenging in Northern Europe, where a humid climate fosters vigorous cover crop development. While considerable efforts must be made to develop appropriate machinery adapted to work with the high biomass remaining on the soil surface, further refinement is needed regarding the optimal cover crop termination practices (number of rollings, date of rolling, etc.), with the aim of improving mulch distribution, prolonging its persistence on the ground until crop harvest, and enhancing the quality of crop sowing (seeds/soil contact). In particular, additional studies are required to quantify the effect of cover crop rolling on the mulch degradation rate until crop harvest under a temperate climate and the effectiveness of the roller crimper on legume species termination.

- **Designing crop rotation:** Integrating MBNT into organic crop rotations in accordance with organic farmers' challenges (e.g., weed management at rotational scale, maintaining crop yields, etc.) while incorporating both production and economic limitations remains a major issue.

Acknowledgments: We acknowledge Paul Mäder (FiBL), F.Xavier Sans (UB), José Manuel Blanco-Moreno (UB), Daniele Antichi (CiRAA), Paolo Bàrberi (SSSA), Annelies Beeckman (INAGRO, department of organic crop production), Federica Bigongiali (SSSA), Julia Cooper (UNEW, NEFG), Hansueli Dierauer (FiBL), Kate Gascoyne (UNEW, NEFG), Meike Grosse (University of Kassel), Juergen Heß (University of Kassel), Andreas Kranzler (FiBL), Anne Luik (EULS), Elen Peetsmann (EULS), Andreas Surböck (FiBL), and Koen Willekens (ILVO), who carried out the European survey within the frame of TILMAN-ORG (www.tilman-org.net) supported by CORE Organic.

Author Contributions: L. Vincent-Caboud reviewed the literature and wrote the initial draft of the paper with assistance from J. Peigné. L. Vincent-Caboud, J. Peigné, M. Casagrande, and E. M. Silva contributed to revising the manuscript.

Conflicts of Interest: The authors declare no conflict of interest.

References

1. Altieri, M.A.; Lana, M.A.; Bittencourt, H.V.; Kieling, A.S.; Comin, J.J.; Lovato, P.E. Enhancing crop productivity via weed suppression in organic no-till cropping systems in santa catarina, Brazil. *J. Sustain. Agric.* **2011**, *35*, 855–869. [CrossRef]
2. Hobbs, P.R.; Sayre, K.; Gupta, R. The role of conservation agriculture in sustainable agriculture. *Philos. Trans. R. Soc. B Biol. Sci.* **2008**, *363*, 543–555. [CrossRef] [PubMed]
3. Kassam, A.; Friedrich, T.; Shaxson, F.; Pretty, J. The spread of conservation agriculture: Justification, sustainability and uptake. *Int. J. Agric. Sustain.* **2009**, *7*, 292–320. [CrossRef]
4. Masutti, C. Action publique et expertise dans la conservation des ressources agricoles aux états-unis dans les années 1930. *Ruralia* **2007**, *25*, 1–25.
5. Horowitz, J. *No-Till Farming is a Growing Practice*; DIANE Publishing: Collingdale, PA, USA, 2011; p. 28.
6. Cannell, R.Q.; Hawes, J.D. Trends in tillage practices in relation to sustainable crop production with special reference to temperate climates. *Soil Tillage Res.* **1994**, *30*, 245–282. [CrossRef]
7. Gohlke, T.; Ingersoll, T.; Roe, R.D.; Oregon, N.; Pullman, W.N. Soil disturbance in no-till and direct seed planting systems. *Nat. Res. Conserv. Serv. Agron. Tech. Note* **2000**, *39*, 1–6.
8. Soane, B.D.; Ball, B.C.; Arvidsson, J.; Basch, G.; Moreno, F.; Roger-Estrade, J. No-till in northern, western and south-western Europe: A review of problems and opportunities for crop production and the environment. *Soil Tillage Res.* **2012**, *118*, 66–87. [CrossRef]
9. Teasdale, J.R.; Coffman, C.B.; Mangum, R.W. Potential long-term benefits of no-tillage and organic cropping systems for grain production and soil improvement. *Agron. J.* **2007**, *99*, 1297–1305. [CrossRef]
10. Triplett, G.B.; Dick, W.A. No-tillage crop production: A revolution in agriculture! *Agron. J.* **2008**, *100*, 153–165. [CrossRef]

11. Favarato, L.; Galvão, J.; Souza, J.; Guarçoni, R.; Souza, C.; Cunha, D. Population density and weed infestation in organic no-tillage corn cropping system under different soil covers. *Planta Daninha* **2014**, *32*, 739–746. [CrossRef]

12. Carrera, L.M.; Abdul-Baki, A.A.; Teasdale, J.R. Cover crop management and weed suppression in no-tillage sweet corn production. *HortScience* **2004**, *39*, 1262–1266.

13. Mäder, P.; Berner, A. Development of reduced tillage systems in organic farming in Europe. *Renew. Agric. Food Syst.* **2012**, *27*, 7–11. [CrossRef]

14. Watson, C.A.; Atkinson, D.; Gosling, P.; Jackson, L.R.; Rayns, F.W. Managing soil fertility in organic farming systems. *Soil Use Manag.* **2002**, *18*, 239–247. [CrossRef]

15. Peigné, J.; Lefevre, V.; Vian, J.F.; Fleury, P. Conservation agriculture in organic farming: Experiences, challenges and opportunities in europe. In *Conservation Agriculture*; Farooq, M., Siddique, K.H.M., Eds.; Springer International Publishing: Cham, Switzerland, 2015; pp. 559–578.

16. Shirtliffe, S.J.; Johnson, E.N. Progress towards no-till organic weed control in western Canada. *Renew. Agric. Food Syst.* **2012**, *27*, 60–67. [CrossRef]

17. Gadermaier, F.; Berner, A.; Fließbach, A.; Friedel, J.K.; Mäder, P. Impact of reduced tillage on soil organic carbon and nutrient budgets under organic farming. *Renew. Agric. Food Syst.* **2012**, *27*, 68–80. [CrossRef]

18. Peigné, J.; Ball, B.C.; Roger-Estrade, J.; David, C. Is conservation tillage suitable for organic farming? A review. *Soil Use Manag.* **2007**, *23*, 129–144. [CrossRef]

19. Casagrande, M.; Peigné, J.; Payet, V.; Mäder, P.; Sans, F.X.; Blanco-Moreno, J.M.; Antichi, D.; Bàrberi, P.; Beeckman, A.; Bigongiali, F.; et al. Organic farmers' motivations and challenges for adopting conservation agriculture in Europe. *Org. Agr.* **2015**. [CrossRef]

20. Peigné, J.; Casagrande, M.; Payet, V.; David, C.; Sans, F.X.; Blanco-Moreno, J.M.; Cooper, J.; Gascoyne, K.; Antichi, D.; Bàrberi, P.; et al. How organic farmers practice conservation agriculture in europe. *Renew. Agric. Food Syst.* **2015**, *31*, 72–85. [CrossRef]

21. Delate, K.; Cwach, D.; Chase, C. Organic no-tillage system effects on soybean, corn and irrigated tomato production and economic performance in Iowa, USA. *Renew. Agric. Food Syst.* **2012**, *27*, 49–59. [CrossRef]

22. Halde, C.; Entz, M.H. Plant species and mulch application rate affected decomposition of cover crop mulches used in organic rotational no-till systems. *Can. J. Plant Sci.* **2016**, *96*, 59–71. [CrossRef]

23. Halde, C.; Gulden, R.H.; Entz, M.H. Selecting cover crop mulches for organic rotational no-till systems in Manitoba, Canada. *Agron. J.* **2014**, *106*, 1193. [CrossRef]

24. Moyer, J. *Organic No-Till Farming: Advancing No-Till Agriculture: Crops, Soil, Equipment*; Acres U.S.A.: Austin, TX, USA, 2011; p. 204.

25. Parr, M.; Grossman, J.M.; Reberg-Horton, S.C.; Brintin, C.; Crozier, C. Nitrogen delivery from legume cover crops in no-till organic corn production. *Crops Soil Mag. Am. Soc. Agron.* **2012**. [CrossRef]

26. Silva, E.M. Screening five fall-sown cover crops for use in organic no-till crop production in the upper midwest. *Agroecol. Sustain. Food Syst.* **2014**, *38*, 748–763. [CrossRef]

27. Davis, A.S. Cover-crop roller-crimper contributes to weed management in no-till soybean. *Weed Sci.* **2010**, *58*, 300–309. [CrossRef]

28. Kornecki, T.S.; Arriaga, F.J.; Price, A.J. Roller type and operating speed effects on rye termination rates, soil moisture, and yield of sweet corn in a no-till system. *HortScience* **2012**, *47*, 217–223.

29. Kornecki, T.S.; Price, A.J. Effects of different roller/crimper designs and rolling speed on rye cover crop termination and seed cotton yield in a no-till system. *J. Cotton Sci.* **2010**, *14*, 212–220.

30. Mirsky, S.B.; Curran, W.S.; Mortenseny, D.M.; Ryany, M.R.; Shumway, D.L. Timing of cover-crop management effects on weed suppression in no-till planted soybean using a roller-crimper. *Weed Sci.* **2011**, *59*, 380–389. [CrossRef]

31. Mischler, R.; Duiker, S.W.; Curran, W.S.; Wilson, D. Hairy vetch management for no-till organic corn production. *Agron. J.* **2010**, *102*, 355–362. [CrossRef]

32. Parr, M.; Grossman, J.M.; Reberg-Horton, S.C.; Brinton, C.; Crozier, C. Roller-crimper termination for legume cover crops in north Carolina: Impacts on nutrient availability to a succeeding corn crop. *Commun. Soil Sci. Plant Anal.* **2014**, *45*, 1106–1119. [CrossRef]

33. Vaisman, I.; Entz, M.H.; Flaten, D.N.; Gulden, R.H. Blade roller–green manure interactions on nitrogen dynamics, weeds, and organic wheat. *Agron. J.* **2011**, *103*, 879–889. [CrossRef]

34. Bernstein, E.R.; Stoltenberg, D.E.; Posner, J.L.; Hedtcke, J.L. Weed community dynamics and suppression in tilled and no-tillage transitional organic winter rye–soybean systems. *Weed Sci.* **2014**, *62*, 125–137. [CrossRef]

35. Carr, P.; Gramig, G.; Liebig, M. Impacts of organic zero tillage systems on crops, weeds, and soil quality. *Sustainability* **2013**, *5*, 3172–3201. [CrossRef]

36. Carr, P.M.; Anderson, R.L.; Lawley, Y.E.; Miller, P.R.; Zwinger, S.F. Organic zero-till in the northern US great plains region: Opportunities and obstacles. *Renew. Agric. Food Syst.* **2012**, *27*, 12–20. [CrossRef]

37. Luna, J.M.; Mitchell, J.P.; Shrestha, A. Conservation tillage for organic agriculture: Evolution toward hybrid systems in the western USA. *Renew. Agric. Food Syst.* **2012**, *27*, 21–30. [CrossRef]

38. Mirsky, S.B.; Ryan, M.R.; Curran, W.S.; Teasdale, J.R.; Maul, J.; Spargo, J.T.; Moyer, J.; Grantham, A.M.; Weber, D.; Way, T.R.; et al. Conservation tillage issues: Cover crop-based organic rotational no-till grain production in the mid-atlantic region, USA. *Renew. Agric. Food Syst.* **2012**, *27*, 31–40. [CrossRef]

39. Mirsky, S.B.; Ryan, M.R.; Teasdale, J.R.; Curran, W.S.; Reberg-Horton, C.S.; Spargo, J.T.; Wells, M.S.; Keene, C.L.; Moyer, J.W. Overcoming weed management challenges in cover crop–based organic rotational no-till soybean production in the eastern United States. *Weed Technol.* **2013**, *27*, 193–203. [CrossRef]

40. Wells, M.S.; Brinton, C.M.; Reberg-Horton, S.C. Weed suppression and soybean yield in a no-till cover-crop mulched system as influenced by six rye cultivars. *Renew. Agric. Food Syst.* **2015**, *31*, 1–12. [CrossRef]

41. Wells, M.S.; Reberg-Horton, S.C.; Mirsky, S.B. Cultural strategies for managing weeds and soil moisture in cover crop based no-till soybean production. *Weed Sci.* **2014**, *62*, 501–511. [CrossRef]

42. Bernstein, E.R.; Posner, J.L.; Stoltenberg, D.E.; Hedtcke, J.L. Organically managed no-tillage rye–soybean systems: Agronomic, economic, and environmental assessment. *Agron. J.* **2011**, *103*, 1169. [CrossRef]

43. Ryan, M.R.; Curran, W.S.; Grantham, A.M.; Hunsberger, L.K.; Mirsky, S.B.; Mortensen, D.A.; Nord, E.A.; Wilson, D.O. Effects of seeding rate and poultry litter on weed suppression from a rolled cereal rye cover crop. *Weed Sci.* **2011**, *59*, 438–444. [CrossRef]

44. Cooper, J.; Baranski, M.; Stewart, G.; Nobel-de Lange, M.; Bàrberi, P.; Fließbach, A.; Peigné, J.; Berner, A.; Brock, C.; Casagrande, M.; et al. Shallow non-inversion tillage in organic farming maintains crop yields and increases soil C stocks: A meta-analysis. *Agron. Sustain. Dev.* **2016**. [CrossRef]

45. Baker, C.J.; Saxton, K.E. *No-Tillage Seeding in Conservation Agriculture*; CABI: Oxfordshire, UK, 2007; p. 340.

46. Derpsch, R.; Friedrich, T.; Kassam, A.; Li, H. Current status of adoption of no-till farming in the world and some of its main benefits. *Int. J. Agric. Biol. Eng.* **2010**, *3*, 1–25.

47. Armengot, L.; Berner, A.; Blanco-Moreno, J.M.; Mäder, P.; Sans, F.X. Long-term feasibility of reduced tillage in organic farming. *Agron. Sustain. Dev.* **2014**, *35*, 339–346. [CrossRef]

48. Emmerling, C. Reduced and conservation tillage effects on soil ecological properties in an organic farming system. *Biol. Agric. Hortic.* **2007**, *24*, 363–377. [CrossRef]

49. Krauss, M.; Berner, A.; Burger, D.; Wiemken, A.; Niggli, U.; Mäder, P. Reduced tillage in temperate organic farming: Implications for crop management and forage production. *Soil Use Manag.* **2010**, *26*, 12–20. [CrossRef]

50. Bilalis, D.; Kakabouki, I.; Karkanis, A.; Travlos, I.; Triantafyllidis, V.; Dimitra, H. Seed and saponin production of organic quinoa (*Chenopodium quinoa* Willd.) for different tillage and fertilization. *Not. Bot. Horti Agrobot. Cluj-Napoca* **2012**, *40*, 42.

51. Bilalis, D.; Karkanis, A.; Pantelia, A.; Patsiali, S.; Konstantas, A.; Efthimiadou, A. Weed populations are affected by tillage systems and fertilization practices in organic flax (*Linum usitatissimum* L.) crop. *Aust. J. Crop Sci.* **2012**, *6*, 157.

52. Bilalis, D.; Karkanis, A.; Patsiali, S.; Agriogianni, M.; Konstantas, A.; Triantafyllidis, V. Performance of wheat varieties (*Triticum Aestivum* L.) under conservation tillage practices in organic agriculture. *Not. Bot. Horti Agrobot. Cluj-Napoca* **2011**, *39*, 28.

53. Lefèvre, V.; Capitaine, M.; Peigné, J.; Roger-Estrade, J. Soil Conservation Practices in Organic Farming: Overview of French Farmers' Experiences and Contribution to Future Cropping Systems Design. Available online: http://ifsa.boku.ac.at/cms/fileadmin/Proceeding2012/IFSA2012_WS6.3_Lefevre.pdf (accessed on 11 January 2017).

54. Peigné, J.; Cannavaciuolo, M.; Gautronneau, Y.; Aveline, A.; Giteau, J.L.; Cluzeau, D. Earthworm populations under different tillage systems in organic farming. *Soil Tillage Res.* **2009**, *104*, 207–214. [CrossRef]

55. Pelosi, C.; Bertrand, M.; Roger-Estrade, J. Earthworm community in conventional, organic and direct seeding with living mulch cropping systems. *Agron. Sustain. Dev.* **2009**, *29*, 287–295. [CrossRef]

56. Friedrich, T.; Kassam, A.; Corsi, S.; Jat, R.A.; Sahrawat, K.L.; Kassam, A.H. Conservation agriculture in europe. In *Conservation Agriculture: Global Prospects and Challenges*; CABI: Oxfordshire, UK, 2014; pp. 127–170.

57. Bàrberi, P. Weed management in organic agriculture: Are we addressing the right issues? *Weed Res.* **2002**, *42*, 177–193. [CrossRef]

58. Emmerling, C. Response of earthworm communities to different types of soil tillage. *Appl. Soil Ecol.* **2001**, *17*, 91–96. [CrossRef]

59. Sans, F.X.; Berner, A.; Armengot, L.; Mäder, P. Tillage effects on weed communities in an organic winter wheat–sunflower–spelt cropping sequence. *Weed Res.* **2011**, *51*, 413–421. [CrossRef]

60. Morris, N.L.; Miller, P.C.H.; Orson, J.H.; Froud-Williams, R.J. The adoption of non-inversion tillage systems in the United Kingdom and the agronomic impact on soil, crops and the environment—A review. *Soil Tillage Res.* **2010**, *108*, 1–15. [CrossRef]

61. Ball, B.C.; Cheshire, M.V.; Robertson, E.A.G.; Hunter, E.A. Carbohydrate composition in relation to structural stability, compactibility and plasticity of two soils in a long-term experiment. *Soil Tillage Res.* **1996**, *39*, 143–160. [CrossRef]

62. Munkholm, L.J.; Schjønning, P.; Rasmussen, K.J.; Tanderup, K. Spatial and temporal effects of direct drilling on soil structure in the seedling environment. *Soil Tillage Res.* **2003**, *71*, 163–173. [CrossRef]

63. Alakukku, L.; Uusitalo, R.; Särkelä, A.; Lahti, K.; Valkama, P.; Valpasvuo-Jaatinen, P.; Ventelä, A.-M. Phosphorus Stratification in the Ap Horizon of Ploughed and No-Till Soils and Its Effect on P Forms in Surface Runoff. In Proceedings of the ISTRO 18th Triennial Conference Sustainable Agriculture, Izmir, Turkey, 15–19 June 2009; International Soil Tillage Research Organisation: Izmir, Turkey, 2009.

64. Canali, S.; Diacono, M.; Campanelli, G.; Montemurro, F. Organic no-till with roller crimpers: Agro-ecosystem services and applications in organic mediterranean vegetable productions. *Sustain. Agric. Res.* **2015**, *4*, 70. [CrossRef]

65. Berner, A.; Hildermann, I.; Fliesbach, A.; Pfiffner, L.; Niggli, U.; Mader, P. Crop yield and soil fertility response to reduced tillage under organic management. *Soil Tillage Res.* **2008**, *101*, 89–96. [CrossRef]

66. Lehocká, Z.; Klimeková, M.; Bieliková, M.; Mendel, L. The effect of different tillage systems under organic management on soil quality indicators. *Agron. Res.* **2009**, *7*, 369–373.

67. Hiltbrunner, J.; Liedgens, M.; Stamp, P.; Streit, B. Effects of row spacing and liquid manure on directly drilled winter wheat in organic farming. *Eur. J. Agron.* **2005**, *22*, 441–447. [CrossRef]

68. Hiltbrunner, J.; Liedgens, M.; Bloch, L.; Stamp, P.; Streit, B. Legume cover crops as living mulches for winter wheat: Components of biomass and the control of weeds. *Eur. J. Agron.* **2007**, *26*, 21–29. [CrossRef]

69. Hiltbrunner, J.; Streit, B.; Liedgens, M. Are seeding densities an opportunity to increase grain yield of winter wheat in a living mulch of white clover? *Field Crops Res.* **2007**, *102*, 163–171. [CrossRef]

70. Bilalis, D.J.; Karamanos, A.J. Organic maize growth and mycorrhizal root colonization response to tillage and organic fertilization. *J. Sustain. Agric.* **2010**, *34*, 836–849. [CrossRef]

71. Den Hollander, N.G.; Bastiaans, L.; Kropff, M.J. Clover as a cover crop for weed suppression in an intercropping design: I. Characteristics of several clover species. *Eur. J. Agrono.* **2007**, *26*, 92–103. [CrossRef]

72. Hollander, N.G.D. Growth characteristics of several clover species and their suitability for weed suppression in a mixed cropping design. Ph.D. Thesis, Wageningen University, Wageningen, The Netherlands, 2012.

73. Vian, J.F.; Peigne, J.; Chaussod, R.; Roger-Estrade, J. Effects of four tillage systems on soil structure and soil microbial biomass in organic farming. *Soil Use Manag.* **2009**, *25*, 1–10. [CrossRef]

74. Delate, K.; Cwach, D.; Fiscus, M. *Evaluation of an Organic No–Till System for Organic Corn and Soybean Production–Agronomy Farm Trial, 2011*; Organic Ag Program Webpage, Iowa State University: Ames, IA, USA, 2012.

75. Parr, M.; Grossman, J.M.; Reberg-Horton, S.C.; Brinton, C.; Crozier, C. Nitrogen delivery from legume cover crops in no-till organic corn production. *Agron. J.* **2011**, *103*, 1578. [CrossRef]

76. Mirsky, S.B.; Curran, W.S.; Mortensen, D.A.; Ryan, M.R.; Shumway, D.L. Control of cereal rye with a roller/crimper as influenced by cover crop phenology. *Agron. J.* **2009**, *101*, 1589. [CrossRef]

77.	Smith, A.N.; Reberg-Horton, S.C.; Place, G.T.; Meijer, A.D.; Arellano, C.; Mueller, J.P. Rolled rye mulch for weed suppression in organic no-tillage soybeans. *Weed Sci.* **2011**, *59*, 224–231. [CrossRef]

78.	Lefebvre, M.; Leblanc, M.; Gilbert, P.-A.; Estevez, B.; Grenier, M.; Belzile, L. *Semis Direct Sur Paillis De Seigle Roulé En Régie Biologique*; Institut De Recherche Et De Développement En Agroenvironnement: Québec, QC, Canada, 2011; p. 36.

4

A Decade of Progress in Organic Cover Crop-Based Reduced Tillage Practices in the Upper Midwestern USA

Erin M. Silva [1],* and Kathleen Delate [2]

[1] Department of Plant Pathology, University of Wisconsin-Madison, 1630 Linden Dr., Madison, WI 53706, USA
[2] Departments of Agronomy and Horticulture, Iowa State University, 106 Horticulture Hall, Ames, IA 50011, USA; kdelate@iastate.edu
* Correspondence: emsilva@wisc.edu

Academic Editor: Patrick Carr

Abstract: The organic industry continues to expand in the United States (U.S.), with 14,093 organic farms in 2014. The upper Midwestern U.S. has emerged as a hub for organic row crop production; however, the management of these organic row crop hectares heavily relies on tillage and cultivation for weed control. Faced with the soil quality challenges related to these practices, and cognizant of the benefits of conventional no-till practices, organic farmers have shown significant interest in the development of Cover Crop-Based Reduced Tillage (CCBRT) techniques to lessen soil disturbance while achieving successful weed management. To serve this farmer interest, significant research efforts have been conducted in the upper Midwestern U.S., focused on systems-based practices to ensure adequate suppression of weeds, through a combination of agronomic and cover crop species and variety selection. Within this review article, we discuss the agronomic successes that have been achieved in CCBRT using a combination of cereal rye and soybeans, resulting in consistent suppression of weeds while providing fuel and labor savings for farmers, as well as the continued challenges that have persisted with its implementation. Continued investment in research focused on cover crop breeding and management, optimization of CCBRT equipment and fertility management, and a greater understanding of rotation effects will contribute to the further expansion of this technique across organic farms.

Keywords: organic agriculture; cover cropping; reduced tillage; ecosystem services; USA

1. Introduction

The organic industry continues to expand in the United States (U.S.), with 14,093 certified organic farms in 2014 [1–3]. The 2014 National Agricultural Survey of organic production, conducted by the United States Department of Agriculture (USDA), reported 82,328 hectares of organic corn and 39,996 hectares of organic soybean among the 1.4 million organic cropland and vegetable hectares in the U.S. [1].

Management of these organic row crops heavily relies on tillage and cultivation for weed control. While mechanical practices can be effective to manage weeds, these activities prevent organic farmers from fully optimizing the requirement for soil building, as set forth in 7 CFR §205.203 and 205.205 of the National Organic Program (NOP) [4]. In typical organic row crop production in the upper Midwestern U.S., five to six tractor passes with tine weeders, rotary hoes, and/or row cultivators are often necessary for adequate weed control, which can negatively impact soil aggregation and soil organic matter concentrations, while exposing the land to greater risk of erosion. Additionally, the reliance on cultivation as a primary weed management tool poses risks during wet springs, an increasingly common production challenge with heavy rainfalls occurring more frequently as a result of climate change [5]; consistently wet soil conditions can prevent organic producers from implementing timely weed management through cultivation, increasing weed competition and weed seedbanks while negatively impacting yields. Ongoing data from the Wisconsin Integrated Cropping Systems Trial, a long-term management trial begun in 1989, illustrates that while organic management results in comparable yields to conventional management during more moderate production seasons, years exhibiting wet spring conditions result in yields of corn and soybeans falling to approximately 75% of those produced using conventional practices, due to the inability of farmers to conduct timely weed management [6].

In addition to weed management challenges associated with tillage and cultivation, these practices can also increase the risk for soil erosion. Several Midwestern regions of the U.S. which support high concentrations of organic farms, including the Driftless regions of Wisconsin, Iowa, and Minnesota, are characterized by farm fields exceeding 4% or greater slopes. The soils in these regions vary widely, from heavy, poorly drained clay soils to sandy, shallow, droughty soils. These conditions create a landscape susceptible to erosion, negatively impacting both soil and water quality of several key watersheds, including the Mississippi Valley Watershed.

Conventional no-till farming techniques have been promoted for their role in reducing water runoff and soil erosion, as well as maintaining soil carbon [7,8], although the degree and nature to which the conventional no-till systems build soil C, and the degree to which this C is stably stored, can vary [9,10]. No-till systems can also increase infiltration of water into the soil by 25 to 50% compared with conventional tillage systems [11]. In addition, cover crop surface residues can decrease the effect of wind and temperature on soil water evaporation, increase water storage in the soil profile [12,13], scavenge available nitrogen, and prevent soil erosion [14], thus preventing watershed contamination and nutrient losses [14,15].

Faced with the soil quality challenges associated with tillage and cultivation, and cognizant of the benefits of conventional no-till practices, organic farmers have shown significant interest in the development of no- and reduced-tillage techniques suitable for organic production. Within the majority of the reduced tillage systems currently utilized by organic farmers, no-till phases are incorporated throughout the rotation, with tillage limited to establishing the cover crops [16–18]. This technique, often referred to as Cover Crop-Based Reduced Tillage (CCBRT) uses mature cover crop residue as a mulch to smother weeds, replacing the standard organic weed management tactics of tillage and cultivation. Winter annual cover crops (typically cereal rye (*Secale cereale* L.) and hairy vetch (*Vicia villosa* L.)) are seeded in the fall and terminated mechanically without herbicides in the spring, often using a roller-crimper (Figure 1a) which creates an in situ surface mulch that physically suppresses weeds. At the time of cover crop termination, the cash crop (typically soybean (*Glyine max* (L.) Merr.) or field corn (*Zea mays* L.)) is planted directly into the cover crop mulch (Figure 1b), which provides season-long weed suppression without further soil disturbance throughout cash crop production.

(a)   (b)

Figure 1. (a) Photograph of the roller-crimper design commonly used by farmers and researchers in the upper Midwestern U.S.; (b) Photograph of soybeans emerging through the rolled winter cereal rye mulch.

2. Organic Cover Crop-Based Reduced Tillage in the Upper Midwestern U.S.

The upper Midwestern U.S. maintains a long history in organic farming, and still remains a primary center of the U.S. organic industry [1]. Prior to the rise of organic agriculture, regional environmental leaders, such as Aldo Leopold, an ardent environmentalist, and Gaylord Nelson, a Wisconsin politician, inspired a strong land ethic in upper Midwestern culture. Counter to the argument that the maturation of the organic industry will, by default, lead to a concurrent conventionalization of the industry, the production practices used by upper Midwestern organic farmers demonstrate adherence to the soil-building ethic that serves as a foundation of the organic regulation outlined by the NOP [19].

CCBRT research began in the early 2000s in the upper Midwest, with several research programs, including Iowa, Wisconsin, and Michigan, establishing experimental plots both at land-grant university research stations and on working certified organic farms. Research approaches have predominantly centered on techniques and equipment made popular by the Rodale Institute in Kutztown, PA, where a fall-planted cover crop is mechanically terminated in the spring (cereal grains at anthesis at Zadok's growth stage 60, or legumes at 100% bloom). Cover crop termination is achieved with a roller-crimper, a hollow steel cylinder with metal slats arranged in a chevron pattern, welded at uniform spacing along the length of the drum that can be filled with water for added weight [20–22] (Figure 1a). However, research programs have also included cover crop termination treatments that utilize sickle-bar mowing and flail mowing [20,22].

Much of the CCBRT research in the upper Midwestern U.S. has focused on systems-based practices to ensure effective suppression of weeds, primarily through a combination of agronomic practices and cover crop management. Corn and soybean have been the primary commodity crops that have been evaluated within the CCBRT system in this region. However, studies have also addressed other critical aspects of the system, including cash crop yields, economic assessments regarding fuel and labor savings, and reduction in erosion risk.

As research data is generated and practical experiences are reported by organic farmers, CCBRT has also been not only gaining recognition, but is being implemented across the organic agricultural landscape. According to the 2014 USDA Organic Survey, organic farmers in the upper Midwest are increasingly adopting organic no-till management strategies as part of their farming practices. Of the 3319 certified organic farms in of Iowa, Illinois, Michigan, Minnesota, Missouri, Nebraska, and Wisconsin, 957 reported the use of no-till practices [1]. This survey, however, does not provide adequate resolution to determine whether CCBRT practices in row crops are being used, or whether other no-till practices are being reported, such as reseeding grasses and legumes into existing pastures without tillage. On a state level, the reported use of no-till practices ranged from 20 to 37% of organic farms (Iowa: 27% (166 no-till farms, 612 total organic farms); Illinois: 27% (69, 249); Michigan: 20% (67, 332); Minnesota; 37% (189, 512); Missouri: 28% (61, 216); Nebraska: 32% (54, 170); Wisconsin: 29% (351, 1228)) [1].

The remainder of this paper summarizes experiences with CCBRT in the upper Midwestern U.S., incorporating published results from research studies and on-farm observations. Several platforms exist from which data and information are assembled: (1) land grant University research programs, particularly the efforts in Wisconsin and Iowa; and (2) organic farmer networks and participatory research efforts, such as OGRAIN (the Organic Grain Resource and Information Network), led by the University of Wisconsin-Madison. Further, the authors discuss remaining barriers in the CCBRT system which prevent more wide scale adoption in the upper Midwestern US, and further research needs necessary to address those barriers.

3. Summary of CCBRT Research in the Upper Midwestern U.S.

3.1. Organic Soybean Production

Weed management in organic cropping systems remains a significant challenge for organic farmers, particularly in the soybean phase of a typical corn–soybean-winter wheat–alfalfa rotation common to organic grain systems in the upper Midwestern U.S. [21,22]. Unlike corn and winter wheat, which develop above-ground biomass early in the growing season and thus more effectively compete with weeds, soybean crops, often planted on the wider row spacing for cultivation, can be relatively slow to canopy, allowing weeds to establish over a prolonged period during the first half of the production season. As such, the development of CCBRT techniques for this phase of the crop rotation to augment mechanical weed management strategies during soybean production addresses a critical production challenge.

CCBRT research in organic soybean production in the upper Midwestern U.S. have integrated the establishment of a cereal grain cover crop (most often cereal rye) in late summer or early fall, typically seeding at a rate of 180–269 kg·ha^{-1} [20,22] (Table 1). The cereal grain cover crop is then terminated in late spring, with soybean directly seeded through the cereal grain mulch, at rates of 500,000 seeds·ha^{-1}; or more. Other cereal grain cover crops have also been trialed for use in CCBRT, including Winter triticale (*Tritocosecale* Wittm. Ex A Camus), winter barley (*Hordeum vulgare* L.), 'McGregor', and winter wheat (*Triticum aestivum*) [21,22].

While CCBRT has demonstrated its ability to effectively suppress weeds in organic soybean production [20,22], the capacity of the cover crop residue to prevent weeds from establishing varies widely depending on the weed species present in the field. Small-seeded annual broadleaf weeds tend to be suppressed more easily by the mulch and thus are more dominant in tilled organic systems [23–25]; if fields have adequate mulch biomass and complete cover crop termination, up to 80% control of common annual weeds in winter rye stands have been reported [26]. Perennial weeds are less sensitive to suppression by the mulch and readily proliferate in the system over time [25,27].

Weed management using the CCBRT technique is comprised of diversified approaches, with efficacy driven by rye biomass levels, cash crop planting date, seeding rate and placement, and cash crop stand establishment. Under typical cropping conditions, rye can produce biomass levels that range from 4000–11,000 kg·ha^{-1} [22] (Table 1). Research has demonstrated that in order to reliably suppress weeds as a surface mulch, the fall-planted cereal grain cover crop must reach biomass levels at the higher end of this range, ideally exceeding 8000 kg·ha^{-1} [26,28]. This allows the mulch to effectively limit weed populations by not only physically interfering with the emergence process, but also preventing the breaking of seed dormancy and inhibiting weed germination through allelopathy [27].

While CCBRT systems can produce soybean yields that are within state averages for organic soybeans, yields continue to lag behind organic systems that are managed using typical tillage practices. Research in Iowa and Wisconsin has demonstrated that, while during some seasons CCBRT yields are not significantly different than typically managed organic soybean, in many circumstances yield reductions in CCBRT systems as compared to typical organic soybean yields are observed, even with planting dates remaining the same, with up to a 24% or more reduction in yields observed [20,21]. Reasons for these observed yield reductions may be multi-fold: (1) cooler soil temperatures under the rye residue delaying soybean germination; (2) slower or inhibited root growth under the rye mulch; (3) nutrient tie-up under the rye mulch; and (4) allelopathic effects of the cereal rye. Whereas soybean biomass at the end of the production season is similar between CCBRT and typically managed organic soybean, early season growth of CCBRT soybeans has been observed to be slower as compared to typical organic production practices. As the weed pressure experienced within the two systems has been observed to be either equivalent or less in the CCBRT, it is unlikely that weed competition is a primary factor resulting in the observed yield differences.

Table 1. Summary of results of Cover Crop-Based Reduced Tillage (CCBRT) research conducted in the upper Midwestern U.S.A. as reported in peer-reviewed research journals.

Study Location	Cover Crop Variety	Cover Crop Planting Date [1]	Cover Crop Seeding Rate and Row Spacing	Cover Crop Termination Date	Cover Crop Biomass at Termination	Cash Crop Seeding Rate and Row Spacing	Cash Crop Planting Date	Weed Populations/ Biomass	Cash Crop Stand Populations	Soybean Yields	Citation
Soybean ((Glycine max (L.) Merr.))											
Iowa	Cereal rye (*Secale cereale* L.), Variety Not Specified (VNS)/VNS hairy vetch (*Vicia villosa* L.)	12 September and 31 October	72 kg·ha⁻¹ cereal rye, 36 kg·ha⁻¹ hairy vetch	Zadoks growth stage 60	N/A [2]	494,210 seeds·ha⁻¹; 76 cm rows	23–25 May	7–8 weeds·m⁻²	226,719–308,733 plants·ha⁻¹	1067–2724 kg·ha⁻¹	[21]
	Winter wheat (*Triticum aestivum* L.) ('Expedition' and 'Arapahoe'/winter pea (*Pisum sativum* subsp. Arvense) cover crop)	12 September and 31 October	63 kg·ha⁻¹ winter wheat, 21 kg·ha⁻¹ Winter Pea	Zadoks growth stage 60	N/A	494,210 seeds·ha⁻¹; 76 cm rows	23–25 May	9–15 weeds·m⁻²	209,422–275,357 plants·ha⁻¹	628–5668 kg·ha⁻¹	[21]
Wisconsin	Cereal rye, 'Rymin'	5 October and 7 October	180 kg·ha⁻¹, 19 cm row spacing	6–11 June	4.3–10.8 Mg DM·ha⁻¹	625,200 seeds·ha⁻¹, 19 cm row spacing	18–21 May	3–229 kg·ha⁻¹	377,100–505,600 plants·ha⁻¹	2751–2885 kg·ha⁻¹	[20]
	Hairy vetch, 'VNS'	8 September and 13 September	33.6 kg·ha⁻¹, 19 cm row spacing	28 May–8 June	3.7–5.0 Mg DM·ha⁻¹	N/A	N/A	4–47 weeds·m⁻²	N/A	N/A	[22]
	Winter rye, 'VNS'	8 September and 13 September	269 kg·ha⁻¹, 19 cm row spacing	28 May–8 June	10.2–10.3 Mg DM·ha⁻¹	N/A	N/A	1–25 weeds·m⁻²	N/A	N/A	[22]
	Winter triticale (*Triticosecale* Wittm. Ex A Camus), 'Fridge'	8 September and 13 September	269 kg·ha⁻¹, 19 cm row spacing	28 May–8 June	6.4–14.6 Mg DM·ha⁻¹	N/A	N/A	24–26 weeds·m⁻²	N/A	N/A	[22]
	Austrian winter pea (*Pisum sativum* subsp. arvense), 'VNS'	8 September and 13 September	44.6 kg·ha⁻¹, 19 cm row spacing	28 May–8 June	0–6.3 Mg DM·ha⁻¹	N/A	N/A	0–18 weeds·m⁻²	N/A	N/A	[22]
	Winter barley (*Hordeum vulgare* L.), 'McGregor'	8 September and 13 September	269 kg·ha⁻¹, 19 cm row spacing	28 May–8 June	10.3–11.7 Mg DM·ha⁻¹	N/A	N/A	19–43 weeds·m⁻²	N/A	N/A	[22]
Corn (Zea mays L.)											
Iowa	Cereal rye (*Secale cereale* L.), Variety Not Specified (VNS)/VNS hairy vetch (*Vicia villosa* L.)	12 September and 31 October	72 kg·ha⁻¹ cereal rye, 36 kg·ha⁻¹ hairy vetch	Zadoks growth stage 60	N/A	79,073 seeds·ha⁻¹; 76 m rows	23–25 May	N/A	45,302–59,510 plants·ha⁻¹	628–5668 kg·ha⁻¹	[21]
	Winter wheat (*Triticum aestivum* L.) ('Expedition' and 'Arapahoe'/winter pea (*Pisum sativum* subsp. Arvense) cover crop)	12 September and 31 October	63 kg·ha⁻¹ winter wheat, 21 kg·ha⁻¹ winter pea	Zadoks growth stage 60	N/A	79,073 seeds·ha⁻¹; 76 cm rows	23–25 May	N/A	49,824–51,479 plants·ha⁻¹	640–5567 kg·ha⁻¹	[21]

[1] Ranges of values in a given column reflect data from separate years in a given study, due to significant year effects [2] N/A: designates data not included in publication.

3.2. Organic Corn Production

Experimentation with CCBRT in the corn phase of the crop rotation has primarily differed with respect to the cover crop used in the system. Whereas the same planting strategies are used—a winter-hardy cover crop is seeded in the late summer and terminated during the following spring—CCBRT techniques for corn have integrated the leguminous cover crops, occasionally in combination with a cereal grain to enhance mulch biomass [21,22] Hairy vetch (*Vicia villosa* L.) has been the most commonly researched and trialed legume cover crop in the corn CCBRT system, although alternative legumes such as Austrian winter pea and field pea (*Pisum sativum* subsp. arvense) have also been tested [21,22]. Seeding dates of these legume cover crops have been similar to those used in the establishment of cereal grain cover crops (late summer/early fall), with seeding rates ranging from $33.6–44.6$ kg·ha^{-1}.

As with the effective mechanical termination of cereal grain cover crops, legume cover crops must be terminated at specific maturities. To obtain effective termination, roller-crimping or mowing must occur at 100% bloom to early pod set [29,30]. In the upper Midwest, this growth stage varies from late May (Austrian winter pea) to mid-June (hairy vetch) [22]. While a late May planting date can allow for corn grain and silage production. If appropriate short-season varieties of corn are used, mid-June planting dates substantially decrease the yield potential of organic corn, for both silage and grain. Within the Austrian winter pea cultivars that are commercially available, winter-hardiness remains marginal in the region, thus reducing the feasibility of this cover crop into the CCBRT system. With hairy vetch remaining the only reliable option for an overwintering legume cover crop which produces adequate biomass for an effective weed-suppressive mulch, CCBRT organic corn systems remain challenging if not prohibitive.

CCBRT in the corn phase has also proven more challenging than soybean due to increased risk of insect pest interactions. In research trials at the University of Wisconsin Arlington Agricultural research station, CCBRT corn stands have been decimated by army worm (*Mythimna unipuncta* Haworth), particularly when cover crop stands have included a cereal grain. This insect pest, which oviposits on the lower leaves of grasses or the base on grass plants, can be attracted to the cereal grain cover crop during its migration from the southern states of the U.S. Depending on the timing of this migration, newly hatched larvae may emerge from the cover crop residue at the time of corn germination, then feeding upon the corn seedlings. Although it has not been observed, depending on annual conditions and insect life cycles, a similar risk exists with seed corn maggot (*Delia platura* Meigen), a common corn insect pest. which lays eggs on decaying vegetation in late April or early May. Without the option to use chemical insecticide seed treatments, the development of CCBRT strategies must account for ecological-based management solutions, such as avoiding the use of cover crops that are preferred hosts to insect pests and the altering of cash crop planting dates to avoid specific pest cycles, to mitigate the risk of insect damage to the cash crop.

3.3. Economics and Labor/Fuel Savings

Several CCBRT studies conducted in the upper Midwestern U.S. have integrated economic analyses into their data analysis (Table 2). In an analysis conducted by Bernstein et al. [20], returns to labor, capital, and management of the CCBRT treatments using cereal rye were 27% less than in the organic tilled system, primarily due to the 24% yield reduction shown in this particular study. However, the study also documented that labor and fuel inputs were reduced by nearly 50% in the no-till cereal rye treatments. Although the profitability per hectare was greater in the tilled treatment, the return per labor hour was 25% greater in the no-till cereal rye system. These savings not only translate to economic savings, but could have positive impacts on the farmers' quality of life, potentially allowing them to engage in other enterprises or expand their soybean acres. Significant diesel fuel savings were also documented, with 720 L of diesel fuel saved using no-till techniques. Delate et al. [21] found similar results, with significantly less labor costs in the CCBRT systems. As with the Bernstein study, returns to land and management remained less in the CCBRT systems as compared to the typical organic management systems, due to persistent soybean yield reductions.

Table 2. Economic analyses of CCBRT systems in the upper Midwestern U.S.

Study Location	Cash Crop	Cover Crop	Cash Crop Row Spacing	Yields kg·ha^{-1}	Gross Revenue USD·ha^{-1}	Return to Management USD·ha^{-1}	Citation
Iowa	Soybean	Cereal rye/hairy vetch	76 cm	1067–2724 [1]	672–1769	−63–993	[21]
		Winter wheat/winter pea		1042–2862	656–1859	36–1198	
		No cover crop (traditional tillage)		2197–3170	1383–2059	742–1377	
	Corn	Cereal rye/hairy vetch	76 cm	628–5668	217–1394	−660–527	
		Winter wheat/winter pea		640–5567	221–1369	−540–618	
		No cover crop (traditional tillage)		7777–9710	2389–2694	1602–1866	
Wisconsin	Soybean	Cereal rye	19 cm	2751–2885	N/A [2]	1598–1687	[20]
		No cover crop (traditional tillage)	76 cm	3618		2162	

[1] Ranges of values in a given column reflect data from separate years in a given study, due to significant year effects.
[2] N/A: designates data not included in publication.

The CCBRT studies focused on organic corn are much more limited. Delate et al. [21], over two years of investigating CCBRT strategies using hairy vetch/rye and winter wheat/winter pea, found return to management to range from an economic loss (−660 USD·ha^{-1}) to more profitable scenarios (618 USD·ha^{-1}) (Table 2). In both years over both production systems, however, return to management was significantly less than the tilled organic corn systems (1602–1866 USD·ha^{-1}).

3.4. Impact of CCBRT on Soil Quality Parameters

The adoption of reduced-tillage practices such as CCBRT has been promoted as a tool to mitigate soil C loss, build soil organic matter (SOM), and reduce the risk of erosion [31]. However, in large part due to the lack of long-term organic CCBRT experimental sites, the impact of CCBRT techniques on increasing soil C and SOM remains unclear. Clark et al. (2017) [32], investigating the impact of CCBRT techniques on soil parameters in organic row crop production in Missouri, USA, reported no change in soil organic carbon (SOC) under CCBRT management, but concluded this may be due to the short two-year time frame of the experiment. Using the Revised Universal Soil Loss Equation, Version 2 to predict soil loss, Bernstein et al. [20] estimated that the CCBRT soybean plots integrating cereal rye as a cover crop would result in a soil loss of 1.5 Mg·ha^{-1} on a 1.0% slope and 5.6 Mg·ha^{-1} on a 4.5% slope, significantly less than the predicted soil loss on the organic till soybean fields, which ranged from 10.9 Mg·ha^{-1} on a 1.0% slope to 49.3 Mg·ha^{-1} on a 4.5% slope. While not measured directly, these same models predicted changes in soil organic matter (soil conditioning index, on a scale of −2 to +2), were positive in CCBRT rye treatments (+0.4 on 1.0% slope and +0.3 on 4.5% slope) and negative in the tilled rye treatment for both slope grades, (−0.9 on 1.0% slope and >−2.0 on 4.5% slope).

While benefits are predicted with respect to building SOM and minimizing erosion, nutrient availability dynamics may be negatively impacted during the soybean production season. In the aforementioned Bernstein et al. study [20], plant tissue tests demonstrated that while soil nutrient concentrations were similar among CCBRT and tilled soybean treatments, soybean uptake of N, P, and K as measured by tissue tests was several-fold greater in the tilled treatment as compared to the CCBRT treatments. These same results have been demonstrated in subsequent years on other fields at the University of Wisconsin Arlington Agricultural Research Station (data not published).

4. Challenges to Further Adoption of CCBRT in the Upper Midwestern U.S.

4.1. Adapting the CCBRT System to Organic Rotations

While the use of CCBRT has demonstrated success in the soybean phase of the rotation using cereal grain cover crops, strategizing crop rotations that are both agronomically and economically sound remains a challenge for organic farmers in the upper Midwestern U.S. A typical representative rotation may include corn–soybean-winter wheat–alfalfa, with corn harvested either as a silage crop throughout September or as a grain crop from late September into October. Research has demonstrated that cereal rye planting date in the fall significantly impacts ground cover in early spring and biomass at termination at anthesis (Zadoks stage 60), two factors critical to ensuring weed suppression throughout the soybean production season [27]. Thus, an organic crop rotation that includes a CCBRT soybean phase would be limited to a rotation that incorporates silage corn as opposed to grain corn in the more northern areas of the upper Midwestern U.S., if the farm were to follow a typical organic rotation scheme. While this is a viable option for organic livestock farmers utilizing CCBRT techniques, it creates a less economically viable rotation for organic cash grain farmers.

Interactions of CCBRT with the winter wheat is also of concern with respect to rotation management. While cereal rye can be planted after a winter wheat and oat harvest, this rotation may not be ideal as it results in sequential cereal grain plantings, which can increase the risk of certain plant diseases. A CCBRT rye-soybean phase into a typical corn–soybean-winter wheat rotation can also have negative repercussions on the subsequent winter wheat crop. With planting dates delayed 2–3 weeks from typical organic yields to synchronize with the appropriate cereal rye maturity stage, CCBRT delays soybean maturity, with harvest occurring in late-October versus late September or mid-October. With much of the high-carbon cereal rye residue remaining at soybean harvest, this further creates delays of winter wheat planting, which typically occurs between mid and late October. This delay in field preparation may result in the need to shift to spring-planted cereal grains, such as oat or spring wheat, which typically result in lower yields and quality as compared to winter cereal grains grown in the regions of the upper Midwestern U.S. with higher precipitation.

Additionally, some farmers utilizing CCBRT in a rye/soybean phase of the rotation have experienced contamination issues in their subsequent cereal grain crop. If the termination of the cereal rye is not completely effective, cereal rye plants producing viable seed may emerge within the soybean crop. While not having a negative impact on the soybean phase, these seeds may germinate in the subsequent cash crop year, resulting in volunteer cereal rye plants. If another cereal grain crop, such as winter wheat, is sown into this field, unacceptable levels of contamination by rye seed at harvest may cause the product to be rejected. While this is less of an issue for livestock farmers feeding the grain to their own herds, this can be a significant concern for farmers selling into the commodity grain market.

4.2. Fertility Management in CCBRT

The impact of cereal rye on soil inorganic nitrogen (N) availability and the subsequent impacts on both weeds and the soybean cash crop growth have just begun to be investigated. Fertility management can include both the cereal rye cover crop and the soybean cash crop. To achieve a desirable level of biomass of 8000 kg·ha^{-1} or greater, the cereal rye cover crop requires at least 90 to 110 kg of available N·ha^{-1}, if a desired shoot N concentration of 9–11 g N kg^{-1} is used [33,34]. To ensure that the cereal rye produces enough biomass to reliably suppress weeds, in some systems, fertilization of the cereal rye may be necessary.

Fertility management in high-residue environments presents specific challenges that need to be addressed in the CCBRT system. Research suggests that cold soil temperature conditions, such as those typically observed in the spring in CCBRT systems, can reduce nutrient mineralization and consequently yields. Work conducted in Wisconsin by Andraski and Bundy [35] concluded that soil temperature in no-till systems is the main factor affecting net N mineralization in corn systems, as opposed to N immobilization by residues, and recommended increasing N fertilization rates up to 30 kg·ha^{-1} for high residue no-till systems. In the rye/soybean system, CCBRT has been shown to lead to nutrient deficiency in the system as compared to typical organic management. Bernstein et al. (2011) [20] documented N and sulfur (S) deficiency in soybean tissue collected at the R1 growth stage grown under CCBRT in Wisconsin. Similarly, in 2013, Silva documented N deficiency in CCBRT soybean plots in Wisconsin through tissue testing conducted at the initial flowering stage, with tissue N averaging 3.5% and tissue S averaging 0.3% (unpublished data). Furthermore, in the studies of both Bernstein and Silva, visual symptoms of N deficiency were observed throughout the study period in the no-till treatments, from the vegetative cotyledon (VC) soybean stage to the reproductive 2 (R2) stage.

As continued efforts are dedicated to developing CCBRT for organic corn production, the issue of nutrient management in the system becomes even more critical. Unlike soybean, which can overcome some degree of initial N deficiency of the system as nitrogen fixation by rhizobium bacteria begins to occur, corn yields are more affected by insufficient N availability at critical stages of plant growth. With the incorporation of nitrogen-rich above and below ground biomass, legume cover crops can provide substantial N credits to the subsequent crop; however, in CCBRT systems where the aboveground biomass is not incorporated prior to cash crop planting, the N credit may be less than what might be anticipated from typical legume cover crop management. Options for supplemental N fertility may include topdressing or side-dressing with manure during the corn phase. However, in the high residue system where disturbance of the cover crop mulch can lead to exposed soil and subsequent weed establishment, the integration of these strategies may require equipment modification or novel application techniques.

4.3. Equipment Modifications

Even with cereal rye biomass reaching levels sufficient for effective weed suppression, potential yields of the system can be reduced due to poor soybean stands. Adequate seed placement through the thick cereal rye mulch is critical for the success of the system and is dependent on both environmental and physical factors. Equipment selection, modifications, and settings, as well as soil moisture at the time of planting, impact the ability of the planter to penetrate the mulch and achieve good seed-to-soil contact. Both conservation tillage planters and no-till drills have been used in the CCBRT system; planters, as compared to drills, are able to cut through the mulch while providing more precise seed metering and placement. Drills, however, can reduce the time to canopy closure due to narrow seed spacing [27,36]. If cereal rye residue is greater than 5000 kg·ha^{-1}, the planter is the preferred selection due to more effective mulch cutting [27]. Additionally, as the cover crops are typically planted at a high seeding rate, the cereal grains have a propensity to lodge prior to roller-crimping, making consistent seed placement difficult through the thick, uneven mulch using either equipment option.

Additional planter modifications can better ensure success with the CCBRT system. Extra weight can be added to the equipment to help ensure better cutting of the mulch residue and adequate seed placement. Coulters designed specifically for residue cutting, such as straight-edged coulters, can further facilitate planting and adequate seed-to-soil contact through the thick cereal rye residue. Initial evaluations with closing wheels has shown that a single 38-cm spiked closing wheel along with a smooth cast closing wheel on each row unit can help improve closure of the planting row [27]. In Wisconsin, several modifications have been made to the planter dedicated to CCBRT to contribute to more effective seed placement, including frame-mounted no-till coulters, and down-pressure springs, which increase planter weight, equivalent to 91 kg per row unit [37].

4.4. Planting Date Modifications

Research continues to be conducted to explore options beyond the foundational techniques developed by the Rodale Institute, which uses a one-pass operation at cover crop anthesis to roll-crimp the cover crop first and plant the cash crop through the cover crop residue. To improve the planting dates of the cash crop in the upper Midwestern regions where the production season length is limited, further work is being conducted to explore planting the cash crop prior to cover crop termination, providing a two to three-week advantage in cash crop planting. By planting into the standing versus rolled cover crop, seed placement may also be facilitated, mitigating the issue of poor seed to soil contact and lower cash crop stands. Initial work by Bernstein et al. [20] demonstrated yield improvements of over 130 kg·ha^{-1} with earlier versus later planted soybeans in CCBRT systems on narrow row spacing.

5. Future Developments in CCBRT Research

Research data collected over multiple years and multiple sites in the upper Midwestern U.S. indicate that the CCBRT system can be a viable and productive option for organic row crop farmers. Of all the potential cover crop and cash crop combinations, cereal rye and soybean remains the most consistent with respect to weed suppression and yields. While implementing CCBRT in the corn phase currently remains a less viable option, farmers continue to express interest in developing techniques that would allow for acceptable yields. Currently, the most significant barrier to successful CCBRT adoption in this phase is delayed cash crop planting dates. In order to ensure adequate nitrogen availability to the corn crop, as well as to mitigate pest pressure from armyworm or seedcorn maggot, legumes are the preferred cover crop rather than cereal grains. To ensure successful termination of a legume cover crop using mechanical methods, the legume must be rolled-crimped or mowed at 100% bloom or at early pod set. In the upper Midwest, this does not occur until approximately mid-June for hairy vetch, which is one month after typical organic corn planting dates. This results in significantly reduced yields of corn, or in some cases, failure of the crop to fully mature.

Legume cover crop breeding efforts could provide options to allow this system to be more successfully adapted to corn. While hairy vetch is currently the cover crop that has been most frequently tested under CCBRT management due to its ability to overwinter and establish sufficient biomass, other vining legume cover crops, such as Austrian winter pea, could provide more acceptable alternatives. While Austrian winter pea is more variable in its ability to survive upper Midwestern winters, new cultivars with improved cold-hardiness are becoming available. Combined with agronomic practices that enhance overwintering of Austrian winter pea, such as earlier planting dates, deeper planting depth, and sowing with a winterkilled nurse crop such as oat, it may become a more reliable alternative for organic farmers interested in adapting CCBRT to the corn phase of their rotation. Austrian winter pea can offer significant advantages over hairy vetch as a cover crop, including earlier flowering dates and more effective termination, leading to silage corn yields equivalent to those resulting from typical organic management.

In addition to further research and investments in cover crops for suitability for the CCBRT system, cultivar selection and breeding for cash crops adapted to CCBRT remains a component of the system where a dearth of information exists. The value of breeding crops for specific organic environments is increasingly recognized [38]. Similar to the diversity across organic environments imparted by different fertility inputs, pest management strategies, and crop/cover crop rotation, the integration of CCBRT creates yet another unique management aspect to organic crop farms. Soybean and corn cultivars that can withstand cooler soil temperatures, establish through the thick cover crop residue, and withstand a narrow row spacing could provide specific trait advantages to improve performance in the CCBRT system.

Further optimization of equipment adapted to CCBRT can also contribute to the optimization of the system and more consistent, improved results. Manufacturers such as Dawn Biologic in Illinois, USA [39] and Cross Slot®No-Tillage Systems in New Zealand [40] are experimenting with alternatives that may improve proper seed depth and spacing. Further research to optimize existing corn planters and no-till drills is also needed, to provide better recommendations as to the most effective coulters, closing wheels, and other equipment modifications.

Additionally, to more effectively motivate farmers and policymakers to integrate the technique into management and policy decisions, more research must be conducted to understand the long-term impacts of CCBRT on SOC, SOM, and soil microbial communities. While overall organic farming practices have also been shown to increase soil C concentrations [38], the impacts of organic CCBRT on soil C dynamics is less clear. Several European-based studies have estimated gains in soil C stocks with reduction of tillage in organic systems, but have not conducted these estimations in systems where tillage is reduced to the extent of CCBRT [41–44]. While it can be hypothesized that the CCBRT would similarly improve soil quality parameters, this has not yet been well-documented on long-term CCBRT fields under upper Midwestern production and soil conditions.

6. Conclusions

As organic systems continue to expand and mature in the US, organic management strategies to reduce production risk while providing continued ecosystem services will become even more critical to the continued growth and success of the industry. A decade of CCBRT research in the upper Midwestern US has demonstrated that CCBRT can provide a strong management tool for organic farmers aiming to improve their weed management practices while minimizing soil erosion risk, building soil organic matter, and incorporating further crop diversity into their rotations. Particularly in the face of climate change, where extreme weather events will occur with increasing frequency and the need for carbon mitigation tools becomes more imperative, CCBRT provides both management advantages and broader ecosystem services. While research continues to occur at land-grant universities, the continued integration of farmer experiences as adoption increases will provide refinement to the system to further reduce risk and allow for successful implementation of CCBRT over a range of environments and rotation strategies.

Acknowledgments: The authors acknowledge the contributions of all upper Midwestern CCBRT organic farmers and researchers, as well as CCBRT researchers across the U.S.A., for their ongoing dialogue and discussion to further the refinement and adoption of the technique. This material is partially based on work that was supported by the National Institute of Food and Agriculture, USDA, Integrated Organic Program Award No. 2008-51106-19021. Funding was also received from USDA Sustainable Agriculture Research and Educations (SARE) Award No. 2009-28640-19953 and from the Ceres Trust.

Author Contributions: Erin M. Silva and Kathleen Delate equally conceived the concept for the paper and contributed to the writing of all sections.

Conflicts of Interest: The authors declare no conflict of interest.

References

1. U.S. Department of Agriculture, National Agricultural Statistics Survey. *2012 Census of Agriculture: Organic Production Survey (2014)*; U.S. Department of Agriculture, National Agricultural Statistics Survey (USDA NASS): Washington, DC, USA, 2017.

2. U.S. Department of Agriculture, Economic Research Service. Organic Market Overview. Available online: http://www.ers.usda.gov/topics/natural-resources-environment/organic-agriculture/organic-trade.aspx#.U2P7pseAftk (accessed on 4 February 2017).

3. U.S. Department of Agriculture, Economic Research Service. Organic Trade. Available online: http://www.ers.usda.gov/topics/natural-resources-environment/organic-agriculture/organic-trade.aspx#.U2pePceAftk (accessed on 4 February 2017).

4. U.S. Department of Agriculture, Agricultural Marketing Service (USDA-AMS) National Organic Program (NOP). Organic Regulations. Available online: https://www.ams.usda.gov/rules-regulations/organic (accessed on 4 February 2017).

5. U.S. Department of Agriculture (USDA) Midwest Climate Hub. Climate Changes Impacting Midwest Crops. Available online: https://www.climatehubs.oce.usda.gov/sites/default/files/todey_moses_climate_changes_-_ag.pdf (accessed on 4 February 2017).

6. Chavas, J.; Posner, J.L.; Hedtcke, J.L. Organic and Conventional Production Systems in the Wisconsin Integrated Cropping Systems Trial: I. Economic and Risk Analysis 1993–2006. *Agron. J.* **2009**, *101*, 288–295. [CrossRef]

7. McCool, D.K.; Papendick, R.I.; Hammel, J.E. Surface Residue Management. In *Crop Residue Management to Reduce Erosion and Improve Soil Quality*; Papendick, R.I., Moldenhauer, W.C., Eds.; Conservation Report Number 40; USDA-Agricultural Research Service: Washington, DC, USA, 1995; pp. 10–16.

8. Fu, G.; Chen, S.; McCool, D.K. Modeling the Impacts of No-till Practice on Soil Erosion and Sediment Yield with RUSLE, SEDD, and ArcView GIS. *Soil Tillage Res.* **2006**, *85*, 38–49. [CrossRef]

9. Luo, Z.; Wang, E.; Sun, O. Can No-Tillage Stimulate Carbon Sequestration in Agricultural Soils? A Meta-Analysis of Paired Experiments. *Agric. Ecosyst. Environ.* **2010**, *139*, 224–231. [CrossRef]

10. Powlson, D.S.; Stirling, C.M.; Jat, M.L.; Gerard, B.G.; Palm, C.A.; Sanchez, P.A.; Cassman, K.G. Limited Potential of No-Till Agriculture for Climate Change Mitigation. *Nat. Clim. Chang.* **2014**, *4*, 678–683. [CrossRef]

11. Naderman, G.C. Effects of Crop Residue and Tillage Practices on Water Infiltration and Crop Production. In *Cover Crops for Clean Water*; Hargrove, W., Ed.; Soil and Water Conservation Society, West Tennessee Experiment Station: Jackson, TN, USA, 1991; pp. 23–24.

12. Brun, L.J.; Enz, J.W.; Larsen, J.K.; Fanning, C. Springtime Evaporation from Bare and Stubble-Covered Soil. *J. Soil Water Conserv.* **1986**, *41*, 120–122.

13. Smart, J.R.; Bradford, J.M. No-tillage and Reduced Tillage Cotton Production in South Texas. In Proceedings of the Beltwide Cotton Conference, Nashville, TN, USA, 9–12 January 1996; Dugger, P., Richter, D.A., Eds.; The National Cotton Council of America: Memphis, TN, USA, 1996; pp. 1397–1401.

14. Clark, A. *Managing Cover Crops Profitably*; Sustainable Agriculture Network U.S. Department of Agriculture: Beltsville, MD, USA, 2007.

15. Rudisill, A. *2007–2008 Agronomy Guide*; The Pennsylvania State University, University Park: State College, PA, USA, 2008.

16. Mirsky, S.B.; Curran, W.S.; Mortensen, D.A.; Ryan, M.R.; Shumway, D.L. Control of Cereal Rye with a Roller/Crimper as Influenced by Cover Crop Phenology. *Agron. J.* **2009**, *101*, 1589–1596. [CrossRef]

17. Mirsky, S.B.; Ryan, M.R.; Curran, W.S.; Teasdale, J.R.; Maul, J.; Spargo, J.T.; Moyer, J.; Grantham, A.M.; Weber, D.; Way, T.R.; et al. Conservation Tillage Issues: Cover Crop-based Organic Rotational No-till Grain Production in the mid-Atlantic Region, USA. *Renew. Agric. Food Syst.* **2012**, *27*, 31–40. [CrossRef]

18. Mischler, R.A.; Curran, W.S.; Duiker, S.W.; Hyde, J.A. Use of a Rolled-rye Cover Crop for Weed Suppression in No-till Soybeans. *Weed Technol.* **2010**, *24*, 253–261. [CrossRef]

19. Silva, E.M.; Moore, V.M. Cover Crops as an Agroecological Practice on Organic Vegetable Farms in Wisconsin, USA. *Sustainability* **2017**, *9*, 55. [CrossRef]

20. Bernstein, E.R.; Posner, J.L.; Stoltenberg, D.E.; Hedtcke, J.L. Organically Managed No-tillage Rye-Soybean Systems: Agronomic, Economic, and Environmental Assessment. *Agron. J.* **2011**, *103*, 1169–1179. [CrossRef]

21. Delate, K.; Cwach, D.; Chase, C. Organic No-tillage System Effects on Soybean, Corn and Irrigated Tomato Production and Economic Performance in Iowa, USA. *Renew. Agric. Food Syst.* **2012**, *27*, 49–59. [CrossRef]

22. Silva, E.M. Management of Five Fall-sown Cover Crops for Organic No-till Production in the Upper Midwest. *Agroecol. Sustain. Food Syst.* **2014**, *38*, 748–763. [CrossRef]

23. Mohler, C.L.; Teasdale, J.R. Response of Weed Emergence to Rate of *Vicia villosa* Roth and *Secale cereale* L. residue. *Weed Res.* **1993**, *33*, 487–499. [CrossRef]

24. Teasdale, J.R.; Mohler, C.L. The Quantitative Relationship between Weed Emergence and the Physical Properties of Mulches. *Weed Sci.* **2000**, *48*, 385–392. [CrossRef]

25. Bernstein, E.R.; Stoltenberg, D.E.; Posner, J.L.; Hedtcke, J.L. Weed Community Dynamics and Suppression in Tilled and No-tillage Transitional Organic Winter Rye-Soybean Systems. *Weed Sci.* **2014**, *62*, 125–137. [CrossRef]

26. Walters, S.A.; Young, B.G.; Krausz, R.F. Influence of Tillage, Cover Crop, and Pre-emergence Herbicides on Weed Control and Pumpkin Yield. *Int. J. Veg. Sci.* **2008**, *14*, 148–161. [CrossRef]

27. Mirsky, S.B.; Ryan, M.R.; Teasdale, J.R.; Curran, W.S.; Reberg-Horton, C.S.; Spargo, J.T.; Wells, M.S.; Keene, C.L.; Moyer, J.W. Overcoming Weed Management Challenges in Cover Crop-based Organic Rotational No-till Soybean Production in the Eastern United States. *Weed Technol.* **2013**, *27*, 193–203. [CrossRef]

28. Ryan, M.R.; Curran, W.S.; Grantham, A.M.; Hunsberger, L.K.; Mirsky, S.B.; Mortensen, D.A.; Nord, E.A.; Wilson, D.O. Effects of Seeding Rate and Poultry Litter on Weed Suppression from a Rolled Cereal Rye Cover Crop. *Weed Sci.* **2011**, *59*, 438–444. [CrossRef]

29. Creamer, N.G.; Dabney, S.M. Killing Cover Crops Mechanically: Review of Recent Literature and Assessment of New Research Results. *Am. J. Altern. Agric.* **2002**, *17*, 32–40.

30. Mischler, R.; Duiker, S.W.; Curran, W.S.; Wilson, D. Hairy Vetch Management for No-Till Organic Corn Production. *Agron. J.* **2010**, *102*, 355–362. [CrossRef]

31. Hobbs, P.R.; Sayre, K.; Gupta, R. The Role of Conservation Agriculture in Sustainable Agriculture. *Philos. Trans. B* **2008**, *363*, 543–555. [CrossRef] [PubMed]

32. Clark, K.M.; Boardman, D.; Staples, J.S.; Easterby, S.; Reinbott, T.M.; Kremer, R.J.; Kitchen, N.R.; Veum, K.S. Crop Yield and Soil Organic Carbon in Conventional and No-Till Organic Systems on a Claypan Soil. *Agron. J.* **2017**, *109*, 588–599. [CrossRef]

33. Graham, R.; Geytenbeek, P.; Radcliffe, B. Responses of Triticale, Wheat, Rye and Barley to Nitrogen Fertilizer. *Aust. J. Exp. Agric.* **1983**, *23*, 73–79. [CrossRef]

34. Shipley, P.R.; Messinger, J.J.; Decker, A.M. Conserving Residual Corn Fertilizer Nitrogen with Winter Cover Crops. *Agron. J.* **1992**, *84*, 869–876. [CrossRef]

35. Andraski, T.W.; Bundy, L.G. Corn Residue and Nitrogen Source Effects on Nitrogen Availability in No-till Corn. *Agron. J.* **2008**, *100*, 1274–1279. [CrossRef]

36. Hock, S.M.; Lindquist, J.L.; Martin, A.R.; Knezevic, S.Z. Soybean Row Spacing and Weed Emergence Time Influence Weed Competitiveness and Competitive Indices. *Weed Sci.* **2006**, *54*, 38–46. [CrossRef]

37. University of Wisconsin-Madison Integrated Pest and Crop Management. Advances Using the Roller-Crimper for Organic No-Till in Wisconsin. Available online: https://www.youtube.com/watch?v=UtxH4CJa-jk&t=3s (accessed on 4 May 2017).

38. Lyon, A.; Silva, E.M.; Bell, M.; Zystro, J. Seed and Plant Breeding for Wisconsin's Organic Vegetable Sector: Understanding Farmers' Needs and Practices. *Agroecol. Sustain. Food Syst.* **2015**, *39*, 601–624. [CrossRef]

39. Dawn Biologic. ZRX Electro-Hydraulic Roller/Crimper/Row Cleaner. Available online: http://www.dawnbiologic.com/zrx/ (accessed on 4 May 2017).

40. Cross Slot®No-Tillage Systems. Drills. Available online: http://www.crossslot.com/drills (accessed on 4 May 2017).

41. Gattinger, A.; Müller, A.; Haeni, M.; Skinner, C.; Fliessbach, A.; Buchmann, N.; Mäder, P.; Stolze, M.; Smith, P.; Scialabba, N.E.-H.; et al. Enhanced Top Soil Carbon Stocks under Organic Farming. *Proc. Natl. Acad. Sci. USA* **2012**, *109*, 18226–18231. [CrossRef] [PubMed]

42. Cooper, J.; Baranski, M.; Stewart, G.; Nobel-de Lange, M.; Bàrberi, P.; Fließbach, A.; Peigné, J.; Berner, A.; Brock, C.; Casagrande, M.; et al. Shallow Non-Inversion Tillage in Organic Farming Maintains Crop Yields and Increases Soil C Stocks: A Meta-analysis. *Agron. Sustain. Dev.* **2016**, *36*, 22. [CrossRef]

43. Krauss, M.; Ruser, R.; Muller, T.; Hansen, S.; Mader, P.; Gattinger, A. Impact of Reduced Tillage on Greenhouse Gas Emissions and Soil Carbon Stocks in an Organic Grass-Clover Ley-Winter Wheat Cropping Sequence. *Agric. Ecosyst. Environ.* **2017**, *239*, 324–333. [CrossRef] [PubMed]

44. Mäder, P.; Berner, A. Development of Reduced Tillage Systems in Organic Farming in Europe. *Renew. Agric. Food Syst.* **2012**, *27*, 7–11. [CrossRef]

5

Heterogeneous Organizational Arrangements in Agrifood Chains: A Governance Value Analysis Perspective on the Sheep and Goat Meat Sector of Italy

Maria Angela Perito [1,2,*], Marcello De Rosa [3,*], Luca Bartoli [3], Emilio Chiodo [1] and Giuseppe Martino [1]

[1] Faculty of Bioscience and Agro-Food and Environmental Technology, University of Teramo, 64100 Teramo, Italy; echiodo@unite.it (E.C.); gmartino@unite.it (G.M.)

[2] ALISS, UR1303, INRA, F-94205 Ivry-sur-Seine, France

[3] Department of Economics and Law, University of Cassino and Southern Lazio, 03043 Cassino, Italy; bartoli@unicas.it

* Correspondence: maperito@unite.it (M.A.P.); mderosa@unicas.it (M.D.R.)

Academic Editors: Paulina Rytkönen, Javier Sanz Cañada and Giovanni Belletti

Abstract: In the Italian agrifood sector, one observes heterogeneity in the types of quality certification processes. This heterogeneity cannot be explained by standard governance theories like transaction costs economics (TCE). We use the governance value analysis (GVA) perspective that synthesizes TCE and a resources-based view (RBV), to suggest that the observed heterogeneity in organizational forms is a result of heterogeneous differentiating strategies that farms have pursued in the face of competitive pricing pressures. To empirically test GVA, data are obtained using a survey methodology on lamb meat produced by local farms in the Abruzzo region of Italy, challenged by price-costs squeeze. Our empirical test evidences the relevance of the adopted approach, enlightening different organizational arrangements, strictly linked to both the strategic positioning and to the farms' resources and core competencies.

Keywords: governance value analysis; LAFs governance; quality schemes; sheep sector; geographical indications

---

1. Introduction

This paper deals with the governance of agrifood value chains, according to a neoinstitutional perspective. More precisely, it analyses how, within localised agrifood systems, different mechanisms of governance are chosen to reflect different value propositions. Therefore, research questions concern why heterogeneous forms of governance occur within territorialised agrifood systems, where re-anchoring processes of agricultural production are at stake [1]. This topic seems relevant when collective action should characterise valorisation strategies in localised agrifood systems. On the other side, plural forms of governance may emerge, even in cases of a similar transaction carried out in the same territory [2].

Therefore, questioning the way in which governance mechanisms are developed in a localised agrifood system is a relevant field of research. In his seminal work, Stoker [3] (p. 18) defines governance as a *set of institutions and actors that are drawn from but also beyond government*, aiming at identifying *boundaries and responsibilities for tackling social and economic issues*, bringing about something without the need for government authority. Consequently, governance mechanisms aim at specifying institutional arrangements affecting the exchange process [4]. For this purpose, a governance structure *consists of a collection of rules/institutions/constraints structuring the transactions between the various stakeholders* [5] (p. 1).

The governance of a market is a key concept for the agrifood sector, being related to the way in which a supply chain is organized, in addition to the ability to govern existing and create new markets [6]. This is a not simple process, in that governance involves actors with diverse preferences and different incentives [1].

The analysis of governance mechanisms in the agrifood sector is a deeply developed field of neoinstitutional researches, focused on the role of transaction costs in explaining the choice of determined modes of governance [7]. According to us, recent literature has not taken the relevance of resources (resources-based view) into account, jointly with the influence of transaction costs. This paper tries to fill this gap in the literature, with the purpose of describing the organizational arrangement in similar transactions by combining a resources-based view and transaction costs economics. Empirical analysis is carried out in an Italian localised agrifood system specialised in sheep breeding. Here, plural forms of governance emerge, as a response to heterogeneous strategies of value creation. In order to provide an explanation for this territorially concentrated heterogeneity, we put forward an original perspective developed within the neoinstitutional theories. Our theoretical perspective refers to governance value analysis [8], which takes into account both transaction costs and a farm's availability of internal resources. Therefore, it may justify the heterogeneous alignment principle in modes of governance.

The paper is organized as follows: following the introduction (Section 1), in Section 2, we provide an overview of the theoretical background. The aim is not to provide an exhaustive literature review, but to try exploring the links between strategy (qualification or quality certification of agricultural products), resources, and governance, with regards to the relatively recent perspective of governance value analysis. An empirical test is carried out in Sections 3 and 4, with an application to a localised agrifood system specialised in meat lamb production. Section 3 provides methodological insights, while Section 4 discusses the results of the empirical analysis.

2. Theoretical Background

The modernization model of agriculture was effective in pursuing original objectives, linked to the increase in agricultural productivity. However, unexpected outcomes also emerged: first, due to the high use of chemical inputs and to unsustainable methods of production, negative externalities emerged, jointly with a growing loss of biodiversity. From an economic point of view, rising costs and relatively downward trends in farm revenues had characterised farming activity, and the consequent price-costs squeeze called for new strategies of farm development in rural areas [9]. As a consequence, a new European agricultural model is emerging, centred on agronomically sound and sustainable agricultural systems characterized by high-added-value farming and high-quality primary and processed products [10]. The backbones of the new agricultural strategies are quality differentiation and the construction of new, alternative markets. Relevance and credibility are the keywords of this process [11]: relevance means that the commitment of producing high quality products makes sense to the consumers. In order to fulfil the promises and make them credible, mechanisms of guarantees are required. Therefore, the choice of quality strategy may bring about stricter organizational arrangements [12], as in cases of collective quality strategies. Consequently, choosing the appropriate governance structure is a strategic answer to credibly commit to a certain quality.

A neoinstitutional perspective on governance is clear in emphasising how farm strategies may affect governance structures. More precisely, transaction costs economics provides evidence concerning the trade-off between the market and hierarchy in the presence of transaction costs. Higher transaction costs engender a progressive transition towards hierarchic forms of governance [7,13]. As Menard [14] points out, "*contracts represent a focal point in neoinstitutional economics, because of their role in relaxing the constraints of bounded rationality, fixing schemes of references for future actions, and checking on opportunistic behaviour"* (p. 282).

New institutional economics has also inspired a large number of researchers in the agrifood sector, by providing them with new tools to explain various modes of organization of the supply chain. Menard and Valceschini [15] describe various institutional solutions as the outcome of granting food safety, thus generating more centralised organizational arrangements. According to Menard and Valceschini [15], strategies of value creation through product differentiation are associated with an increase in transaction costs, on account of higher commitments with final consumers (a quality promise) being made credible. This may boost stricter coordination mechanisms. Coordination also involves intermediate transactions, through the introduction of compulsory quality standards at each level of the food chain: the respect of quality standard affects transaction attributes, in that choosing different quality standards requires farms to make different levels of investments (transaction attributes) that necessitate corresponding governance forms [16]. In cases of collective brands (for example, geographical indications), the high number of actors may raise these costs, with the risks of either reduction or profit crossing-out. Therefore, the choice of quality product may be adopted in cases of net positive revenues for farmers. Thus, how to match quality strategies and forms of governance becomes a relevant field of analysis [6]. Against this background, transaction costs theories have recently faced new challenges, with reference to the evidence of plural forms of governance, not always explainable only through the lens of transactions costs economics. Menard's [2] (p. 125) key question: "why are different forms of governance often adopted for organising similar transaction?", has brought about a more extended framework which does not properly consider the availability of a firm's internal resources.

Ghosh and John put forward an alternative perspective. The starting point is a similar question pointed out by Ghosh and John in their seminal paper [8]: why do firms working in the same industry choose different positioning strategies and governance forms? To answer this question, we intend to complement TCE through a RBV perspective. If transaction costs analysis seems effective in depicting organizational arrangements, some limits underlined in recent literature suggest integrating it with new analytical perspectives: one of them, the resource-based view, takes into account a firm's resources. As pointed out by Rantamäki-Lahtinen [17], the analysis of a firm's resources and governance mechanisms brings together transaction costs economics and a resource-based view.

A resource-based view focuses on the relevance of a firm's available physical and human resources in the ability of building up competitive advantages [18,19]. The need for considering internal capabilities in transaction costs economics is well known by neoinstitutional theorists [20]. A recent analysis carried out by Rindfleisch et al. [21] brings together transaction costs economics and GVA. We agree with the authors in underlying how the "classic" perspective of transaction costs economics underestimates the role of resources and capability, by emphasizing the relevance of three key dimensions in the explanation of governance (asset specificity, uncertainty, and frequency). Therefore, to efficiently design governance forms that exploit resources and capabilities leads to the existence of heterogeneous/plural governance forms. As pointed out by Bradach and Eccles [22], mixed governance modes may minimize transaction costs. Therefore, the *coexistence of different arrangements for operating similar transactions* [23] (p. 575) is not surprising. As a matter of fact, Rindfleisch et al. [21] specify four contextual conditions influencing the nature of governance: the type of transaction, its level of exchange, the capabilities and resources surrounding this exchange, and the mixture of mechanisms used to govern this exchange.

As far as empirical analysis is concerned, a lot of work remains to be done to excavate the complexity and the variety of institutional arrangements [12,23]. This paper posits that GVA may be of help in evaluating strategic positioning within localised agrifood systems (LAFs). Resource-based perspectives, in combination with transaction costs analysis, are at the basis of Ghosh and John's [8] GVA. According to them, the economizing calculus of transaction costs analysis should be replaced by strategizing calculus, through including strategic choices (for example, quality strategies), specific investments, and modes of governance. In this context, value creation and value claiming are central [24]. Consequently, the strategizing calculus argues for a simultaneous, three-way choice of resources, investments, and governance that yields the highest expected outcomes [25] (p. 146). More precisely, in order to explain the differences in the contractual relationships among firms operating in the same industry and hence the use of plural forms observed in industry sectors, the authors hypothesise an alignment between resources, specific investments, and governance, bringing about the highest expected incomes. Therefore, according to the authors, strategic positioning leads the analysis of exchange attributes and, consequently, the alternative governance modes. However, they underline that modes of governance must fit with a firm's resources. In their empirical test of GVA, Ghosh and John [25] (p. 146) affirm that the strategizing calculus argues for a simultaneous, three-way choice of resources, investments, and governance that yields the highest expected outcomes. Consequently, this perspective maintains the idea of specific resources in the firm and the objective of maintaining a skill difference among the firms in order to create value. The interesting aspect of the approach is the consideration of transaction costs as endogenous variables, in that alignment may be obtained through the modification of attributes, for example, by intervening in the specificity of resources.

Therefore, modes of governance are the result of a more complex strategizing calculus, bringing about different types of governance, well explained within the framework of transaction costs economics [26–28]. The Table 1 illustrates the typologies of governance.

Table 1. Typologies of governance.

The Spot Market Contract	A Contract for the Immediate Exchange of Goods or Services at Current Prices. The Identity of the Parties is Irrelevant
The relational bilateral governance (also implicit contract).	A non-written (not legally enforceable) contract that specifies only the general terms and objectives of the relationship. This governance introduce the idea of repeated relations with the same agents
The relational bilateral governance with "qualified partner(s)".	This structure is similar to the previous one. However, agents are not free to choose their partners, but have to select a "qualified" transactor (accredited for example by a collective organization)
The formal (written) bilateral contract.	A legally enforceable set of promises that defines all or part of each party obligations
The financial participation in the ownership of the partner(s).	In this case, buyer (respectively seller) is a stockholder of the other but stay legally independent from the seller (respectively buyer). Joint-venture is a canonical example of this type of governance structure
Vertical integration.	Bringing two or more successive stages of the supply chain under common ownership and management

Source: processed from Raynaud et al. [11].

Furthermore, through the analysis of contractual arrangements, GVA sheds light on "*comprehensive models that incorporate both value creation and value claiming*" [24] (p. 59).

Empirical tests of GVA have concerned various sectors and, recently, have been applied to agrifood analysis: Raynaud and Valceschini [11] underline GVA's utility in describing a quality strategy in agrifood supply, by emphasising the eventual needs for acquiring specific resources. More recently, Nazzaro et al. [29] put forward a Value Portfolio and Multifunctional Governance Analysis (VPMGA), aiming at integrating the value chain analysis, transaction costs, and value creation within the framework of multifunctional agriculture. Nonetheless, these analyses do not provide answers to the question concerning the "*coexistence of different organizational arrangements for operating similar transactions*" [23] (p. 575). Therefore, we believe a gap in the literature needs to be filled, related to the analysis of coordination mechanisms which take into account both economic and strategizing calculus. In what follows, we will try to fill this gap by putting forward an empirical analysis in a territorial agrifood system of Abruzzo, Italy, where different modes of governance have been found in the same food industry.

3. Methodology

3.1. Sample and Methodology Description

The sector under study is the sheep and goat meat produced in the Region of Abruzzo (Italy). This is a specialised area, even though it has been witnessing a continuous decline. As a matter of fact, a real collapse in the number of farms throughout the region (−64%), in addition to a significant decrease in the number of heads (−25%), has been observed in the last two censuses of Italian agriculture (referring to years 2000 and 2010). This trend characterizes the smallest and less specialized farms and marks the entire region, differently from the national situation where the number of heads seems to remain constant. The average size of the farms (32 to 66 animals per farm) is still much lower than that at the national level (132.7 animals per farm) or in Central Italy (165.5).

The Abruzzo Region boasts a long tradition in the field of pastoralism, and has a high appreciation by the consumers. Nevertheless, the region is a net importer of live sheep and goats and represents nearly 20% of total Italian purchases of this product.

The census data of 2010 reveal a dichotomous structure of sheep breeding: on the one hand, small and unspecialized farms (with very few animals per holding), and on the other hand, large farms with a number of animals in line with the national context.

In order to analyse the organizational arrangements, data have been collected through a direct and structured survey (developed in 2014, see Appendix A). The survey was conducted through a questionnaire organized into four parts:

-    general information (number of heads, rearing, autochthonous breeds, use of transhumance, etc.);
-    supplier and upstream relationships;
-    quality management (quality of product, labels, geographical indication, process quality management);
-    farm diversification.

We submitted the questionnaire through face-to-face interviews. Moreover, in order to acquire further information, qualitative interviews with regional experts and with the Regional Association of Breeders (ARA) were realized. The survey was carried out in cooperation with ARA. After conducting a pilot survey and in-depth interviews with some stakeholders in the Abruzzo Region, we made some minor adjustments to the questionnaire.

Through this direct investigation, it has been possible to collect data on 101 farms in the region, for a total of 35,980 heads (2013), representing 3.1% of the farms in the Abruzzo Region and more than 17% of the animals at a regional level. Farms were casually extracted from the list of the those registered in the regional association of breeders and 50% of the farms registered were included. Consequently, our sample may be considered as representative of farms under the control of the regional body in charge of the control over ovine and goat breeding. The sample has been processed through post-stratification methods and strata have been ascribed by this sample, which sensibly reduces the variability of the estimators.

The farms extracted were professional, on account of their average dimension, which is about 350 heads per farm [30].

In order to acquire further information, our sample was matched with the latest Italian agriculture census. This permitted the inclusion of a significant number of issues: the breeding system, type of farming (wild, semi-wild, extensive rearing, etc.), animal feed and supply methods, organic certification, adoption of technological innovation (as mechanization, milking mechanization and milk refrigeration), development of multifunctional activities, management skills, quality controls, and market channels.

Data collected represent the basis for the application of the GVA approach (Figure 1).

As far as resources and competencies are concerned, by adapting Ghosh and John's indication, three main categories of resources were considered. They are further synthesised in Table 2. As far as tangible resources are concerned, they take into account the structural characteristics of the farms, represented by the number of heads, types of farming (professional activity), type and composition of the family farm, standard output, and diversification in agricultural and non-agricultural activities. With respect to family composition, family farms have been distinguished according to the phase of life cycle of the manager (young, mature, old), the presence of children, and the presence of farmers' assistants (either young or not) [31].

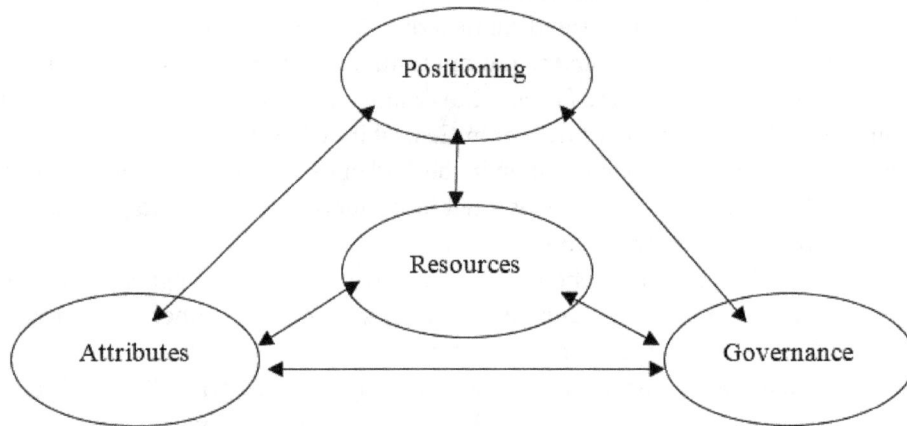

Figure 1. GVA model, processed from Ghosh and John [24] (p. 56).

Table 2. Active variables.

Active Variables	Categories of Variables	Category Description
Adoption of rural development policies	2	Yes/no
CUF-MEAT certification	2	Yes/no
SATA certification	2	Yes/no
Organic farming/transition	2	Yes/no
Supply chain governance	8	- Wholesalers managed by the regional breeding association. - Specialised wholesalers for PGI products. - Other wholesalers - Short food supply chain
Label	4	Buongusto, Protected Geographical Indication (PGI), Buongusto and PGI, no quality sign
Family type	7	**Young phase** Young farmer and a not young assistant Young farmers with other (assistant may be young or may be not) **Mature phase** Mature farmer and a young assistant (assistant may be old, mature or may be not) **Old phase** Older farmer and a young assistant Older farmers with other (assistant may be mature, old or may be not)

Intangible resources refer to the technology, quality, and management control system. Therefore, aspects concerning the presence of the management control system and the adoption of quality processes control have been analysed through the presence of a specific certification scheme, the CUF-meat, a control system employed to monitor the quality of meat at a farm level. Members must adapt to traditional farming, with typical breeds reared on pasture, feeding based on breast milk until weaning, and feeding with fodder and other typical cereals in the fattening phase.

Furthermore, human resources are divided up into basic skills and entrepreneurial skills, according to Rudmann et al.'s classification [32]:

○   basic skills refer to the farmers' competencies on farm management (professional and managerial skills). Accordingly, the presence of autochthonous breeds in the farm; to this end, the practice of wild, semi-wild, extensive, or intensive breeding, and the presence of the transhumance system, have been considered. To evaluate managerial competencies, we have considered the adoption of the S.A.T.A. system (Technical Assistance Service to Breeders), which provides breeders with specific experts with the purpose of encouraging the business organization and the achievement of management efficiency.

○   entrepreneurial skills include: (a) opportunity skills (capability to exploit opportunities), (b) cooperation and networking skills, and (c) strategy skills. For the purpose of our paper, opportunity skills are synthesised by the capability of gaining access to a rural development policy (revealed by the farm's application for rural policy funds). As far as strategy skills are concerned, investments in quality schemes (geographical indications, quality labels, and organic farming) and diversification into agricultural and non-agricultural activities are considered.

According to Ghosh and John's perspective depicted in Figure 1, positioning, that is the value proposition, is synthesized by the presence of differentiation strategies through the participation in quality systems. Four types of differentiation strategies have been identified:

-   adhesion to the "Buongusto" brand (private and local brand of Regional Association of Breeders, ARA) label; it is a collective mark associated with a quality code of practices;
-   adhesion to the protected geographical indication (PGI "Agnello del Centro Italia");
-   adhesion to both Buongusto and PGI labels;
-   no quality marks.

According to our hypothesis, positioning influences the attributes of transactions, resulting in different modes of governance [14]. The positioning strategies may result in a 'commitment intensive' as, for example, in the qualification of agricultural products.

As said before, Raynaud and Valceschini [11] point out that relevance and credibility are the main characteristics of these strategies, probably raising governance costs and boosting discriminant alignment mechanisms aiming at value creation. Consequently, diversified supply chain governance may emerge as a response to diversified attributes, characterising each strategic positioning. As a matter of fact, adhesion to certification (either CUF-meat and SATA, or PGI certification) implies different transaction attributes to be respected. Therefore, various organizational adjustments and, consequently, a different supply chain governance, may emerge.

In order to test the mode of governance, specific questions on distribution channels (wholesalers, retailer, short supply, etc.) and the type of governance structures (ownership of quality signal and coordination vertical relations) were submitted.

In order to verify the modes of governance, a multivariate analysis (cluster analysis and correspondence analysis) has been carried out, in order to identify homogeneous farms with reference to positioning, available resources, and the mode of governance. Cluster analysis is a method of farms' classification that is able to aggregate farms with homogeneous strategies related to positioning/mode of governance. Through multiple correspondence analysis (MCA), a reduced set of factors is extracted in order to reduce the complexity of the variables. MCA is a way to apply the typical Correspondence analysis singular value decomposition algorithm to the so-called "super-indicator matrix", the boolean table in which the rows denote the categories of the survey variables [33]. The Chi-squared distances between the rows of that matrix are based on the variation of the counts of the set of responses of the units, as the sum of the squared differences between their own row data [34].

Clustering procedure has been carried out through the SPAD 3.21 program, by following a hierarchic procedure according to the Ward methodology of aggregation (with 10 iterations with mobile centres). The choice of the three clusters is drawn on the dendrograph, the main graphical tool for looking at a hierarchical cluster solution [35]. This clustering method reveals the groups with the lowest internal inertia and the highest external inertia.

A description of each cluster only takes into account statistically significant variables ($p \leq 0.05$). Active variables are listed below with possible categories.

Illustrative variables concern the average dimension of the farm (physical and economic), diversification of farming activity, type of breeding, and annual family work.

3.2. Characteristics of the Sample

The farms under investigation hold different quality positions. The sample is described in Table 3 and includes four categories of quality schemes: 22.8% are in a PGI scheme (PGI "Agnello del Centro Italia"), 10.9% adhere to the Buongusto brand, 14.9% participate in both schemes, and 51.5% are not taking part in any quality scheme. It has to be noted that the interviewed farms of the PGI category account for 40.4% of all Abruzzo's farms included in the PGI "Agnello del Centro Italia".

The farms without any quality brand have a lower propensity to implement and use systems leading to improvements in the control process, like CUF-Meat, whereas the propensity is higher for the farms with Buongusto certification. Among the quality management systems, the SATA certification is highly adopted, above all in farms with double certification (PGI and Buongusto), even though non-certified farms present a relatively good level of this SATA certification.

The Buongusto members have a larger share of organic farms (27.3%), even if the share is quite similar to farms producing PGI meat (26.1%). The farms that display more traditional attributes are those with the Buongusto label, with a higher rate of autochthonous breeds (47.5%), while practices of transhumance are mainly realized by farms with both Buongusto and PGI certifications.

Farms with Buongusto certification mainly sell their production through wholesalers; nonetheless, they also sell to local butchers, restaurants, and directly front door farms. Wholesalers are linked to ARA (Regional Association of Breeders) through a specific private company (named SCA—*Servizi Commerciali Allevatori*), specifically created to sell the product to large retailers. Nearly 20% of Buongusto farms directly sell to large retailer buyers. The "only PGI certificated" farms sell the product to specialised wholesalers in the PGI supply chain; moreover, they verify whether each supplier adheres to the standard.

Table 3. Percentage distribution of the sample.

	No Certification	PGI + Buongusto	PGI	Buongusto	TOT.
Quality certification (%)	51.5	14.8	22.8	10.9	100
Number of heads per farm (mean)	299	329	453	331	
Std. Deviation	284.5	296.4	244.1	172.9	
Min	30	90	115	160	
Max	1600	1152	910	800	
Organic (%)	13.5 (100%)	20.0 (100%)	26.1 (100%)	27.3 (100%)	
Autochthonous breeds (%)	32.8 (100%)	67.7 (100%)	29.4 (100%)	47.5 (100%)	
Transhumance system (%)	42.3 (100%)	86.7 (100%)	82.6 (100%)	72.7 (100%)	
Average weight of animals (kg)	21	16	19	13	
Average age of animal (days)	65	55	56	57	
Transhumance system (km)					
Mean	11.9	13.1	29.0	15.5	
Std. Deviation	25.6	11.2	31.8	11.3	
Max	120	30	130	35	
CUF-MEAT (%)	50.0 (100%)	60.0 (100%)	78.3 (100%)	81.8 (100%)	
SATA (%)	63.5 (100%)	93.3 (100%)	60.8 (100%)	90.9 (100%)	
Average UAA (hectare)	70	171	105	132	97
Average standard output	108,170	190,398	145,811	117,001	128,787
Average use of policy instruments	0.4	0.6	0.7	0.8	0.6
Supply chain *					
Short foods supply chain (%)	36.5	13.3	4.4	36.4	
Retailer (directly) (%)			100	18.2	
Wholesaler (%)	76.9	93.3		54.5	

* Total is not 100%.

4. Results

Multivariate Analysis

Through a multiple correspondence analysis (MCA), we have taken into account 22 variables; eight active and 14 illustrative (eight qualitative and six quantitative). The choice of multiple correspondence analysis is due to the preponderance of qualitative variables, with respect to quantitative variables. On the basis of the factors extracted, a cluster analysis has been carried out, with the aim of aggregating homogeneous farms, whilst at the same time obtaining the highest heterogeneity among the groups. Eighteen factorial axes have been created with a total sum of eigenvalues equal to 2.25 (Table 4). First, three factors have been considered to build up the following cluster analysis. These factors explain a total inertia of 32.16%. In order to correct this value of inertia and make it more reliable for the first factors, we have made reference to Benzècri [36]:

$$\lambda^* = \left(\frac{p}{p-1}\right)^2 * \left(\lambda - \frac{1}{p}\right)^2 ; \lambda > \frac{1}{p}$$

By applying the formula, the first three factors may explain a total inertia of 80.6%, thus confirming a minimum loss of information.

Table 4. Eigen values.

Factors Extracted	Eigen Values	Inertia %	Cumulated %	Reassessed Inertia %	Cumulated %
	Total Inertia: 2.25000				
1	0.2671	11.87	11.87	38.4577	38.4577
2	0.2483	11.03	22.90	28.9436	67.4014
3	0.2084	9.26	32.16	13.2409	80.6423
4	0.2012	8.94	41.11	11.0741	91.7163
5	0.1772	7.88	48.98	5.1914	96.9078
6	0.1595	7.09	56.07	2.2645	99.1722
7	0.1456	6.47	62.54	0.8084	99.9806
8	**0.1282**	**5.70**	**68.24**	**0.0194**	**100.0000**
9	0.1110	4.93	73.17		
10	0.1053	4.68	77.85		
11	0.0977	4.34	82.19		
12	0.0872	3.88	86.07		
13	0.0856	3.81	89.88		
14	0.0750	3.33	93.21		
15	0.0633	2.81	96.02		
16	0.0521	2.31	98.34		
17	0.0258	1.15	99.48		
18	0.0116	0.52	100.00		

Following this, cluster analysis aggregates homogeneous farms in relation to the active variables used in the analysis. A clear misalignment in the modes of governance emerges, attributable to both the presence of quality marks and to the resources available at the farm level, which discriminates organizational arrangements.

Table 5 points out the test value of the three factors extracted to create clusters, which are significant for each cluster of homogeneous farms. By taking into account the test values, a description of the cluster on the basis of the three factors is statistically robust.

Table 5. Factors' value tests per each cluster.

	First Factor	Second Factor	Third Factor
		Test Values	
I cluster	6.1	3.6	1.5
II cluster	2.5	−7.0	1.2
III cluster	−6.9	3.0	1.5

The first factor (10.9%) points out the positioning/resources couple. It demonstrates, on the one hand, the presence or absence of value proposition, strictly linked to the endowment of resources and competencies. Consistent with the GVA perspective, the absence of value proposition, that is of a quality mark, characterises the smallest farms (less than 100 heads per farm) that are equipped with a very low level of either tangible or intangible human resources. On the other hand, the presence of resources and competencies in the farm typifies the use of quality brands, more specifically farms with either Buongusto or both Buongusto and PGI brands. The second factor (10.2%) is coherent with the theoretical perspective accepted here: it couples positioning decisions and modes of governance. More precisely, this factor highlights dedicated modes of governance strictly associated with specific brand strategies. The third factor (8.4%) describes the presence of intangible resources in the farm. More precisely, it evidences the adoption of quality process control tools (CUF-MEAT scheme).

On the basis of the previous factors, the following cluster analysis has aggregated farms into three homogeneous clusters. Neither outliers nor highly skewed data have been found; accordingly, it has been possible to extract robust clusters. An interesting element of our empirical analysis concerns the homogeneity between a farm's positioning and consequent supply chain governance.

I cluster—Local quality oriented farms with coordinated mechanisms of governance

The first cluster includes 17.9% of farms, with an average size (301–500 heads) and prevailingly managed by young farmers without young helpers. These farms' value proposition consists of strategies of specific quality, based on geographical indications, by using the Buongusto label (local label − value test = 2.7), and sometimes joined with the PGI marks as tools to qualify their products (value test = 5.2). To accomplish credible commitments of the strategy, a coordinated governance mechanism is evident, in the form of relational bilateral governance, in which farms have to trade their products within a supply chain coordinated by the regional association of breeders, through a specified type of wholesaler (value test = 7.1).

To better comprehend this form of governance, a resource-based perspective is of help: the farms of this cluster raise only autochthonous livestock (value test = 1.8) and they are provided with a SATA certification (value test = 1.6) demonstrating adequate managerial skills and competencies. Moreover, specific quality is granted through quality control systems, based on specific rearing techniques, making use of traditional and local feeding systems (CUF-MEAT certification − value tests = 1.6).

II cluster—farms with specific mark and dedicated mechanisms of governance

In the second cluster, 21.8% of farms have been extracted, characterized by the presence of both big farms (>500 heads (value test = 2.2) and, to a lower extent, farms of an average size (value test = 0.6). A clear combination of value proposition and mode of governance emerges: as confirmed by the high value test (7.9), farms of the cluster adopt a specific differentiation strategy, based on protected geographical indication (PGI). This strategic positioning choice brings about a diverse and dedicated mode of governance. Moreover, to qualify their products further, farmers have adhered to organic methods of production. As a consequence, coherently with the GVA model, a specific mode of governance fits with this value proposition. More precisely, as said in the methodological note, the PGI farms trade the product through specialised wholesalers in the PGI supply chain (value test = 5.9), by checking out the respect of specific quality standards. The analysis of key resources evidences the presence of the quality product certification, but, differently to the first cluster, SATA has not been identified as a key factor.

III cluster—farms without either quality strategies or specific governance mechanisms

The third cluster includes the majority of farms, comprising 60.3% of the total. They are mainly small farms, managed by elderly people and in a few cases, by young farmers. In these farms, no value proposition has been put forward: as a matter of fact, these farms make use of no label (value test = 8.0) to market their products and make use of either a generic mode of governance (value test = 3.7) or alternative food networks. Consequently, they sell meat through "generic" wholesalers and/or through direct selling (value test = 2.21). The analysis of internal resources provides possible explanations concerning difficulties adopting specific quality strategies. The location in the elderly phases of the life cycle, above all in the absence of generational renewal, limits the propensity to adopt differentiation strategies. Moreover, the scarcity of both tangible and intangible resources hampers other value propositions (no CUF meat or SATA certifications).

In order to synthetize previous information and to size up the role of each variable, we have positioned the clusters and included other key-variables on the basis of factorial coordinates. To this end, in Figure 2, the horizontal axis reports the factorial coordinates of the first factors (label/resources) extracted from the correspondence analysis, while, on the vertical axis, factorial coordinates of the second factor (label/mode of governance) are calculated. Factorial coordinates are obtained on the basis of chi-squared distances, delivered by the Burt matrix (Escoutier*).

A differentiation among clusters, explainable according to our theoretical perspective, emerges. Clusters 1 and 2 are clearly positioned in the area where value proposition is "supported" by the presence of core competencies and internal resources. Differences between the two clusters emerge in terms of strategizing calculus: farms in the first cluster pursuit a regional brand strategy, with the aim of serving local-regional markets. Farms of the second cluster aim at valorising their product through the adhesion to a geographical indication (PGI). This implies the choice of a larger market, thanks to the availability of large cattle and to the entrepreneurial profile of the manager, usually the young manager of a family farm.

On the contrary, the third cluster evidences a reduced level of resources. Consequently, it is positioned in the area where no brand strategies are adopted.

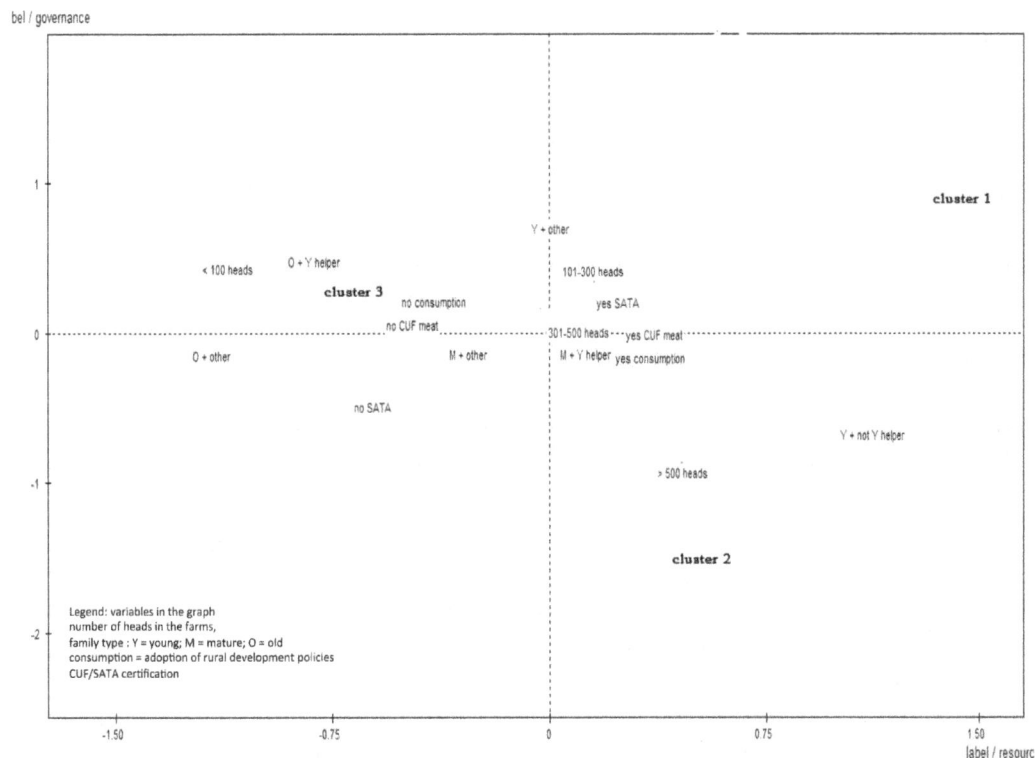

Figure 2. Distribution of clusters and key variables on the basis of the first two factorial coordinates.

5. Discussion and Conclusions

Rindfleisch et al. [21] 's stimulus "to continue to explore the relationship between resources and capabilities upon both transaction costs and governance mechanisms" has been acknowledged in this paper. GVA seems particularly effective in describing links between positioning, the availability of resources, and modes of governance. Actually, three different value propositions [8] characterize our farms, generating three different modes of governance. Moreover, if resources may be defined "as the scarce and imperfectly mobile skills, assets, or capabilities owned by a party to an exchange" [8] (p. 14), a resource-based view permits the cluster to be clearly differentiated. Therefore, if, on the one hand, GVA retains the key elements of transaction costs theory, on the other hand, through the concept of strategizing calculus, it revisits the positioning process by taking into account firms' core competencies and motivations.

Our empirical analysis demonstrates how farms have coherently chosen couples (positioning/governance) fitting best with their resources and competencies. Three different trajectories have been pointed out, where, coherently with GVA, farms design modes of governance with the aim of joint value maximization (value creation and value claiming) [24]:

1. The first one concerns the majority of farms with limited available resources: value creation and value claiming are a consequence of this. If the farm has no key resources (for example, in cases of farms located in the elderly phases of the life cycle), it has to opt out of differentiation strategies. As a matter of fact, value proposition is based on generic quality and on generic supply chains, where the product is sold through wholesaler or informal food networks. Accordingly, price becomes the most dominant and the most relevant variable in a competitive arena where imported (low price) meat is the reference.

2. A second trajectory involves farms with a clear differentiation strategy based on a local regional brand (Buongusto). In this case, key resources can be put into play and imply a more coordinated mechanism of governance, which calls for resources to be adequate: to support this strategy, tangible, intangible, and human resources are involved. Accordingly, value creating and claiming implies the involvement of vertical systems of governance where the institutional role of the regional breeders' association is fundamental. As a matter of fact, this association acts as a meso-institution [23], with the purpose of facilitating organizational arrangements in the meat value chain.

3. Finally, a third trajectory adpots a deeper market strategy, based on the recognition of a geographical indication of central Italy, entailing a dedicated form of governance. Value creation is targeted to a deeper market and value claiming leads to specific governance.

Relational bilateral governance characterizes the farms of the clusters, even though two clusters provide qualified partners, while the third one works in generic circuits with non-written contracts and generic conditions to be respected.

Previously described strategies lead to different implications for the modes of governance along the food chain. As a matter of fact, to become credible, a quality signal raises the costs of the governance of intermediate transactions [15]; this is particularly true in cases of complex food chains, enhancing dedicated mechanisms of governance. Nonetheless, the core competencies perspective has permitted a more integrated approach to the individuation of strategies and of "coherent" modes of governance. Therefore, we think that this paper provides a contribution to the literature by integrating transaction costs economics with resources-based perspectives and, thus, clarifies modes of governance as the outcome of this double perspective strategizing calculus.

Finally, the alignment between positioning and the mode of governance feeds the value creation, put into effect thanks to a premium price achieved by farms within quality circuits [37]. However, considering the premium price in detail, the interviews with stakeholders, and our analyses of the main Italian markets, we present evidence that the price differential between PGI and Buongusto farms and not certified farms is more or less than 10%. This aspect may imply a misalignment between value creation and value claiming [24] and, as a consequence, this may partly explain the 60% of our sample not involved in the quality schemes. Consequently, creating value may be an insufficient strategy in cases of difficulty of value claimed by farms [38]. On the other hand, these farms prefer to diversify agricultural production, in order to escape foreign competition.

Set against this background, GVA has revealed its effectiveness in discriminating three different strategizing calculi in the same territorialized agrifood system, thus confirming Menard's [11] idea of the diversity of institutional rules.

Of course, our paper suffers limits related to the restricted sample investigated and to the regional level of the analysis. Therefore, further analyses are necessary to consolidate the approach and above all to evaluate the economic performance of the various marketing channels. Nonetheless, we think that GVA could be supportive to enlarge the transaction costs perspective, thus taking into consideration resources and competencies, as suggested by Rindfleisch et al. [21]. Moreover, normative implications can be drawn from our analysis, starting from Menard's idea that different organizational arrangements are not to be considered as optimal solutions, but they *leave room for adaptation and evolution* [23] (p. 574). In this trajectory, the consideration of a farm's core competencies and resources may address policy implication, within the framework of the new rural development policies for the period 2014–2020. The large set of measures aiming at boosting collective action (funds for either collective investments, or collective brands, or innovations, etc.), in combination with the classic instruments for a farm's structural adjustments (generational renewal, measure to upgrade entrepreneurial skills of farmers, etc.), may be at the basis of future strategies. The role of rural policies in acquiring specific resources seems remarkable and may involve all farms classified in the three clusters. This could redefine future trajectories of value proposition and, consequently, new organizational arrangements based on new couples and higher expected incomes.

Acknowledgments: Authors would like to thank Mrinal Ghosh for very useful suggestions. This work is part of the project "Caratterizzazione e miglioramento degli indici salutistici e sicurezza alimentari delle produzioni lattiero ovine tipiche abruzzesi a marchio di origine," supported by a grant from Rural Development Plan 2007–2013 (MISURA 1.2.4) Regione Abruzzo. Project manager Giuseppe Martino. The authors are grateful to "Associazione Regionale Allevatori d'Abruzzo" for their kind cooperation.

Author Contributions: The presented research was conjointly designed and elaborated. The discussion was realized conjointly by all authors and all authors contributed equally to the writing of this paper. All authors have read and approved the final manuscript.

Conflicts of Interest: The authors declare no conflict of interest. The founding sponsors had no role in the design of the study; in the collection, analyses, or interpretation of data; in the writing of the manuscript, and in the decision to publish the results.

Abbreviations

ARA	Regional Association of Breeders
CUF	Control system quality of meat
LAFs	Localised agrifood systems
GVA	Governance value analysis
PGI	Protected Geographical Indication
RBV	Resources-based view
SATA	Technical Assistance Service to Breeders
VPMGA	Value Portfolio and Multifunctional Governance Analysis
TCE	Transaction costs economics

Appendix A

Date _____   Data collector _____
Farm name
Municipality   Address
Identification Code
Number of heads:.......
Race...

    Type of tenure: Family farming   ☐   No. of workers...........
    With salaried workers   ☐   No. of workers.........
        Cooperative   ☐   No. of workers.............
        Other   ☐   No. of workers............
    Certifications: CUF Milk   ☐   CUF Meat   ☐   SATA   ☐
    "Buongusto Agnello d'Abruzzo" brand   ☐
       "Agnello del Centro Italia" PGI   ☐
Organic farming certification (related to breeding crops)   ☐
In transition towards organic farming certification (related to breeding crops)   ☐
Organic livestock certification   ☐
In transition towards organic livestock certification   ☐

Breeding system
Wild (without any building)   ☐   Semi-wild (with buildings but not stables)   ☐
Extensive (stabled and on pasture)   ☐   Grazing months per year
Intensive (only stabled)   ☐

Transhumance towards mountain pastures: NO ☐   YES ☐
Municipality ....
Place ....
Distance ....

    Meat marketing
    Average weight of selling/slaughtering lambs _____   Average age of the lambs (days)
    _____
    Selling to dealers   ☐   __ % of heads
    Direct selling within the farm shop   ☐   __ % of heads
    Other types of direct selling: ☐   __ % of heads
    - *Ethical Purchasing Groups*   ☐
    - *Farmers' markets*   ☐
    - *Personal delivery*   ☐
    - *Internet*   ☐
    - *Other (specify)*   ☐
Butchers' shops   ☐   __ %
Retailers   ☐   __ %
Restaurants   ☐   __ %
Other (specify) _____   ☐ __ %
Milk production and selling
Milking: NO ☐   YES   ☐
Milk selling   NO ☐   YES   ☐
Dairy within the farm: NO ☐   YES ☐

Wool selling   NO ☐   YES ☐

Distribution of farm land used for breeding
Crops..........
Ha..........

References

1.   Torre, A.; Traversac, J.B. *Territorial Governance. Local Development, Rural Areas and Agrofood Systems*; Springer: Heidelberg, Germany; New York, NY, USA, 2011.

2.   Menard, C. Plural forms of organization: Where do we stand. *Manag. Decis. Econ.* **2013**, *34*, 124–139. [CrossRef]

3.   Stoker, G. Governance as theory: Five proposition. *Int. Soc. Sci. J.* **1998**, *50*, 17–28. [CrossRef]

4.   Hesterly, W.S.; Liebeskind, J.; Zenger, T.R. Organizational economics: An impending revolution in organization theory? *Acad. Manag. Rev.* **1990**, *15*, 402–420. [CrossRef]

5.   Hendrikse, G.W.J. Governance of chains and networks. A research agenda. *J. Chain Netw. Sci.* **2008**, *3*, 1–6. [CrossRef]

6.   Vihinen, H.; Kroger, L. The governance of markets. In *Unfolding Webs*; van der Ploeg, J.D., Marsden, T., Eds.; van Gorcum: Assen, The Netherlands, 2008.

7.   Williamson, O.E. The new institutional economics: Taking stock, looking ahead. *J. Econ. Lit.* **2000**, *38*, 595–613. [CrossRef]

8.   Ghosh, M.; John, G. Governance value analysis and marketing strategy. *J. Mark.* **1999**, *63*, 131–145. [CrossRef]

9.   Renting, H.; Vogelenzang, L.; Roep, D.; Oostindie, H.; van der Ploeg, J.D. Going backwards to find a way forward: The Netherland. In *Driving Rural Development: Policy and Practice in Seven EU Countries*; O'Connor, D., Renting, H., Gorman, M., Kinsella, J., Eds.; van Gorcum: Assen, The Netherlands, 2006; pp. 51–81.

10. Van der Ploeg, J.D. Rural development and territorial cohesion in the new CAP. Detailed briefing note. In *2010 European Parliament, Directorate General for Internal Policies, Policy Department B: Structural and Cohesion Policies, Agriculture and Rural Development*; European Parliament: London, UK, 2010.

11. Raynaud, E.; Valceschini, E. Creation and capture of value in sectors of the agrifood industry: Strategies and governance. In *Working Party on Agricultural Policies and Markets*; OECD: Paris, France, 2007.

12. Menard, C. The diversity of institutional rules as engine of change. *J. Bioecon.* **2014**, *16*, 83–90. [CrossRef]

13. Williamson, O.E. *The Economic Institutions of Capitalism*; Free Press: New York, NY, USA, 1985.

14. Menard, C. A new institutional approach to organization. In *Handbook of New Institutional Economics*; Menard, C., Shirley, M.M., Eds.; Springer: Berlin, Germany, 2008.

15. Menard, C.; Valceschini, E. New institutions for governing the agri-food industry. *Eur. Rev. Agric. Econ.* **2005**, *32*, 421–440. [CrossRef]

16. Raynaud, E.; Sauvée, L.; Valceschini, E. Aligning branding strategies and governance of vertical transactions in agri-food chains. *Ind. Corp. Chang. Adv.* **2009**, *18*, 1–34. [CrossRef]

17. Rantamäki-Lahtinen, L. The success of the diversified farm—Resource-based view. *Agric. Food Sci.* **2009**, *18*, 1–134.

18. Barney, J.; Wright, M.; Ketchen, D. The resource-based view of the firm: Ten years after 119. *J. Manag.* **2001**, *27*, 625–641.

19. Eghtedari, N.; Hosseini, M.; Malek Mohammadi, I.; Chizari, M. Resource-Based View, Innovative Orientation and Performance In Iran's Agricultural Advisory Services Corporations. *J. Appl. Sci. Agric.* **2014**, *9*, 68–76.

20. Williamson, O.E. Strategy research: Governance and competence perspectives. *Strateg. Manag. J.* **1999**, *20*, 1087–1108. [CrossRef]

21. Rindfleisch, A.; Antia, K.; Bercovitz, J.; Brown, J.R.; Cannon, J.; Carson, S.J.; Ghosh, M.; Helper, S.; Robertson, D.C.; Wathne, K.H. Transaction costs, opportunism, and governance: Contextual considerations and future research opportunities. *Mark. Lett.* **2010**, *21*, 211–222. [CrossRef]

22. Bradach, J.L.; Eccles, R.G. Price, authority, and trust: From ideal types to plural forms. *Annu. Rev. Sociol.* **1989**, *15*, 97–118. [CrossRef]

23. Menard, C. Embedding organizational arrangements: Towards a general model. *J. Inst. Econ.* **2014**, *16*, 567–589. [CrossRef]

24. Ghosh, M.; John, G. Progress and prospects for governance value analysis in marketing: When Porter meets Williamson. In *Handbook of Business-to-Business Marketing*; Lilien, G.L., Grewa, R., Eds.; Edward Elgar: Northampton, MA, USA, 2012; pp. 54–726.

25. Ghosh, M.; John, G. Strategic Fit in Industrial Alliances: An Empirical Test of Governance Value Analysis. *J. Mark. Res.* **2005**, *42*, 346–357. [CrossRef]

26. Williamson, O.E. *The Mechanisms of Governance*; Oxford University Press: Oxford, UK, 1996.

27. Raynaud, E.; Sauvée, L.; Valceschini, E. Governance of the agrifood chains as vector of credibility for quality signalisation in Europe. In Proceedings of the 10th EAAE Congress, Saragoza, Spain, 28–31 August 2002.

28. Milgrom, P.; Roberts, J. *Economics, Organization and Management*; Prentice Hall: Upper Saddle River, NJ, USA, 1992.

29. Nazzaro, C.; Marotta, G.; Pascucci, S. Creation and governance of value in agricultural cooperation: The role of policies. In Proceedings of the 126th EAAE Seminar, Capri, Italy, 27–29 June 2012.

30. Sotte, F.; Arzeni, A. Imprese e non-imprese nell'agricoltura Italiana. *Agriregionieuropa* **2013**, *32*, 65.

31. Bartoli, L.; De Rosa, M. Family farm business and access to rural development policies: A demographic perspective. *Agric. Food Econ.* **2013**, *1*, 12. [CrossRef]

32. Rudman, C.; Vesala, K.M.; Jäckel, J. Synthesis and recommendations. In *Entrepreneurial Skills and their Role in Enhancing the Relative Independence of Farmers. Results and Recommendations from the Research Project Developing Entrepreneurial Skills of Farmers*; Rudmann, C., Ed.; Research Institute of Organic Agriculture: Frick, Switzerland, 2008.

33. Greenacre, M.J. Canonical correspondence analysis in social science research. In *Studies in Classification, Data Analysis, and Knowledge Organization*; Locarek-Junge, H., Weihs, C., Eds.; Springer: Berlin/Heidelberg, Germany, 2010.

34. Lebart, L.; Piron, M.; Morineau, A. *Statistique Exploratoire Multidimensionnelle*; Dunod: Paris, France, 2006.

35. Maimon, O.; Rokach, L. Clustering methods. In *Data Mining and Knowledge Discovery Handbook*; Maimon, O., Rokach, L., Eds.; Springer: New York, NY, USA, 2010.

36. Benzècri, J.P. *L'analyse des Données—Leçons sur L'analyse Factorielle et Reconnaissance des Formes et Travaux*; Dunod: Malakoff, France, 1982.

37. Quiñones-Ruiz, X.F.; Penker, M.; Belletti, G.; Marescotti, A.; Scaramuzzi, S. Why early collective action pays off: Evidence from setting Protected Geographical Indications. *Renew. Agric. Food Syst.* **2016**, *32*, 179–192. [CrossRef]

38. Verwaal, E.; Commandeur, H.; Verbeke, W.; Wilem, J.M.I. Value creation and value claiming in strategic outsourcing decisions: A resource contingency perspective. *J. Manag.* **2009**, *35*, 420–444. [CrossRef]

6

Net Greenhouse Gas Budget and Soil Carbon Storage in a Field with Paddy–Upland Rotation with Different History of Manure Application

Fumiaki Takakai [1,*], Shinpei Nakagawa [2], Kensuke Sato [2], Kazuhiro Kon [2], Takashi Sato [1] and Yoshihiro Kaneta [1]

[1]   Faculty of Bioresource Sciences, Akita Prefectural University, 241-438 Aza Kaidobata-Nishi, Shimoshinjo Nakano, Akita 010-0195, Japan; t_sato@akita-pu.ac.jp (T.S.); ykaneta@akita-pu.ac.jp (Y.K.)

[2]   Akita Prefectural Agricultural Experiment Station, 34-1, Aza Genpachizawa, Yuwa Aikawa, Akita 010-1231, Japan; Nakagawa-Shinpei@pref.akita.lg.jp (S.N.); Sa-Ke@pref.akita.lg.jp (K.S.); kon-kazuhiro-085@pref.akita.lg.jp (K.K.)

*   Correspondence: takakai@akita-pu.ac.jp or ftakakai@gmail.com

Academic Editor: Ryusuke Hatano

Abstract: Methane (CH_4) and nitrous oxide (N_2O) fluxes were measured from paddy–upland rotation (three years for soybean and three years for rice) with different soil fertility due to preceding compost application for four years (i.e., 3 kg FW m^{-2} $year^{-1}$ of immature or mature compost application plots and a control plot without compost). Net greenhouse gas (GHG) balance was evaluated by integrating CH_4 and N_2O emissions and carbon dioxide (CO_2) emissions calculated from a decline in soil carbon storage. N_2O emissions from the soybean upland tended to be higher in the immature compost plot. CH_4 emissions from the rice paddy increased every year and tended to be higher in the mature compost plot. Fifty-two to 68% of the increased soil carbon by preceding compost application was estimated to be lost during soybean cultivation. The major component of net GHG emission was CO_2 (82–94%) and CH_4 (72–84%) during the soybean and rice cultivations, respectively. Net GHG emissions during the soybean and rice cultivations were comparable. Consequently, the effects of compost application on the net GHG balance from the paddy–upland rotation should be carefully evaluated with regards to both advantages (initial input to the soil) and disadvantages (following increases in GHG).

Keywords: carbon dioxide; methane; nitrous oxide; paddy–upland rotation; preceding compost application; rice; soybean

1. Introduction

"Paddy–upland rotation" involves alternating between rice cultivation in paddy fields and upland crops cultivation in drained paddy fields every few years. Recently, this has become a popular practice for adjusting rice production to decreasing demands in Japan. Currently, soybean is a major crop cultivated in the converted uplands in Japan [1]. In the northern part of Japan that has heavy snow in winter, cultivation of rice or upland crops generally occurs once a year.

Soil conditions in such rotated paddy fields could change drastically along with the cycle of flooding and drainage. Furthermore, the change in soil conditions influences greenhouse gas (GHG) dynamics greatly. Flooded paddy fields are a major source of methane (CH_4, the second most major GHG) due to anaerobic decomposition of organic matter under reductive conditions [2]. In comparison, upland fields emit carbon dioxide (CO_2, the most major GHG) derived from aerobic decomposition of organic matter as well as nitrous oxide (N_2O, the third major GHG) derived from nitrogen in fertilizer, plant residue, and soil organic matter [2–4]. In particular, drained paddy fields can cause depletion of soil carbon accumulated during use as a paddy field [5,6]. It is reported that N_2O emissions from upland fields tend to be higher in poorly drained soils [7], which are major soils for paddy–upland rotation.

It was reported that CH_4 emission from a paddy field converted from an upland field decreased significantly compared to a continuous paddy field in the first year after conversion in Japan [8,9]. Furthermore, the suppressing effect of CH_4 emission was found notably in the field incorporated with rice straw [10]. However, the length of this effect has not been well clarified yet. In an upland field converted from a rice paddy field, although N_2O emission is increased compared to a continuous paddy field, the decrease in CH_4 emission exceeds the increase based on CO_2 equivalent evaluation [9,10].

However, to our knowledge, reports on GHG balance are sparse, including CO_2 emission in paddy–upland rotation systems. In addition, there have been no reports on the GHG balance in paddy rice–upland soybean rotation, which is the major system in northern Japan.

Recently, depletion in available soil nitrogen has been reported in fields with paddy (rice)–upland (soybean) rotation in northern Japan [11,12]. To maintain the soil nitrogen availability of rotated paddy fields, application of organic matter (e.g., compost and green manure) is considered to be an efficient practice. Organic matter application, such as manure compost to paddy fields, could increase soil carbon storage [11,13]. In comparison, organic matter application could enhance CH_4 emissions from paddy fields [14] and N_2O emissions from upland fields [15,16]. However, reports on the effects of organic matter application are also sparse, including preceding application on GHG balance in paddy–upland rotation system.

Consequently, the objective of this study was to evaluate the effect of preceding manure compost application on GHG balance in a paddy–upland rotation field in northern Japan.

2. Materials and Methods

2.1. Site Description and Plant Cultivation

The experiment was conducted at the lysimeter plots of the Akita Prefectural Agricultural Experiment Station (39°35' N, 140°12' E), Akita, Japan for six years (June 2008 to May 2014). Three lysimeter plots (15 m^2 in area and 2 m in depth for each plot) were filled with gray lowland soil (Eutric Fluvisols; Food and Agriculture Organization/UNESCO) with subsurface drainage at 60 cm depth in each plot. Soil fertilities differed among the plots due to preceding compost application to forage rice (*Oryza sativa* L. cv. Bekoaoba) cultivation for four consecutive years (2004–2007) before this study. During the cultivation, 3.0 kg m^{-2} (as fresh matter) of immature or mature compost made of livestock manure (mixture of poultry/swine/cattle = 2:3:7, C/N ratios: 18.2–24.3 and 10.0–16.9, respectively) was applied to the plots (i.e., immature compost or mature compost plots, respectively) each year. In the control plot, forage rice was cultivated without compost application. Chemical fertilizers were applied to all plots equally. All treatments were conducted with one replication (lysimeter). More detailed information is provided in related papers [17,18]. The chemical properties of studied soils (0–10 cm) in each plot are provided in detail by Takakai et al. [17]. Briefly, the soil pH and cation exchange capacity (CEC) ranged from 5.6 to 6.0 and 21.8 to 23.9 $cmol_C$ kg^{-1}, respectively. The total nitrogen contents of the immature and mature compost plots (2.03 and 2.14 g kg^{-1}, respectively) were higher than that of the control plot (1.67 g kg^{-1}) at the beginning of the experiment. The total carbon contents of the immature and mature compost plots (25.0 and 26.4 g kg^{-1}, respectively) were also higher than that of the control plot (18.0 g kg^{-1}) (Table S1).

The mean annual temperature and precipitation recorded by the automated meteorological data acquisition system (AMeDAS) at a location 5 km from the experimental field are 10.9 °C and 1775 mm, respectively (Table S2). In this region, snowfall is generally observed from December to March.

Soybean (upland) and staple rice (paddy) was cultivated for the three consecutive years (2008–2010 and 2011–2013), respectively. Plant cultivation was conducted based on the guidelines of Akita

Prefecture [19,20]. All plots did not have applications of any organic materials throughout the study period. All agricultural practices (i.e., chemical fertilizers and agrochemicals) were applied to all plots equally. Detailed information about plant cultivation was described in our previous paper [18].

Soybean (*Glycine max* (L.) Merr. cv. Ryuho) was cultivated with applying chemical fertilizer at the rate of 0 or 2 g N m^{-2} (ammonium sulfate) as a basal fertilizer to all plots in 2008 or 2009 and 2010, respectively. In all years, top-dressing of fertilizer was not conducted. Soybean seeds were stripe-sown (10.9 plants m^{-2}) in early June. Ridging was conducted once at the end of June or July. Plant residue after harvesting (early October) was subsequently scattered to each plot. After the harvesting, ridges were leveled with a hoe.

Thereafter, staple rice cultivar (*Oryza sativa* L. cv. Yumeobako or Akitakomachi for 2011 or 2012–2013, respectively) was cultivated for three years. In the middle of May, plowing and basal fertilizer application (chemical fertilizer with the rate of 0 or 6 g N m^{-2} (ammonium sulfate) was applied to all plots in 2011 or 2012 and 2013, respectively), with puddling conducted. In late May, transplanting was carried out at a density of 20.8 hills m^{-2}. In late July, a total of 3 or 2 g N m^{-2} of chemical fertilizer (2011 or 2012–2013, respectively) was top-dressed. During the flooding period before mid-season drainage, flooding water depth was kept around 3–5 cm by irrigation and surface drainage. Mid-season drainage was conducted from late June to the middle of July according to plant growth and field moisture condition. Intermittent drainage was carried out during the end of mid-season drainage and final drainage in late August. Harvest was conducted in late September. Rice straw after harvesting was subsequently scattered to each plot and was left until plowing in the next spring.

For the first year of soybean and rice cultivation (2008 and 2011), to avoid over-luxuriant growth of crops due to increased soil nitrogen supply caused by paddy–upland rotation, basal fertilizers were not applied based on the guidelines of Akita Prefecture [19,20]. The plant nitrogen accumulations in 2008 (soybean) and 2011 (rice) were similar to or higher than those in the corresponding other two years despite no application of basal fertilizer [18].

2.2. CH$_4$ and N$_2$O Fluxes

CH$_4$ and N$_2$O fluxes were measured using a closed-chamber method based on the method described in Takakai et al. (for soybean upland [17,18] and rice paddy [18,21]).

For soybean uplands, cylindrical stainless-steel chambers (18.5–21.0 cm in diameter and 25 cm in height) were used for the measurements. Two stainless-steel bases equipped with a groove for sealing by water were installed into the soil between the rows and on the rows of each plot. In total, the gas flux measurement was conducted for each plot with four replicates. After ridging, the difference in height of measurement points between "inter-rows" and "on the rows" was approximately 20 cm. During the snow period, gas fluxes from the snow surface were measured by inserting the chamber into snow directly. Measurements were conducted almost once a week during the growing period (June–October) and one to three times per month during the fallow season (December–May, including the snow period). The frequency of measurement increased during one month after fertilization, plowing and sowing. Gas samples were taken at 0 and 20 min after the chamber was closed. Soil temperature at a depth of 5 cm and volumetric soil water content at a depth of 0–6 cm was measured simultaneously with each gas flux measurement by thermometer and amplitude domain reflectometry (ADR, ML2 Theta Probe Delta-Y Devices, Cambridge, UK), respectively. The volumetric soil water content was converted into a value of water filled pore space (WFPS) by soil porosity measured using soil core samples.

For rice paddies, rectangular transparent acryl chambers (30 × 60 × 50 or 100 cm in length × width × height) were used for the measurements during the rice growing period (end of May to late September). The flux measurement was conducted with three replicates per plot. Gas samples were taken at 1, 11 and 21 min after the chamber was closed. During the fallow period, gas fluxes from soil surface were measured in the same manner with soybean upland. Measurements were conducted almost once a week during the growing period and one to three times per month during the fallow season. Soil redox potential (Eh) at a depth of 5 cm was measured using platinum-tipped electrodes and a portable Eh meter (PRN-41, Fujiwara Scientific Company Co. Ltd., Tokyo, Japan) simultaneously with each measurement of gas fluxes. After transplanting rice, three electrodes were inserted into the soil at a depth of 5 cm per plot and kept in place throughout the rice growing period. Soil temperature was also measured. To avoid any disturbance during the flux measurements, all operations were performed from a boardwalk.

The CH_4 and N_2O concentrations were analyzed using a gas chromatograph (GC-14B, Shimadzu, Kyoto, Japan) equipped with a flame ionization detector and an electron capture detector, respectively. CH_4 and N_2O fluxes were calculated using a linear regression method. Annual emissions of CH_4 and N_2O were calculated by integrating the daily fluxes by linear interpolation.

2.3. Soil Carbon Storage and Decrease Rate

Changes in soil carbon storage were calculated by using the soil samples obtained by Takakai et al. [18]. Briefly, bulk soil samples were taken at three different depths (0–10, 10–20, and 20–30 cm), with three replicates conducted at each plot before the start of this study (November 2007), after 3 years of soybean cultivation (May 2011) and after 3 years of rice cultivation (April 2014). The total carbon content of air-dried, sieved (2-mm mesh) and finely ground samples were measured using an N/C analyzer (NC-900 and NC-22F, Sumika Chemical Analysis Service, Ltd., Osaka, Japan). Soil carbon storage (0–30 cm) was calculated based on soil mass [3,22], using the value of bulk density in May 2008 (before plowing) as described in Takakai et al. [18].

In this study, assuming that all carbon losses from soil contributed to CO_2 and CH_4 emissions, annual CO_2 emissions from soils were calculated by subtracting annual CH_4 emissions from the annual rate of carbon loss.

2.4. Net Greenhouse Gas Balance

Net GHG balance in the field was calculated by integrating CH_4, N_2O and CO_2 emissions described above as equivalent to CO_2. Global warming potentials for CH_4 and N_2O were 34 and 298, respectively [2].

2.5. Statistical Analyses

Differences in cumulative GHG emissions among the plots were compared by two-way analysis of variance (ANOVA, Year × Plot) followed by a Tukey's test. In this study, differences with $p < 0.10$ were considered significant. For all statistical analyses, Excel Statistics 2012 for Windows (SSRI, Tokyo, Japan) was used.

3. Results

3.1. GHG Emissions from the Upland Soybean Field

In all years, there were no differences in soil temperature and WFPS among the plots (Figure 1). The WFPS mainly ranged from 40% to 60% during the growing period, and tended to increase after harvesting with an increase in precipitation. Episodic CH_4 emissions were randomly found irrespective of the plots. Significant N_2O fluxes from all plots were found after sowing (June–July) and after harvesting (October). On the other hand, there was no significant N_2O flux during the snow and snow-melting period. In the first year (2008), N_2O flux from the mature compost plot was lower than those in the other two plots. In comparison, in the second (2009) and third years (2010), high N_2O flux was sometimes found even in the mature compost plot.

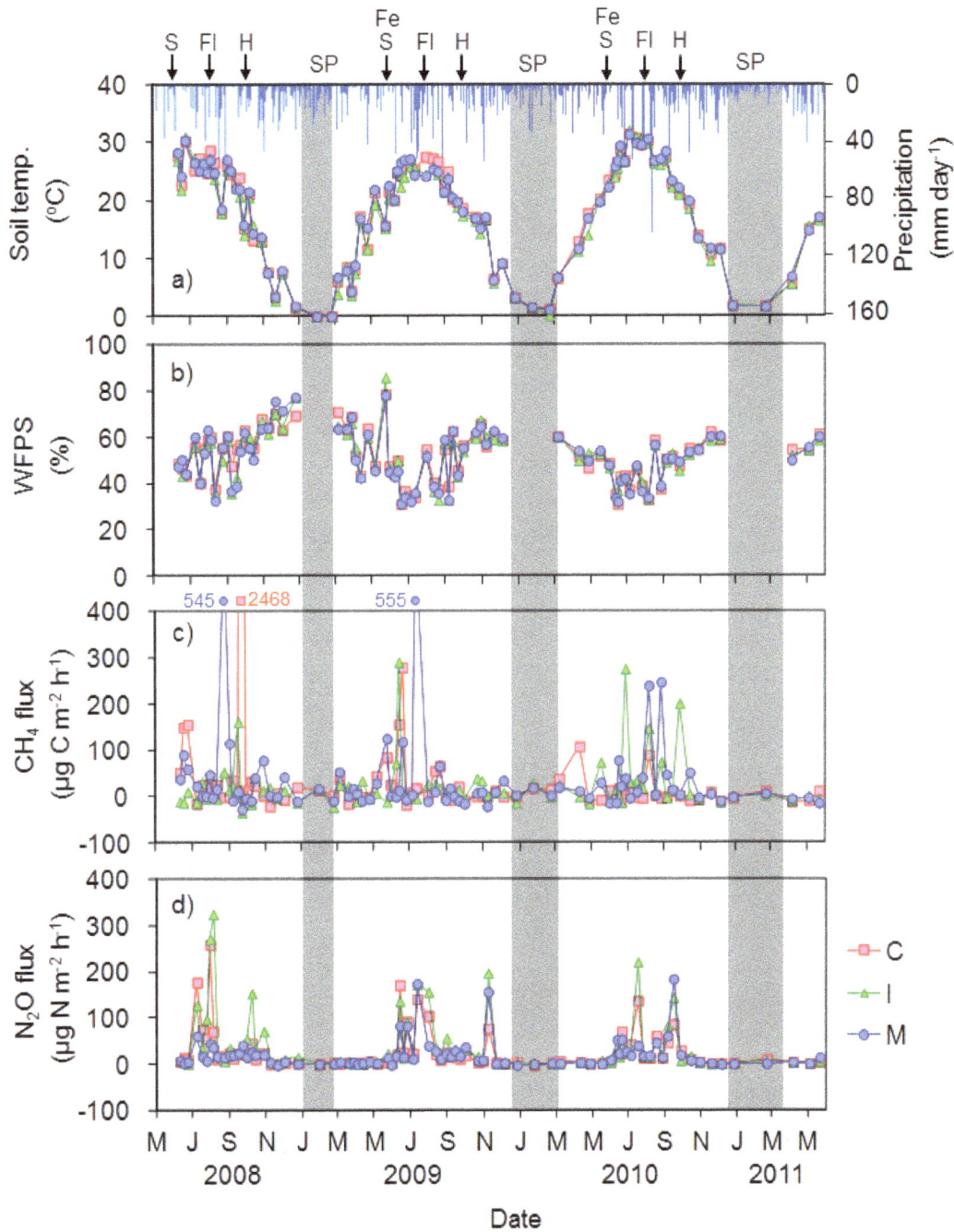

Figure 1. Seasonal changes in (**a**) soil temperature and precipitation (bars); (**b**) water filled pore space (WFPS); as well as (**c**) methane (CH_4) and (**d**) nitrous oxide (N_2O) fluxes during the soybean cultivation period (upland). Positive flux values indicate emission to the atmosphere and negative indicate uptake from the atmosphere. C: control; Fe; Fertilizer application; Fl: flowering stage; H: harvesting; I: immature compost; M: mature compost; S: sowing; SP: snow period (gray area).

During the soybean growing period after ridging, there was no significant difference in N_2O fluxes from inter-row and from on the rows, except for the immature compost plot, which had a tendency to have higher flux from those located on the rows (Figure 2, $p > 0.05$, paired t-test, statistical data not shown).

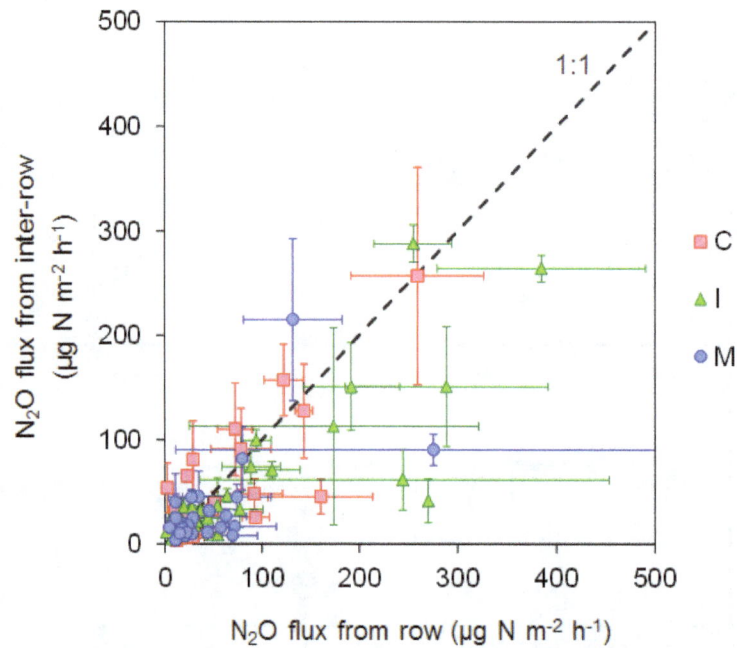

Figure 2. Comparison of N_2O flux from row and inter-low during the soybean cultivation period after ridging. Error bars indicate standard error. C: control; I: immature compost; M: mature compost.

The annual CH_4 emissions did not differ significantly among the years and the plots (Table 1). The annual N_2O emissions from soybean upland did not have obvious differences among the years. For the three-year average, the annual N_2O emissions increased in the following order: mature compost plot < control plot < immature compost plot. A significant difference was found between the mature and immature compost plots. GWP (CO_2 equivalent) of N_2O emissions were 6 to 19 times higher than those of CH_4 emissions.

Table 1. Annual methane (CH_4) and nitrous oxide (N_2O) emissions from the soybean cultivated field (upland) for three years (2008–2010).

Year	Plot	CH_4 (kg C ha^{-1} year^{-1})	N_2O (kg N ha^{-1} year^{-1})	GWP (Mg CO_2-eq ha^{-1} year^{-1}) [†] CH_4	N_2O	Total	Measurement Period
2008	C	5.51	1.73	0.25	0.81	1.06	
	I	0.89	2.70	0.04	1.26	1.30	10 Jun 2008 to
	M	3.14	0.88	0.14	0.41	0.56	8 Jun 2009
	Average [‡]	3.18 ± 1.33 [a]	1.77 ± 0.52 [a]	0.14 ± 0.06	0.83 ± 0.25	0.97 ± 0.22 [a]	
2009	C	1.96	1.85	0.09	0.87	0.96	
	I	1.35	2.73	0.06	1.28	1.34	8 June 2009 to
	M	2.49	1.90	0.11	0.89	1.00	11 Jun 2010
	Average [‡]	1.93 ± 0.33 [a]	2.16 ± 0.28 [a]	0.09 ± 0.01	1.01 ± 0.13	1.10 ± 0.12 [a]	
2010	C	0.15	1.44	0.01	0.67	0.68	
	I	1.94	1.75	0.09	0.82	0.91	11 Jun 2010 to
	M	1.73	1.29	0.08	0.61	0.68	22 May 2011
	Average [‡]	1.28 ± 0.57 [a]	1.49 ± 0.13 [a]	0.06 ± 0.03	0.70 ± 0.06	0.75 ± 0.07 [a]	
Three years average [‡]	C	2.54 ± 1.57 [A]	1.68 ± 0.12 [A,B]	0.12 ± 0.07	0.78 ± 0.06	0.90 ± 0.11 [A,B]	-
	I	1.40 ± 0.30 [A]	2.39 ± 0.32 [B]	0.06 ± 0.01	1.12 ± 0.15	1.18 ± 0.14 [B]	
	M	2.45 ± 0.41 [A]	1.36 ± 0.30 [A]	0.11 ± 0.02	0.64 ± 0.14	0.75 ± 0.13 [A]	

Positive values indicate net emissions to the atmosphere. [†] Calculated using the GWP of CO_2:CH_4:N_2O = 1:34:298 [2]. [‡] Values represent average $\pm$ standard error for three plots in each year or three years and numbers of CH_4 and N_2O emission and total GWP within a column followed by different letters differ significantly among the years (lowercase) or plots (uppercase) (Two-way ANOVA (Year $\times$ Plot) followed by Tukey's test, $p < 0.10$). C: control; I: immature compost; M: mature compost.

3.2. GHG Emissions from the Rice Cultivated Paddy Field

In all years, there were no differences in soil temperature and soil Eh among the plots (Figure 3). The soil Eh in all plots began to decrease after transplanting and increased to positive values during the mid-season drainage. The decrease in the third year (2013) tended to be faster than those in the first (2011) and second year (2012). The soil Eh after mid-season drainage in the first year tended to be higher than those in the other two years. In the first year, CH_4 fluxes before mid-season drainage from both compost application plots were higher than that from the control plot. In the second year, significant CH_4 fluxes after mid-season drainage were observed. In the third year, CH_4 fluxes were higher than those in the previous two years. In addition, the CH_4 fluxes in both compost application plots tended to be higher than that in the control plot until the rapid decrease in the immature compost plot caused by accidental drainage with drainage system trouble in late August. Throughout the three-year rice cultivation period, N_2O fluxes in all plots indicated slight uptake or emissions, aside for several episodic emissions. High N_2O emissions from the immature compost plot in late August of the third year could be caused by the accidental drainage.

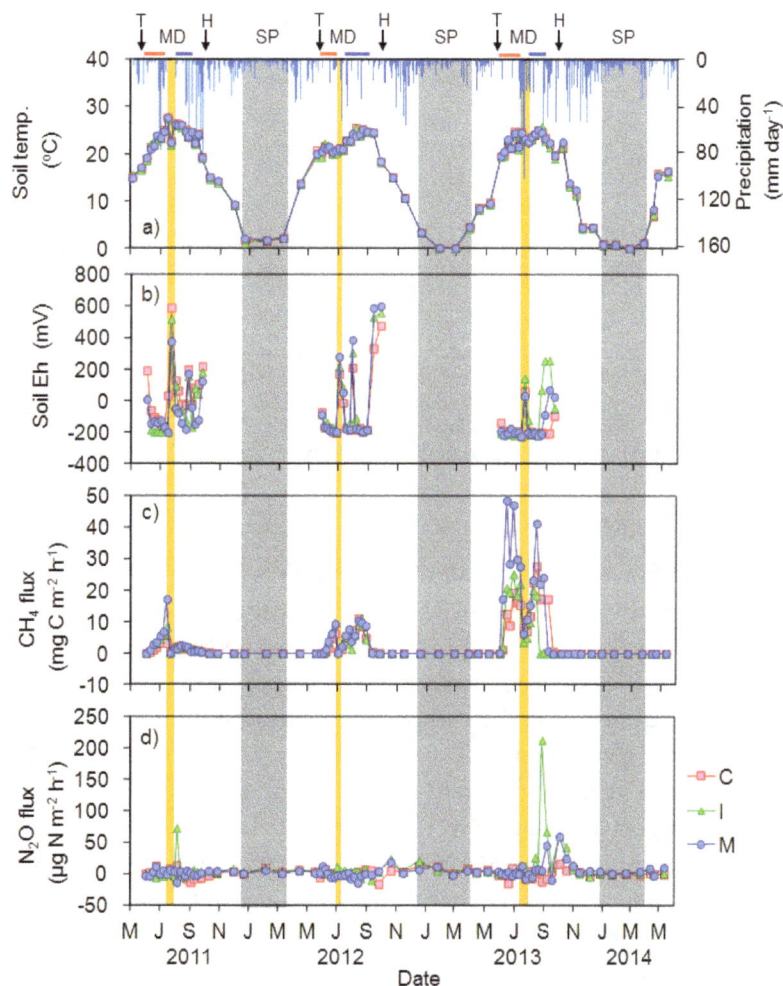

Figure 3. Seasonal changes in: (**a**) soil temperature and precipitation (bars); (**b**) soil Eh; as well as (**c**) methane (CH_4) and (**d**) nitrous oxide (N_2O) fluxes during the rice cultivation period (paddy). Positive flux values indicate emission to the atmosphere and negative indicate uptake from the atmosphere. In the late August 2013, accidental drainage with drainage system trouble occurred in the immature compost plot. C: control; H: harvesting; I: immature compost; M: mature compost; MD: Mid-season drainage (orange area); SP: snow period (gray area). Red lines before MD and blue lines after MD indicate continuous flooding and intermittent drainage, respectively.

The annual CH_4 emissions from all plots increased year by year and were significantly higher in the third year compared to those in previous two years (Table 2). For the three-year average, the annual CH_4 emissions increased in the following order: immature compost plot < control plot < mature compost plot, although there was no significant difference among the plots. The lowest CH_4 emission from the immature compost plot could be attributed to the accidental drainage in the third year. The annual N_2O emissions from all plots did not differ obviously among the years, and tended to be higher in both compost application plots compared to the control plot. GWP (CO_2 equivalent) of CH_4 emissions were 22 to 122 times higher than those of N_2O emissions.

Table 2. Annual methane (CH_4) and nitrous oxide (N_2O) emissions from the rice cultivated field (paddy) for three years (2011–2013).

Year	Plot	CH_4 (kg C ha^{-1} year^{-1})	N_2O (kg N ha^{-1} year^{-1})	GWP (Mg CO_2-eq ha^{-1} year^{-1}) [†]			Measurement Period
				CH_4	N_2O	Total	
2011	C	48.7	0.11	2.21	0.05	2.26	2 Jun 2011 to 17 May 2012
	I	71.5	0.37	3.24	0.17	3.41	
	M	83.5	0.18	3.78	0.08	3.87	
	Average [‡]	67.9 ± 10.2 [a]	0.22 ± 0.08 [a]	3.08 ± 0.46	0.10 ± 0.04	3.18 ± 0.48 [a]	
2012	C	106.9	0.28	4.84	0.13	4.97	27 May 2012 to 8 May 2013
	I	113.7	0.50	5.15	0.23	5.39	
	M	133.9	0.25	6.07	0.12	6.19	
	Average [‡]	118.1 ± 8.1 [a]	0.34 ± 0.08 [a]	5.36 ± 0.37	0.16 ± 0.04	5.52 ± 0.36 [a]	
2013	C	364.0	0.03	16.50	0.01	16.52	30 May 2013 to 12 May 2014
	I [§]	274.1	1.15	12.43	0.54	12.96	
	M	580.0	0.59	26.29	0.27	26.57	
	Average [‡]	406.1 ± 90.8 [b]	0.59 ± 0.32 [a]	18.41 ± 4.12	0.28 ± 0.15	18.68 ± 4.07 [b]	
Three years average [‡]	C	173.2 ± 96.9 [A]	0.14 ± 0.07 [A]	7.85 ± 4.39	0.07 ± 0.03	7.92 ± 4.37 [A]	-
	I	153.1 ± 61.7 [A]	0.67 ± 0.24 [A]	6.94 ± 2.80	0.32 ± 0.11	7.25 ± 2.91 [A]	
	M	265.8 ± 157.8 [A]	0.34 ± 0.13 [A]	12.05 ± 7.15	0.16 ± 0.06	12.21 ± 7.21 [A]	

Positive values indicate net emissions to the atmosphere. [†] Calculated using the GWP of CO_2:CH_4:N_2O = 1:34:298 [2]. [‡] Values represent average ± standard error for three plots in each year or three years and numbers of CH_4 and N_2O emission and total GWP within a column followed by different letters differ significantly among the years (lowercase) or plots (uppercase) (Two-way ANOVA (Year × Plot) followed by Tukey's test, $p < 0.10$). [§] Influenced by accidental drainage with drainage system trouble in the late August. C: control; I: immature compost; M: mature compost.

3.3. Changes in Soil Carbon Storage

Assuming that soil carbon storage was similar for all plots at a depth of 0–30 cm before compost application for forage rice cultivation (2004–2007), the application of immature and mature compost caused carbon storage to increase by 1.16 and 1.32 kg C m^{-2}, respectively, compared to the control plot over four years (Figure 4, Table S3). The increase in soil carbon storage occurred in the surface soil at a depth of 0–10 cm. The decreases in the carbon storage during the 3 years of soybean cultivation in the immature and mature compost plots (1.05 and 0.96 kg C m^{-2}, respectively) were higher than that in the control plot (0.33 kg C m^{-2}). Sixty-eight percent and 52% of the carbon increase from immature and mature compost application were estimated to be lost during the soybean cultivation period. During the soybean cultivation, soil carbon storage in the surface soil decreased remarkably in both compost application plots, with no change found in the control plot. On the other hand, during the rice cultivation period, the decreases in the carbon storage did not differ among the plots (0.17 to 0.25 kg C m^{-2}) and were lower than those during the soybean cultivation period. During the rice cultivation, soil carbon storage in surface soil (0–10 cm) tended to increase slightly in all plots, while soil carbon storage in subsoil (10–30 cm) decreased in all plots.

During the soybean cultivation period, the annual CO_2 emissions from soil in the control, immature and mature compost plots were 110, 350 and 315 g C m^{-2} year^{-1}, respectively. On the other hand, during the rice cultivation period, the CO_2 emission in the control, immature and mature compost plots were 58, 79 and 83 g C m^{-2} year^{-1}, respectively.

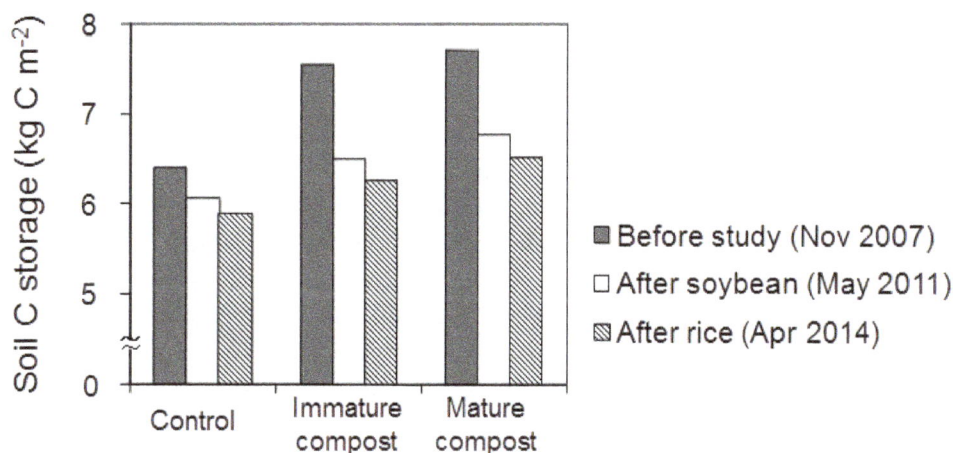

Figure 4. Decrease in soil carbon (C) storage (0–30 cm) under the soybean–rice cultivation.

3.4. Net GHG Balance

During the three years of soybean cultivation, the major component of net GHG balance was CO_2 emission (82–94% of the total; Figure 5). The net GHG balances in both compost application plots (12.3–14.0 Mg CO_2-eq ha^{-1} year^{-1}) were higher than that in the control plot (4.9 Mg CO_2-eq ha^{-1} year^{-1}) along with the difference in CO_2 emission. During the three years of rice cultivation, the major component of net GHG balance was CH_4 emission (72–84% of the total). The net GHG balance in the mature compost plot (14.3 Mg CO_2-eq ha^{-1} year^{-1}) was higher than those in the other two plots (9.6–9.9 Mg CO_2-eq ha^{-1} year^{-1}) along with the difference in CH_4 emission. Consequently, the net GHG balances during the rice cultivation period were higher than that in the control plot and were similar to or lower than those in both compost application plots during soybean cultivation period.

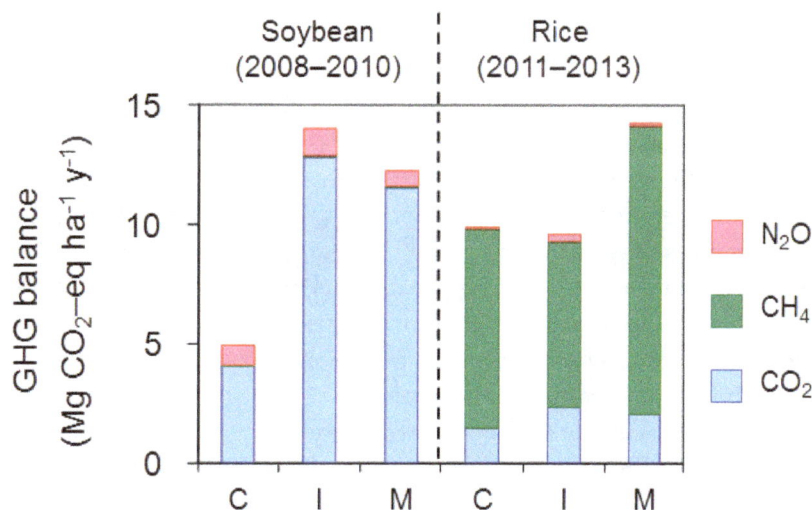

Figure 5. Comparison of net greenhouse gas (GHG) balance in soybean (upland) and rice (paddy) cultivation field. Positive values indicate net emission to the atmosphere. Carbon dioxide (CO_2) was estimated from soil carbon loss. Methane (CH_4) and nitrous oxide (N_2O) fluxes were obtained by the closed chamber method. C: control; I: immature compost; M: mature compost.

4. Discussion

4.1. N₂O Emissions from the Upland Soybean Field

Increase in N_2O flux in June and July (Figure 1) was consistent with the results from a soybean-cultivated converted paddy field in Yamagata, northern Japan [10]. It could be caused by increases in soil temperature and moisture for the period. The increase in N_2O flux after harvesting (October) could be derived from decomposition of plant residue and nodule [23]. There was no clear tendency between the N_2O fluxes from inter-row and from those on the rows with different conditions of the surface soil such as soil moisture across the plots (Figure 2). It suggested that there is only a minor contribution of N_2O production in the surface soil on N_2O flux. Therefore, it is considered that there is a significant contribution of N_2O production via denitrification near ground water table [24] fluctuating around drainage pipe buried at a depth of 60 cm.

Any clear trend in annual N_2O emissions from soybean upland was not found across the years (Table 1). N_2O production in soil is affected greatly by soil moisture condition [25]. In addition, N_2O emissions from upland fields tend to be higher in poorly drained soils [7]. Improvement of soil drainage properties in converted (drained) paddy fields with increasing year after paddy condition [6] may influence N_2O production and emission. In this study, there was a possibility that subsurface drainage pipe buried in the soil affected the soil drainage properties even in the first year after paddy condition.

Although there was a significant difference in N_2O emissions among the plots, the difference did not correlate with the available nitrogen in the soil (Tables 1 and S1). Therefore, the reason for the difference did not become clear in this study. The annual N_2O emissions in this study ranged from 0.88 to 2.73 kg N ha^{-1} year^{-1}. These values were similar to or slightly higher than the mean value of N_2O emissions from unfertilized uplands in Japan (0.36 and 1.4 kg N ha^{-1} year^{-1} for the well-drained and poorly-drained soils, respectively [7]). Furthermore, these values were lower than those from an onion-cultivated and unfertilized upland on gray lowland soil in Hokkaido (4.88 kg N ha^{-1} year^{-1}; [26]) in addition to those from a soybean-cultivated upland converted from a paddy field on gray lowland soil in Yamagata, northern Japan (3.3–4.4 kg N ha^{-1} year^{-1}; [10]).

4.2. CH₄ and N₂O Emissions from the Rice Paddy Field

The annual CH_4 emission from the rice paddy was the lowest in the first year after conversion and increased year by year (Table 2). In this study, the suppressing effect of the paddy–upland rotation on CH_4 emission [8,9] could not be evaluated quantitatively due to the absence of a continuous paddy field as a control. However, at least, the suppressing effect was considered to continue until the second year based on the comparison with the third year. The suppressing effect could be caused by the absence in rice straw, a major substrate for CH_4 production [10] and changes in availability of electron donors, redox status of soil Fe and activity of methanogens [27]. In the third year (2013) CH_4 fluxes before the accidental drainage at the immature compost plot increased in the following order: control plot < immature compost plot < mature compost plot (Figure 3). The order was consistent with the order of available nitrogen in surface soil shown in our previous paper [18]. A positive relationship between available nitrogen content and CH_4 production rate was found in Japanese paddy soils [28]. Therefore, increased mineralizable soil organic matter caused by preceding manure application could enhance CH_4 emission from rice paddy fields even in six years after application.

The annual N_2O emission from the rice paddy ranged from 0.03 to 0.59 kg N ha^{-1} year^{-1}, with the exception of high emission caused by the accidental drainage in the late August (the immature compost plot in the third year). These values were lower than the mean value of N_2O emissions from fertilized paddy fields with mid-season drainage in the world (0.99 kg N ha^{-1} season^{-1}; [29]).

4.3. CO₂ Emission and Net GHG Balance

The decreases in soil carbon storage during the soybean cultivation period were higher than those during the rice cultivation period (Figure 4). The result was consistent with previous reports [5,6,11].

It could be mainly attributed to the difference in oxidative–reductive status between the upland and paddy areas. Carbon loss from upland and grassland fields on brown lowland soil, brown forest soil and gray lowland soil in central Hokkaido estimated by the carbon budget were found to range from 193 to 410 $g\,C\,m^{-2}\,season^{-1}$ [30]. Carbon loss from an upland plot on andosol in Hokkaido estimated by change in soil carbon storage was 134 $g\,C\,m^{-2}\,year^{-1}$ [22]. Carbon loss from uplands converted from rice paddy estimated by continuous measurement of CO_2 flux were 275–343 and 256–361 $g\,C\,m^{-2}\,year^{-1}$ for upland rice and soybean-wheat, respectively [31]. The carbon losses estimated in this study were considered to be comparable with those values in previous reports.

At upland fields converted from paddy fields, trade-off relationships between decrease in CH_4 emission and increase in N_2O emission have been reported [9,10]. Shiono et al. [10] indicated that the suppressing effects of the paddy-upland rotation on GHG ($CH_4 + N_2O$) increased in the field with rice straw incorporation, because the decrease in CH_4 emission exceeded the increase in N_2O emission greatly. The results in this study agreed with the trend. However, significant CO_2 emission from soybean-cultivated upland occurred in this study (Figure 5). Taking the increased CO_2 emission into account, the suppressing effect of paddy–upland rotation on CH_4 emission may be canceled to some extent. A significant amount of accumulated soil carbon by the preceding compost application had been released during the soybean cultivation period. Thus, quantitative evaluation of change in soil carbon storage related to organic matter application is considered to be required in future.

5. Conclusions

Although compost application to the paddy–upland rotation system increased soil carbon storage, it also increased net GHG emission after application, including CO_2 emission during the soybean cultivation period under upland conditions. Therefore, to evaluate the effect of compost application to the net GHG balance from the paddy–upland rotation system, integration of both advantages (the initial input to the soil) and disadvantages (the following increase in GHG) should be conducted in the future.

Acknowledgments: We are deeply grateful to the staff members of the Akita Prefectural Agricultural Experiment Station (Kazuki Sekiguchi, Keiji Sasaki, and others) for their support in the management of the experimental field. We would also like to thank Keiko Hatakeyama, Emiko Sato, and Tomoko Suzuki for their great help on field survey and laboratory analyses.

Author Contributions: Kazuhiro Kon and Yoshihiro Kaneta conceived and designed the experiments; Fumiaki Takakai performed the experiments and analyzed the data; Kensuke Sato, Shinpei Nakagawa, and Kazuhiro Kon contributed cultivation and measurements in the lysimeter; Fumiaki Takakai wrote the paper; Takashi Sato and Yoshihiro Kaneta provided many constructive comments on this manuscript.

Conflicts of Interest: The authors declare no conflict of interest.

References

1. Takahashi, T.; Sumida, H.; Nira, R. A new framework for study of irrigated paddy rice and upland crops rotation farming and its relation to soil and plant nutrition science. 1. Advances and perspectives in irrigated paddy rice and upland crops rotation farming. *Jpn. J. Soil Sci. Plant Nutr.* **2013**, *84*, 202–207. (In Japanese).
2. Intergovernmental Panel on Climate Change (IPCC). The physical science basis: Anthropogenic and natural radiative forcing. In *Climate Change 2013*; Myhre, G., Shindell, D., Eds.; Cambridge University Press: Cambridge, UK, 2013.
3. Paustian, K.; Andrén, O.; Janzen, H.; Lal, R.; Smith, P.; Tian, G.; Tiessen, H.; van Noordwijk, M.; Woomer, P. Agricultural soil as a C sink to offset CO_2 emissions. *Soil Use Manag.* **1997**, *13*, 230–244. [CrossRef]
4. Bouwman, A.F. Exchange of greenhouse gases between terrestrial ecosystems and the atmosphere. In *Soils and the Greenhouse Effect*; Bouwman, A.F., Ed.; John Wiley: New York, NY, USA, 1990; pp. 61–127.

5. Mitsuchi, M. Characters of humus formed under rice cultivation. *Soil Sci. Plant Nutr.* **1974**, *20*, 249–259. [CrossRef]

6. Moroyu, H. Changes of physical and chemical properties of paddy soils under the cultivation of upland crops. *Jpn. J. Soil Sci. Plant Nutr.* **1983**, *54*, 434–441. (In Japanese)

7. Akiyama, H.; Yan, X.; Yagi, K. Estimations of emission factors for fertilizer-induced direct N_2O emissions from agricultural soils in Japan: Summary of available data. *Soil Sci. Plant Nutr.* **2006**, *52*, 774–787. [CrossRef]

8. Kumagai, K.; Konno, Y. Methane emission from rice paddy fields after upland farming. *Jpn. J. Soil Sci. Plant Nutr.* **1998**, *69*, 333–339. (In Japanese with English Summary)

9. Nishimura, S.; Akiyama, H.; Sudo, S.; Fumoto, T.; Cheng, W.; Yagi, K. Combined emission of CH_4 and N_2O from a paddy field was reduced by preceding upland crop cultivation. *Soil Sci. Plant Nutr.* **2011**, *57*, 167–178. [CrossRef]

10. Shiono, H.; Saito, H.; Nakagawa, F.; Nishimura, S.; Kumagai, K. Effects of crop rotation and rice straw incorporation in spring on methane and nitrous oxide emissions from an upland paddy field in a cold region of Japan. *Jpn. J. Soil Sci. Plant Nutr.* **2014**, *85*, 420–430. (In Japanese with English Summary)

11. Sumida, H.; Kato, N.; Nishida, M. Depletion of soil fertility and crop productivity in succession of paddy rice-soybean rotation. *Bull. Natl. Agric. Res. Cent. Tohoku Reg.* **2005**, *103*, 39–52. (In Japanese with English Summary)

12. Nishida, M.; Sekiya, H.; Yoshida, K. Status of paddy soils as affected by paddy rice and upland soybean rotation in northeast Japan, with special reference to nitrogen fertility. *Soil Sci. Plant Nutr.* **2013**, *59*, 208–217. [CrossRef]

13. Shirato, Y.; Yagasaki, Y.; Nishida, M. Using different versions of the Rothamsted Carbon model to simulate soil carbon in long-term experimental plots subjected to paddy–upland rotation in Japan. *Soil Sci. Plant Nutr.* **2011**, *57*, 597–606. [CrossRef]

14. Kumagai, K.; Shiono, H.; Morioka, M.; Nagasawa, K.; Nakagawa, F. Effect of application of livestock dung compost instead of spring incorporation of rice straw on methane emission from paddy fields in Yamagata. *Bull. Agric. Res. Yamagata Prefect.* **2010**, *2*, 1–18.

15. Toma, Y.; Hatano, R. Effect of crop residue C:N ratio on N_2O emissions from Gray Lowland soil in Mikasa, Hokkaido, Japan. *Soil Sci. Plant Nutr.* **2007**, *53*, 198–205. [CrossRef]

16. Akiyama, H.; Tsuruta, H. Nitrous oxide, nitric oxide, and nitrogen dioxide fluxes from soils after manure and urea application. *J. Environ. Qual.* **2003**, *32*, 423–431. [CrossRef] [PubMed]

17. Takakai, F.; Takeda, M.; Kon, K.; Inoue, K.; Nakagawa, S.; Sasaki, K.; Chida, A.; Sekiguchi, K.; Takahashi, T.; Sato, T.; et al. Effects of preceding compost application on the nitrogen budget in an upland soybean field converted from a rice paddy field on gray lowland soil in Akita, Japan. *Soil Sci. Plant Nutr.* **2010**, *56*, 760–772. [CrossRef]

18. Takakai, F.; Kikuchi, T.; Sato, T.; Takeda, M.; Sato, K.; Nakagawa, S.; Kon, K.; Sato, T.; Kaneta, Y. Changes in the nitrogen budget and soil nitrogen in a field with paddy–upland rotation with different histories of manure application. *Agriculture* **2017**, *7*, 39. [CrossRef]

19. Department of Agriculture, Forestry and Fisheries, Akita Prefecture. *Guidelines for Soybean Cultivation*; Department of Agriculture, Forestry and Fisheries, Akita Prefecture: Akita, Japan, 2015; pp. 26–29. (In Japanese)

20. Department of Agriculture, Forestry and Fisheries, Akita Prefecture. *Guidelines for Rice Cultivation*; Department of Agriculture, Forestry and Fisheries, Akita Prefecture: Akita, Japan, 2014; pp. 43–51. (In Japanese)

21. Takakai, F.; Ichikawa, J.; Ogawa, M.; Ogaya, S.; Yasuda, K.; Kobayashi, Y.; Sato, T.; Kaneta, Y.; Nagahama, K. Suppression of CH_4 emission by rice straw removal and application of bio-ethanol production residue in a paddy field in Akita, Japan. *Agriculture* **2017**, *7*, 21. [CrossRef]

22. Koga, N.; Sawamoto, T.; Tsuruta, H. Life cycle inventory-based analysis of greenhouse gas emissions from arable land farming systems in Hokkaido, northern Japan. *Soil Sci. Plant Nutr.* **2006**, *52*, 564–574. [CrossRef]

23. Uchida, Y.; Akiyama, H. Mitigation of postharvest nitrous oxide emissions from soybean ecosystems: A review. *Soil Sci. Plant Nutr.* **2013**, *59*, 477–487. [CrossRef]

24. Minamikawa, K.; Nishimura, S.; Nakajima, Y.; Osaka, K.; Sawamoto, T.; Yagi, K. Upward diffusion of nitrous oxide produced by denitrification near shallow groundwater table in the summer: A lysimeter experiment. *Soil Sci. Plant Nutr.* **2011**, *57*, 719–732. [CrossRef]

25. Davidson, E.A.; Keller, M.; Erickson, H.E.; Verchot, L.V.; Veldkamp, E. Testing a conceptual model of soil emissions of nitrous and nitric oxides. *Bioscience* **2000**, *50*, 667–680. [CrossRef]

26. Toma, Y.; Kimura, S.D.; Yamada, H.; Hirose, Y.; Fujiwara, K.; Kusa, K.; Hatano, R. Effects of environmental factors on temporal variation in annual carbon dioxide and nitrous oxide emissions from an unfertilized bare field on Gray Lowland soil in Mikasa, Hokkaido, Japan. *Soil Sci. Plant Nutr.* **2010**, *56*, 663–675. [CrossRef]

27. Eusufzai, M.K.; Tokida, T.; Okada, M.; Sugiyama, S.; Liu, G.C.; Nakajima, M.; Sameshima, R. Methane emission from rice fields as affected by land use change. *Agric. Ecosyst. Environ.* **2010**, *139*, 742–748. [CrossRef]

28. Cheng, W.; Yagi, K.; Akiyama, H.; Nishimura, S.; Sudo, S.; Fumoto, T.; Hasegawa, T.; Hartley, A.E.; Megoniga, J.P. An empirical model of soil chemical properties that regulate methane production in Japanese rice paddy soils. *J. Environ. Qual.* **2007**, *36*, 1920–1925. [CrossRef] [PubMed]

29. Akiyama, H.; Yagi, K.; Yan, X.Y. Direct N$_2$O emissions from rice paddy fields: Summary of available data. *Glob. Biogeochem. Cycles* **2005**, *19*. [CrossRef]

30. Mu, Z.J.; Kimura, S.D.; Hatano, R. Estimation of global warming potential from upland cropping systems in central Hokkaido, Japan. *Soil Sci. Plant Nutr.* **2006**, *52*, 371–377. [CrossRef]

31. Nishimura, S.; Yonemura, S.; Sawamoto, T.; Shirato, Y.; Akiyama, H.; Sudo, S.; Yagi, K. Effect of land use change from paddy rice cultivation to upland crop cultivation on soil carbon budget of a cropland in Japan. *Agric. Ecosyst. Environ.* **2008**, *125*, 9–20. [CrossRef]

7

Phenotypic Variability Assessment of Sugarcane Germplasm (*Saccharum officinarum* L.) and Extraction of an Applied Mini-Core Collection

Atena Shadmehr [1], Hossein Ramshini [1,*], Mehrshad Zeinalabedini [2,*] (ID),
Masoud Parvizi Almani [3], Mohammad Reza Ghaffari [2] (ID), Ali Izadi Darbandi [1] and
Mahmoud Fooladvand [3]

[1] Department of Agronomy and Plant Breeding Sciences, Agricultural College of Aburaihan,
University of Tehran, Emam reza Blvd, Pakdasht, Tehran 3391653755, Iran; at.shadmehr@gmail.com (A.S.);
aizady@ut.ac.ir (A.I.D.)

[2] Systems Biology Department, Agricultural Biotechnology Research Institute of Iran, Agricultural Research,
Education and Extension Organization (AREEO), 31359-33151 Karaj, Iran; ghaffari@abrii.ac.ir

[3] Department of Biotechnology, Cane Development and Sidelong Industrial Research and Education Institute,
Golestan Blvd, Khuzestan 1465834581, Iran; mparvizi_almani@yahoo.com (M.P.A.);
foolad594@yahoo.com (M.F.)

* Correspondence: ramshini_h@ut.ac.ir (H.R.); mzeinolabedini@abrii.ac.ir (M.Z.)

Academic Editors: Eva Johansson and Les Copeland

Abstract: The sugarcane germplasm collection located in Khuzestan, Iran, is one of the most important genetic resources with valuable accessions from different continents. However, this collection has not been properly used by breeders due to the extremely large population. The aim of this study was to phenotypically characterize the sugarcane germplasm and form a mini-core collection. Hence, 13 morphological traits were evaluated on 253 accessions. The primary germplasm was grouped into 10 clusters based on partial repeated bisection (RB) data, where the smallest cluster contained three accessions from two breeding centres (USA and Cuba). Using principal component analysis (PCA), the first two PCs (principal component) explained 59.5% of the total variation. A mini-core of 21 accessions was created by using the maximization strategy, with a low mean difference percentage (MD = 2.31%) and large coincidence rate of range (CR = 93.96%). The sugarcane mini-core represented the major diversity of the primary collection. The means and medians between the mini-core and the primary collection did not differ significantly. Accessions with high sugar and cane yield, originating from the USA, Cuba, Argentina, and South Africa, were in the mini-core collection. In this paper, we established, for the first time, an applied mini-core collection in sugarcane germplasm. The mini-core collection, as a breeding collection, is a highly suitable, manageable, and efficient subset for the enhanced use of sugarcane germplasm in breeding programs.

Keywords: breeding collection; cluster analysis; coincidence ratio; diversity; mini-core collection

1. Introduction

Sugarcane belongs to the genus *Saccharum* L., of the tribe *Andropogoneae* in the family of *Poaceae* (*Gramineae*), which is grown widely in tropical and subtropical regions across the world [1].

In Iran, there is a Cane Development and Sidelong Industrial Research and Education Institute located in Khuzestan province that was officially founded in 1981. The sugarcane germplasm collection located at this institute includes approximately 340 accessions from *Saccharum officinarum* L. and interspecific hybrids. This set has been collected from different continents, such as the USA, Cuba,

Brazil, Argentina, South Africa, India, and Australia, and has been maintained over the years in Iran. It seems that present accessions are well adapted to the climate of Iran, and the collection probably contains genotypes carrying genes for cold tolerance. The sugarcane collection in Khuzestan (primary collection), which has been selected manually from whole accessions across the world, is actually a core collection (CC). Due to the large size of the collection (340 accessions), there are problems concerning the management and use of this genetic resource. Hence, a diverse mini-core collection or breeding collection could be a beneficial resource for sugarcane breeders [2]. Formation of a reduced subset of sugarcane genetic resources in Khuzestan is very important for assessing the population structure and diversity and for identifying sources of variation in breeding programs. So far, there has been no research to assess the morphological diversity of the entire sugarcane germplasm in Iran, indicating the necessity of a comprehensive evaluation of the accessions in the mentioned collection.

Despite the diversity of resources and genotypes, one of the major problems faced during plant breeding programs is insufficient use of the whole germplasm, because it is difficult to manage and utilize a large number of genotypes in germplasm collections. It is accepted that recognition and use of diversity in germplasm is very important prior to starting a breeding program [3]. There is a growing body of literature that recognizes the importance of constructing a core or mini-core collection for conservation of novel variations in genetic resources [2,4,5]. A core collection is defined as a limited set of accessions representing the minimum of repetition and the maximum of the genetic diversity of a crop species and its wild relatives [6]. Similar to the definition of a core collection [6], a mini-core collection is defined as a reduced subset of a core collection with minimum repetitiveness and maximum genetic diversity of the source germplasm [7]. Previous studies have reported that a core collection would be better if it contains 10% of the entire collection and a mini-core collection contains 10% of the core collection or 1% of the entire collection [8]. A mini-core collection is important in order to identify sources of resistance to biotic and/or abiotic stresses and agronomic and nutritional traits. In sugarcane, it is arguable that the accessions with high sucrose content can be selected for sugar production. More importantly, such a subset of germplasm is also useful for recognition of superior energy cane cultivars, i.e., with high biomass and lignocellulosic compounds and fiber, which are appropriate for lignocellulosic ethanol production [9]. Additionally, a core or mini-core collection plays an important role in assessing allelic richness and association genetics, especially for identifying important QTLs (quantitative trait locus) controlling desirable traits, such as yield and fiber content.

Over the years, phenotypic traits have played an important role in genetic diversity analysis and in discriminating genotypes in order to form a core and/or mini-core collection [10–12]. Similarly, previous studies have reported establishing a core collection of germplasm in sugarcane based on morphological traits [2,13–15]. Phenotypic characterization of a germplasm collection is an increasingly important step in plant breeding programs since breeders can evaluate variations and select high performance accessions more efficiently. Although several studies have been carried out on core collection construction in sugarcane [2], no single study has established the breeding collection (BC), actually named mini-core collection, of sugarcane germplasm.

The goals of our work were (1) comprehensive evaluation of sugarcane genetic resources in Iran and measuring important phenotypic traits on four dates over two growing seasons and (2) to create a mini-core collection or breeding collection that captures most of the phenotypic diversity of the primary collection. The present research attempts, for the first time, to establish a mini-core collection in sugarcane germplasm. This study provides an exciting opportunity to advance our knowledge of germplasm characterization for discovering new sources of variations in order to enhance the use of germplasm for cultivar improvement.

2. Materials and Methods

2.1. Plant Materials

The materials comprised 253 sugarcane accessions collected from different continents. All accessions in the collection belong to the *Saccharum officinarum* L. and all of them are hybrids.

The accessions' names have been derived from their origin or breeding centre. For example, C85-102 is a variety of Cuban sugarcane, and the breeding centre of the FGO2-250 variety is France. These accessions were evaluated in a randomized complete block design with four replications at the experimental station of the Cane Development and Sidelong Industrial Research and Education Institute of Khuzestan, Iran in two successive seasons (2012 and 2013). The plot size was 100 m^2 including five rows, 20 m long with a between-row spacing of 1 m.

2.1.1. Measurements of Phenotypic Traits

Sugarcane planting was done in June 2010, and morphological traits and sugar yield factors were measured during two separate seasons. The first season was January 2012, and coincided with 18-month-old plants. The second season was comprised of three successive time points, namely November 2012, January 2013, and February 2013, and coincided with ratoon plants. The plan was to record data in the abovementioned time points in order to gain a detailed understanding of growth periods, and to assess the difference between them and, consequently, to recognize critical stages for gaining the optimum yield. Thirteen traits (Table 1) were measured for 253 accessions grown in the field, across two separate seasons. First, the whole collection (comprising 340 accessions) was chosen, but because of the expected difficulty of obtaining phenotypic traits for all accessions, and also because of missing data in sugar yield measurements such as Brix% or sugar%, measurements were restricted to 253 accessions. Stalk height (cm), internode length (cm), and stalk diameter (mm) were measured on five stalks at harvest according to Dillewijn [16] and were averaged subsequently. The canes were cut and the average weight of 10 canes was recorded. After peeling the selected canes, they were crushed in a mill to calculate the percentage and weight of juice. The Brix value (total soluble solids) was determined with a refractometer, and Pol (apparent sucrose) in sugarcane juice was determined with saccharimeter (a saccharimeter is simply a polarimeter specially designed for measuring the polarization of sugars). Purity is the percentage of pure sucrose in dry matter that was calculated as [Pol/Brix] × 100. In order to calculate the recoverable sucrose (%) (RS), the following formula was used:

$$\text{Recoverable sucrose \%} = \left\{ \text{Pol\%} - \left(\frac{\text{Brix\%} - \text{Pol\%}}{2} \right) \right\} \times \text{juice extraction\%} \tag{1}$$

Cane yield was estimated as the number of stalk (m^2) × stalk weight. For calculation of sugar yield (kg sugar ha^{-1}), RS was multiplied with cane yield per hectare [17].

Table 1. Descriptors used for characterizing the sugarcane accessions of entire collection (253 accessions) in order to build a mini-core collection.

Traits	Description
Stalk Height (cm)	Stalk height from ground level to the insertion of the top visible dewlap leaf (TVD)
Stalk Internode Number	Number of internode
Stalk Internode Length (cm)	Length of internode
Stalk Diameter (mm)	Diameter of stalk
Stalk Weight (kg)	Ten-stalk weight in kilogram
Juice Weight (kg)	Weight of juice extracted after crushing the stems
Juice %	Percentage of juice extracted after crushing the stems
Brix %	Brix is the total soluble solids in the aqueous solution from the stalk as a percentage by weight (% w/w)
POL%	The apparent sucrose content expressed as a mass percent measured by the optical rotation of polarized light passing through a sugar solution
PTY%	Purity is the sucrose content as a percent of the dry substance or dissolved solids content.
RS%	Recoverable sucrose: The percentage of white sugar
Cane Yield (tone/ha)	Yield of cane stalk dependent on two factors: number of stalk and stalk weight.
Sugar Yield (tone/ha)	Yield of sugar dependent on two factors: cane yield and RS.

2.2. Statistical Analysis

2.2.1. Descriptive Statistics

Data management and analysis were carried out using SAS [18]. Descriptive statistics including mean, median, range, and variance were generated for all variables for both the primary germplasm and the mini-core collection. A normality test was performed with the Shapiro–Wilk W method. In this study, the analysis of variance (ANOVA) was conducted and the experimental design was considered as a randomized complete block design with four replications. Each time point was mentioned as a replication.

2.2.2. Correlation and Regression Analysis

Pairwise correlations were obtained by using the Pearson method in Proc Corr of SAS [18]. Stepwise multiple linear regression analysis was carried out using SAS program using Proc Reg [18]. In this analysis, recoverable sucrose (RS) was considered as dependent variable and the remaining traits as independent variables. Prior to analyzing the regression, in order to identify important characteristics affecting the amount of white sugar (RS), and moreover to avoid the loss of fit due to multicollinearity phenomenon (correlation between predictors that one can be linearly predicted from the others with a substantial degree of accuracy), the level of multicollinearity was estimated with the most widely used criterion of the Variance Inflation Factor (VIF) as suggested by Hair et al. [19]. High VIF values (above 10) indicate a high collinearity [19].

2.2.3. Clustering

To reduce data dimensions for better visualization of accessions and traits, principal component analysis (PCA) was conducted for 13 traits and 253 accessions using statistical software SPSS (version 19, IBM Corp, Armonk, NY, USA). Clustering analysis of accessions in the primary collection based on morphological traits was performed using gCLUTO software (version 1.0, University of Minnesota, Twin Cities, MI, USA) based on the RB (repeated bisection) method. The advantage of Graphical Clustering Toolkit (gCLUTO) [20] is the better visualization of the clusters.

2.3. Establishment of Mini-Core Collection

PowerCore software v 1.1 [21] was used to construct the mini-core collection by analysing phenotypic data using maximization strategy (M strategy). The M strategy was used to select entries of subset collection with highest diversity through a modified heuristic algorithm [22]. In order to compare means or medians of traits between primary and mini-core collection, t-test and Wilcoxon signed-rank method was used in Proc t-test and in Proc NPAR1WAY, respectively [18]. For traits with normal distribution, the t-test was performed, and for traits with non-normal distribution, the non-parametric Wilcoxon signed-rank method was used. The Ansari-Bradley test was performed using Proc NPAR1WAY to compare the variance of traits with non-normal distributions between the mini-core and primary collection. Moreover, comparison of the variances of traits with normal distributions, between two collections, was carried out using F test. The Shannon–Weaver diversity index [23] was estimated using PowerCore software based on all the traits to measure the diversity of accessions in the primary and mini-core collection, and finally this index was compared between entire collection and mini-core collection with pairwise t-test in SAS Proc means.

3. Results

3.1. Phenotypic Characterization & Trait Distributions

3.1.1. Morphological Traits

The average stalk height in the second season was 168.2 cm, which was significantly ($p \leq 0.05$) higher than the stalk height in the first year (125.6 cm). There were no significant differences in stalk height among the three time points of the second season (Figure 1a). The same trend was found for internode length, as the first season mean (8.8 cm) was lower than that of the second season (10.1 cm) (Figure 1b). Contrary to these findings, the average stalk diameter in the first season (23.4 mm) was higher than that of the second season (22.6 mm), and this difference was found to be significant ($p \leq 0.05$) (Figure 1c).

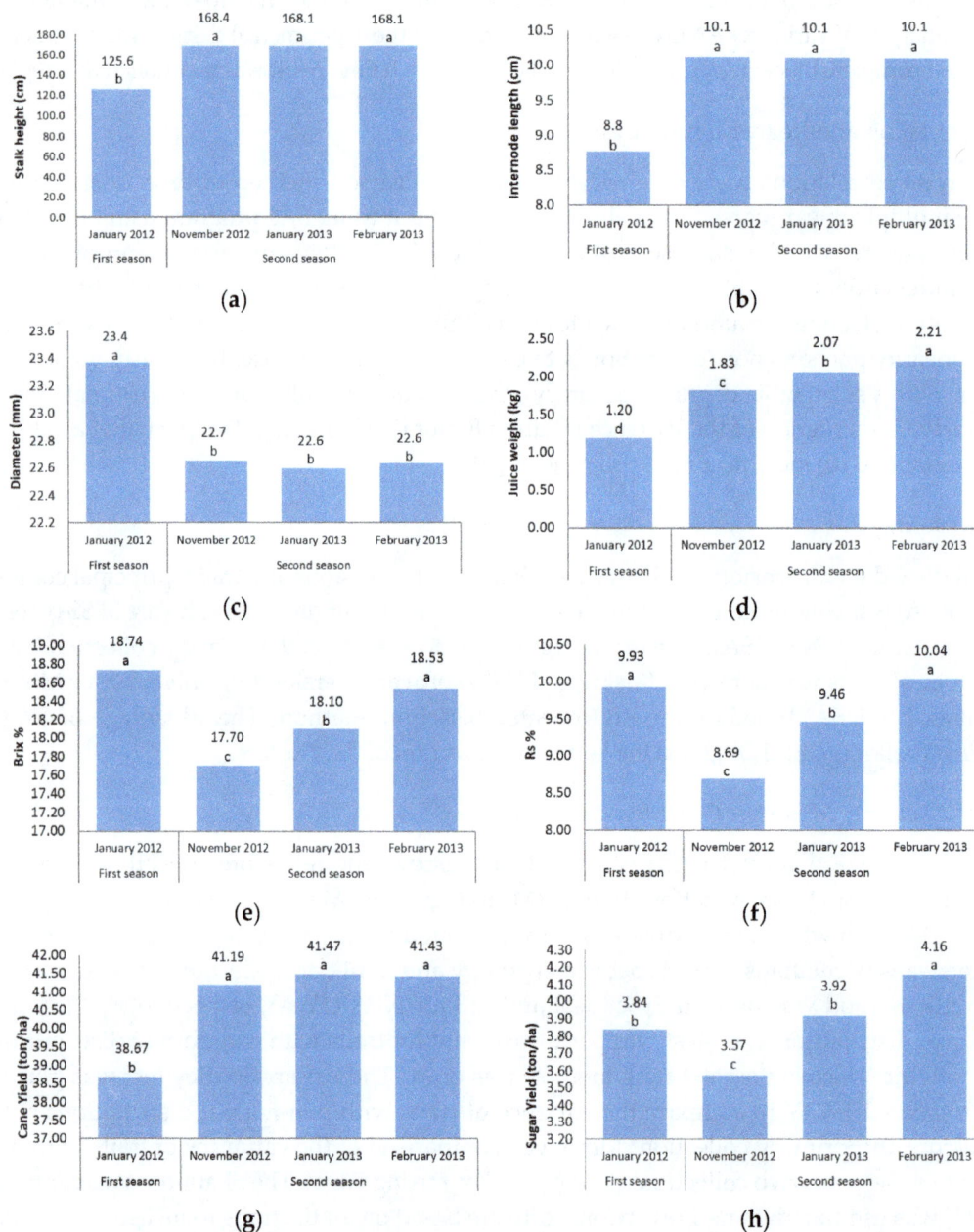

Figure 1. (**a**) Mean stalk height (cm); (**b**) Mean internode length (cm); (**c**) Mean diameter (mm); (**d**) Mean juice weight (kg); (**e**) Mean percentage Brix of juice extracted from stalks; (**f**) Mean recoverable sucrose; (**g**) Mean cane yield (ton/ha); and (**h**) Mean sugar yield (ton/ha) of primary collection of sugarcane accessions measured in two seasons (in four time points). In all Figures (**a–h**), means followed with the same letters are not significantly different at $p \leq 0.05$.

3.1.2. Sugar-Related Traits and Cane Yield

There was a clear trend of an increase in in juice weight during two seasons (Figure 1d). In February 2013, juice weight was significantly higher than those in other time points. The results obtained from the measurement of mean percentage of Brix are presented in Figure 1e. The mean percentage of Brix (18.74%) was at a maximum in January 2012. In November 2012, the value obtained was minimum (17.70%), but it increased in subsequent time points. It was 18.10% and 18.53% at harvest times in January 2013 and February 2013, respectively. There were significant differences ($p \leq 0.05$) between the three harvest times in the second season for this trait (Figure 1f). The trend for recoverable sucrose (RS %) was the same as that of Brix, i.e., the RS percentage was significantly higher in the first season than that of the second season. As shown in Figure 1g, the average cane yield in the second season was much higher than that in the first season. Cane yield in November 2012 was 25% higher than January 2012, but no significant differences were found between the three time points in the second season. Figure 1h provides an overview of mean sugar yield of the primary collection sugarcane accessions measured in four harvest time points. In the third time point in the second season (February 2013), the mean sugar yield was significantly higher ($p \leq 0.05$) than other time points.

The normality test was carried out according to the Shapiro–Wilk test. The results revealed that the distributions of all traits in the first season were non-normal, but in the second season, distributions for stalk height, stalk diameter, stalk internode number, cane weight, RS percentage, and sugar yield were normal. The distribution of the means of two seasons for stalk height, stalk internode length, stalk diameter, POL%, and RS% was normal, while for other traits it was non-normal.

3.2. Correlations Analysis of Traits Measured in Different Time Points

The pair-wise Pearson product-moment correlation analysis was used to determine the relationship between different harvest time points in two seasons in the primary collection (Table 2). The analysis showed medium R-value for most traits among different time points. Nevertheless, in several traits there were significant positive correlations among different harvest times. For example, a significant positive correlation ($p \leq 0.01$) was found between January 2013 and February 2013 for stalk height, internode length, stalk diameter and cane yield with R-values of 0.98, 0.99, 0.98 and 0.99, respectively. In contrast, the lowest correlations between time points were between January 2012 and November 2012 for Brix ($R = 0.21$) and RS ($R = 0.15$). Generally, as Table 2 shows, there are modest correlations among different time points for different traits.

Table 2. Correlation coefficients among months for the different traits measured in two seasons of the collection of sugarcane.

Traits	November 2014:January 2015	November 2014:February 2015	January 2015:February 2015	January 2014:November 2014	January 2014:January 2015	January 2014:February 2015
Stalk height	0.69 **	0.69 **	0.98 **	0.42 **	0.46 **	0.46 **
Internode length	0.70 **	0.70 **	0.99 **	0.28 **	0.32 **	0.32 **
Stalk diameter	0.67 **	0.68 **	0.98 **	0.33 **	0.38 **	0.38 **
Juice weight	0.35 **	0.38 **	0.60 **	0.28 **	0.40 **	0.23 **
Brix	0.23 **	0.24 **	0.53 **	0.21 **	0.27 **	0.23 **
RS	0.27 **	0.27 **	0.56 **	0.15 *	0.28 **	0.25 **
Cane yield	0.59 **	0.60 **	0.99 **	0.59 **	0.97 **	0.98 **
Sugar yield	0.50 **	0.49 **	0.89 **	0.83 **	0.83 **	0.43 **

* and **, significant at $p < 0.05$ and 0.01, respectively. RS% = Recoverable sugar percent.

Further correlation analysis among traits (Table 3) revealed the strong positive correlation between POL% and RS% ($R = 0.96$, $p \leq 0.01$) and also between Brix% and POL% ($R = 0.95$, $p \leq 0.01$). It is apparent from this table that there is a significant positive correlation between cane yield and sugar yield ($R = 0.92$). Also, RS% was correlated significantly with Brix%; likewise, juice weight with cane weight, both of them with $R = 0.90$ ($p \leq 0.01$). A significant negative correlation was detected between stalk diameter and cane yield (Table 3). In this investigation, the correlation between sugar yield with internode number and POL% was not significant (Table 3). Among the traits, RS%, Pol%, Brix%, PTY%, stalk height, and internode length had significant correlation with sugar yield, suggesting that these traits perhaps can be manipulated for sugar yield improvement in sugarcane.

Table 3. Correlation coefficients among thirteen traits measured in 253 sugarcane accessions for mean measurement of two seasons.

Traits	SH	InN	InL	SD	CW	JW	J%	Br%	Po%	PT%	RS%	CY	SY
SH	1												
InN	0.46 **	1											
InL	0.71 **	0.01	1										
SD	−0.16 **	0.13 *	−0.26 **	1									
SW	0.52 **	0.40 **	0.25 **	0.50 **	1								
JW	0.45 **	0.33 **	0.24 **	0.48 **	0.90 **	1							
J%	0.07	0.04	0.09	0.28 **	0.35 **	0.62 **	1						
Br%	−0.10	−0.02	−0.14 *	0.13 *	0.04	0.01	−0.11	1					
Po%	−0.11	−0.05	−0.13 *	0.12 *	0.04	0.01	−0.11	0.95 **	1				
PT%	−0.07	−0.07	−0.06	0.05	0.04	0.03	−0.04	0.64 **	0.80 **	1			
RS%	−0.12	−0.06	−0.13 *	0.11	0.01	−0.01	−0.09	0.90 **	0.96 **	0.87 **	1		
CY	0.33 **	0.09	0.31 **	−0.31 **	−0.001	−0.04	−0.07	−0.004	−0.01	−0.03	−0.03	1	
SY	0.27 **	0.06	0.25 **	−0.25 **	−0.002	−0.05	−0.10	0.32 **	0.33 **	0.29 **	0.33 **	0.92 **	1

* and **, significant at the 5% and 1% levels probability, respectively. SH = Stalk height, InN = internode number, InL = internode length, SD = stalk diameter, SW = 10-stalk weight, JW = juice weight, J% = juice percent, Br% = Brix percent, Po% = POL percent (sucrose content in the cane), PT% = PTY (the purity of juice), RS% = recoverable sugar percent, CY = cane yield, SY = sugar yield.

3.3. Multiple Stepwise Regression Analysis

During genetic improvement, breeders are interested in those traits that have causative effects on economical traits. In this research, to do multiple stepwise regressions, the percentage of recoverable sucrose (RS%) was considered as the dependent variable and other remaining characters as independent variables. The result of the stepwise regression analysis is summarized in Table 4. Accordingly, 96% of total variation for the RS% could be explained by two characters; Brix percent (82%) and PTY percent (14%). The most positive effective traits on RS% were Brix and PTY. Juice weight had a significant negative effect on RS%.

Table 4. Stepwise regression model for RS% as dependent variable and the remaining characters as independent variables.

Variable	Parameter Estimate	Standardized Estimate	Partial R-Square	t Value	pr > |t|
Intercept	−16.32534	0	0	−39.00	<0.0001
Br%	0.48689	0.58710	0.8218	41.94	<0.0001
PT%	0.19221	0.50455	0.1478	36.78	<0.0001
JW	−0.19206	−0.10041	0.0015	−4.94	<0.0001
J%	0.01104	0.04620	0.0008	3.12	0.0020
SD	0.01568	0.04265	0.0007	2.99	0.0030
SH	0.00107	0.02461	0.0003	1.68	0.0952

Br% = Brix percent, PT% = PTY (the purity of juice), JW = juice weight, J% = juice percent, SD = stalk diameter, SH = stalk height.

3.4. Principal Component Analysis

Figure 2 exhibits the distribution of accessions and traits in groups according to PCA. The first two principal components (PCs) provided a reasonable summary of the data and explained 59.51% of the total variation. PC1 explained 32.36% of the total variation. The most effective traits on this component were sugar yield, cane yield, stalk height and internode length. The second PC, which represented 27.15% of the variation, mainly represented RS% and Brix%. According to PCA result, CL61-620 and CL73-239 (from the USA) were two accessions with the highest content of RS and Brix percent. Likewise, CP73-1547, CP44-101 (from the USA) and TUC68-19 (from Argentina), were accessions with the highest sugar yield and cane yield. From this data, it can be seen that F134, Q138 and CL54-336 accessions had the lowest content of Brix and RS percent. Also, some accessions like CP75-1353, CP81-1254 and CP73-21 (from the USA) had high values for both Brix and sugar yield.

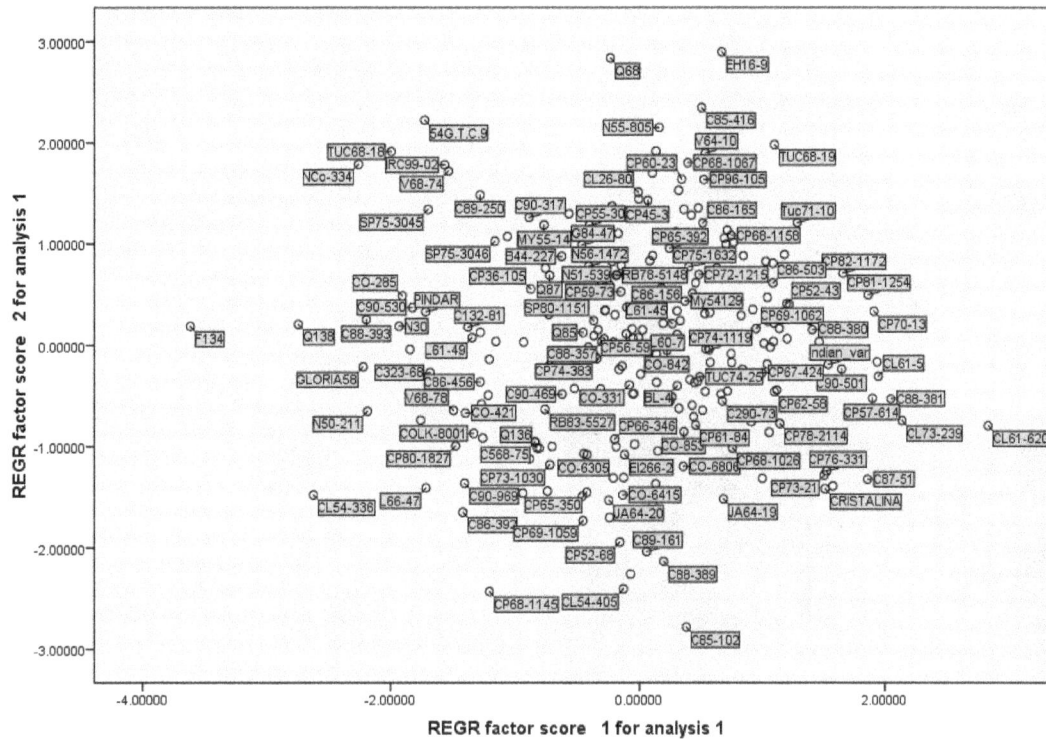

Figure 2. Two-dimensional scatter plot of principal component analysis for the first two principal components created from phenotypic traits from the sugarcane collection (*Saccharum* spp.) in Khuzestan province, Iran.

3.5. Graphical Clustering Analysis

The results obtained from the clustering analysis are summarized in Table 5. As can be seen, all the accessions were grouped in 10 segregated clusters based on the RB method. The highest and lowest numbers of accessions were observed in Clusters 8 and 1, respectively. This data showed that Clusters 7 and 9 resulted in the highest value of average internal similarities (Isim) (0.999), and Clusters 2 and 6 with Isim of 0.997 had the lowest value of internal similarities. Data from this table can be compared and completed with Figure 3, which shows the mountain visualization of relationships between the 10 clusters. Considering the distances between peaks, cluster 1 is the farthest group from other clusters and thus has the lowest value of average external similarities (Esim) (Table 5). Moreover, the lowest peak height was seen in Clusters 2 and 6, which represents the lowest value of internal similarity in these clusters. As Figure 3 shows, accessions from Cuba, the USA, South Africa, India, and Iran were grouped in the same cluster (Cluster 3). The data in Figure 3 and Table 5 makes it apparent that the smallest group is Cluster 1, which contains only three accessions from two breeding centres in the USA and Cuba, while other clusters, which had accessions from several breeding centres, were grouped in the same cluster.

Table 5. Clustering analysis of 253 sugarcane accessions with morphological traits based on repeated bisection (RB) method.

Cluster Number	Number of Accessions inside the Cluster	Isim	Isdev	Esim	Esdev
1	3	0.998	0.000	0.970	0.003
2	15	0.997	0.001	0.983	0.003
3	9	0.998	0.001	0.988	0.004
4	23	0.998	0.001	0.988	0.004
5	28	0.998	0.001	0.989	0.003
6	32	0.997	0.001	0.990	0.003
7	29	0.999	0.001	0.982	0.002
8	44	0.998	0.001	0.993	0.002
9	32	0.999	0.001	0.993	0.001
10	38	0.998	0.001	0.993	0.001

Isim = internal similarity, Isdev = internal standard deviation, Esim = external similarity, Esdev = external standard deviation.

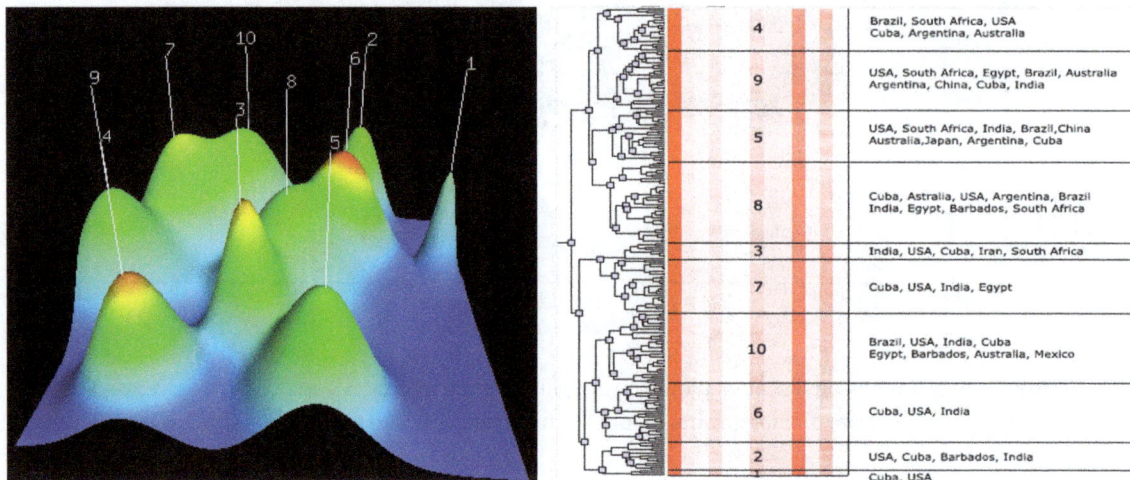

Figure 3. The mountain (**left**) and matrix (**right**) visualization of relationships among 10 clusters constructed in collection of sugarcane (*Saccharum* spp.) in Khuzestan. The clustering analysis and visualization were performed using gCLUTO.

3.6. Constructing a Mini-Core Collection

Based on the PowerCore program output, conducted on the entire collection (253 accessions), a mini-core collection was generated that included 21 accessions using M strategy and heuristic search. Descriptions of the 21 selected accessions with important morphological and sugar-related traits have been shown in Table 6.

To compare the primary collection with the mini-core collection, the variance difference (VD = 53.5%), mean difference (MD = 2.31%), coincidence rate (CR = 93.96%), and variable rate (VR = 147.28%) were calculated. In traits with normal distribution, there were no significant differences between the means of two collections (Table 7). Similarly, in non-normal traits, there were no significant differences between the medians of two collections (Table 7). The *F* test showed significant differences between variances of two collections for normally distributed traits, while, for non-normally distributed traits, except for juice weight, no significant differences were found between two collections (Table 7). Interestingly, for those traits with significant difference in variation, the mini-core collection was more diverse than the whole collection.

Table 6. Description of 21 sugarcane accessions created by M-strategy and heuristic algorithm.

Clone Name	Origin	Parents	Maturity Group	Height	InterL	Diameter	Juice %	Brix %	RS %	Cane Yield	Sugar Yield
BL-4	Barbados	POJ2878 × Polycross	Medium-Late	112.90	7.05	28.85	29.35	18.99	9.73	37.00	3.60
C85-102	Cuba	Unknown	Early-Medium	97.10	6.27	20.55	25.73	18.28	9.39	34.50	3.24
C86-12	Cuba	Unknown	Early-Medium	135.95	8.54	27.00	35.03	19.46	10.37	14.50	1.50
C87-51	Cuba	Unknown	Early-Medium	106.25	8.20	21.80	29.78	20.72	11.18	46.00	5.14
CL54-336	USA	Unknown	Late	126.30	8.15	19.95	34.76	13.96	7.51	27.00	2.03
CL61-620	USA	CP52-68 × Polycross	Early	137.25	6.81	24.40	23.49	21.78	12.22	32.50	3.97
CL73-239	USA	Unknown	Early-Medium	116.15	8.22	25.30	25.49	20.63	11.39	51.50	5.87
CO-407	India	Unknown	Late	150.73	9.16	19.53	21.53	17.80	8.63	35.33	3.05
CP45-3	USA	Unknown	Medium-Late	169.25	11.30	21.15	33.50	18.50	9.62	66.50	6.40
CP65-315	USA	Unknown	Medium-Late	143.15	16.00	20.66	29.49	18.07	9.32	47.50	4.42
CP68-1067	USA	Unknown	Medium-Late	132.65	6.94	30.95	36.53	19.38	10.01	42.50	4.26
CP73-21	USA	CP56-63 × CP66-1043	Early-Medium	143.75	8.69	18.95	15.07	19.65	10.76	53.50	5.75
EH16-9	Unknown	LCP81-325 × LCP81-30	Early-Medium	190.00	11.92	25.37	26.61	19.71	10.32	33.00	3.41
F134	China	Unknown	Late	183.10	11.50	22.70	29.17	14.88	6.46	27.50	1.78
L61-49	USA	Unknown	Medium-Late	171.42	11.38	19.87	29.66	16.65	8.33	53.00	4.41
MY55-14	Cuba	CP34-79 × B45181	Medium-Late	163.72	9.57	25.93	32.55	18.37	9.16	26.17	2.40
N30	South Africa	77F0637 × 78F1025	Medium-Late	214.47	12.68	17.82	23.68	16.55	7.72	62.00	4.79
NCo-376	South Africa	Co421 × Co312	Early-Medium	163.98	11.00	23.47	31.63	17.76	9.38	45.17	4.24
Q68	Australia	POJ2878 × Co290	Early-Medium	175.82	9.96	28.93	39.02	18.29	9.88	29.50	2.91
TRITON	Unknown	Co270 × Eros	Late	139.93	10.10	22.97	28.82	18.91	7.67	36.50	2.80
TUC68-18	Argentina	CP50-28 × CB38-79	Medium-Late	181.77	9.79	24.88	35.95	16.18	8.01	39.00	3.12

InterL = internode length, RS % = recoverable sugar percent.

Table 7. Comparison of descriptive statistics of the accessions growing in the primary germplasm of the sugarcane collection (means of two season), and the accessions from this collection selected for the mini-core collection (breeding collection), table (b) is continue of table (a).

(a)

	Height		Inter N		Inter L		Diameter		10 Cane W		Juice W		Juice %	
	CC	MCC	CC	MCC	CC	MCC	CC	MCC	CC	MCC	CC	MCC	CC	MCC
Mean	146.87	150.26	15.39	15.60	9.44	9.67	22.99	23.38	5.25	5.52	1.63	1.74	29.78	29.37
Median	146.97	143.75	15.50	15.50	9.49	9.57	22.95	22.97	5.14	4.98	1.58	1.47	29.78	29.49
Range	129.17	117.37	13.72	13.12	10.92	9.73	15.46	13.13	6.14	5.85	3.46	3.33	25.51	23.95
Max	214.47	214.47	22.77	22.77	16	16	32.35	30.95	8.57	8.57	3.99	3.99	39.68	39.02
Min	85.30	97.10	9.05	9.65	5.08	6.27	16.89	17.82	2.43	2.72	0.53	0.66	14.17	15.07
Variance	496.07	919.24 *	3.47	8.68	2.41	5.41 **	6.89	13.07 *	1.43	3.09	0.25	0.84 *	16.32	32.52
Shannon-Weaver	0.18	0.20 **	0.15	0.19 **	0.16	0.19 **	0.17	0.20 **	0.19	0.20 **	0.16	0.19 **	0.17	0.20 **

(b)

	Brix %		Pol %		PTY %		RS %		Cane Yield		Sugar Yield	
	CC	MCC	CC	MCC	CC	MCC	CC	MCC	CC	MCC	CC	MCC
Mean	18.42	18.31	15.85	15.64	85.75	84.32	9.96	9.38	40.02	40.00	3.86	3.76
Median	18.49	18.37	15.95	15.83	85.99	85.35	9.72	9.39	37.83	37.00	3.76	3.60
Range	7.82	7.82	8.54	8.54	15.83	15.20	5.76	5.76	58.00	52.00	5.45	4.90
Max	21.78	21.78	20.06	20.06	91.10	90.47	12.22	12.22	72.50	65.50	6.95	6.40
Min	13.96	13.96	11.52	11.52	75.27	75.27	6.46	6.46	14.50	14.50	1.50	1.50
Variance	1.35	3.60	1.80	4.37 **	6.41	14.54	0.93	2.06 **	102.94	162.57	1.05	1.82
Shannon-Weaver	0.16	0.20 **	0.16	0.19 **	0.16	0.19 **	0.17	0.20 **	0.18	0.19 **	0.19	0.20 **

Inter N = internode number, Inter L = internode length, 10 SW = 10-stalk weight, JW = juice weight, Pol = sucrose content in the cane, PTY = the purity of juice, RS% = recoverable sugar percent. * and **, significant at $p < 0.05$ and 0.01, respectively. CC: core collection: Primary collection. MCC: mini core collection.

To compare the variability of two collections, mini-core collection and primary collection, the Shannon–Weaver diversity index was also used for each trait in each collection (Table 7). The mean of the Shannon–Weaver diversity index scores for primary and mini-core collections were 17.50% and 20.11%, respectively. Pairwise t-test for comparing this index between two collections revealed significant difference, i.e., two collections have different variability. Altogether, these results confirmed that the mini-core collection was a good representation of the primary collection and captured much of the diversity present in the primary collection.

4. Discussion

The first question of this study sought to characterize the sugarcane germplasm with the assessment of phenotypic traits. The mean analysis showed that no differences were found among three harvest time points of the second season for stalk height, internode length, stalk diameter, and cane yield (Figure 1a–c,g). A possible explanation for this result might be that there is cold weather and short daylight in November, January, and February in the second season in Iran, which may have led to a lack of adequate photosynthesis and no differences between the time points were found. Another reason could probably be due to the short time intervals between three harvests; hence, plants did not have growth opportunities. However, for other traits there were significant differences between the three time points in the second season (Figure 1d–f,h). In fact, for morphological traits, the harvest time differences were not significant; while for sugar-related traits significant differences between the three time points of the second season were observed. A significant increase in height from one season to the next was observed (Figure 1a). These results are consistent with those reported by Todd et al. [2], who showed a significant increase in height in the *Saccharum officinarum* species at different measurement times.

The present experiment was conducted during four harvest times (in two seasons) to determine the effect of growth periods to select the best harvest time in order to achieve the highest cane and sugar yield. It is interesting to note that in all sugar-related traits, significant differences between two seasons and/or different time points were observed (Figure 1d–f,h). This finding showed that the best time point with the maximum sugar yield is February in the second season. The same trend was observed for cane yield in which second season values were significantly ($p \leq 0.05$) higher than first season records. Due to the significant increase in cane yield in the second season, these findings suggest the best time for sugarcane harvest to be the second season. Also, it could conceivably be hypothesized that energy-cane cultivars/accessions are preferably harvested in the second year (ratoon stage) due to the importance of cane yield in the production of high biomass and lingo-cellulosic compounds (Figure 1g).

As mentioned earlier, the greatest magnitudes of correlations were observed between January 2013 and February 2013 for stalk height, internode length, stalk diameter and cane yield (Table 2). The observed correlation may be due to the short time interval between January and February harvest times and subsequently, a lack of plant growth. These results are in agreement with those reported by Todd et al., [2] in which there were high correlations for stalk height, stalk diameter, and internode length between two time points. In correlation analysis among traits, positive association was seen between cane yield with stalk height and internode length, at $\alpha = 0.01$. Cane yield is one of the major parameters for assessment of sugarcane performance [24]. Hence, stalk height and internode length are two effective traits to improve and to recognize the high cane yield genotype. These results are in agreement with recent studies indicating that cane yield was associated positively with stalk height, internode length and cane thickness [25]. Another important significant positive correlation was found between sugar-related factors such as RS%, Pol%, Brix%, PTY%, with sugar yield. The present results are significant in at least two major respects. On the one hand, for sugar yield improvement in sugarcane accessions, it is suggested to focus and to manipulate the factors such as RS%, Pol%, Brix%, PTY%. On the other hand, for increase in cane yield, it is recommended to focus on stalk height and internode length.

In current research, Brix% and PTY% were identified as effective traits on RS%, according to multiple stepwise regression analysis results. Positive regression coefficients of these variables indicated that these variables can be mentioned as logical and reasonable indices for selection in order to improve recoverable sucrose in sugarcane. This result confirms the strong association between RS% with two variables of Brix% and PTY%, (Table 3). The question that arises here is that in the study of correlation analysis (Table 3), positive correlation between RS% and POL% was found while it is missing in the model. In fact, due to high multicollinearity among independent variables scaled by VIF, POL% was omitted from the analysis.

The PCA scatterplot of two principal components axes provided the distribution of primary collection accessions in a biplot. The PCA finding has important implications for introducing accessions with desirable traits, such as accessions with high content of RS% and Brix% or accessions with high/low sugar yield or cane yield (Figure 2). Arguably, depending on the approach of sugarcane breeding, these traits can be utilized by breeders. Some of the clear findings in the biplot are that CL61-620 and CL73-239 are accessions with the highest content of RS% and Brix% percent, and that CP45-3, Q68 and EH16-9 are accessions with the highest cane yield (Figure 2). Interestingly these accessions are present in the constructed mini-core collection in this research (Table 6).

The clustering analysis revealed the morphologic similarity of different accessions from different breeding centres. For example, grouping of the USA and Cuba accessions in the same clusters indicated that there are morphologic similarities between these accessions (Figure 3). Likewise, the grouping of the Iran–Cuba hybrid (Iran) with the USA, India, Cuba, and South Africa varieties in Cluster 3 demonstrate the morphologic similarities between them (Figure 3).

The range values of all traits in both collections are shown in Table 7. As can be seen from the table (above), in the core collection (primary collection), the range value of stalk height and stalk diameter obtained 129.17 (cm) and 15.46 (mm) respectively. Todd et al. showed that range value of stalk height and stalk diameter 304.13 (cm) and 37.79 (mm), respectively, in World Collection of Sugarcane (*Saccharum* spp.) and Related Grasses of Sugarcane in Miami, Florida. This is different from the findings presented here about the range value of traits, but the mean value of the mentioned traits in our collection are consistent with data obtained with Todd et al [2].

With respect to the second research goal, PowerCore software was used to construct a mini-core collection. PowerCore software had been used previously to create core/mini collections for rice (*Oryza sativa* L.) [26,27], sesame (*Sesamum indicum* L.) [28], and chickpea (*Cicer arietinum* L.) [29]. Using this software, 21 accessions were selected to build a mini-core collection out of 253 accessions of sugarcane. These results confirm the appropriate number of subset collections (10% of the primary collection), as was mentioned in the literature [8]. A core/mini-core collection with a coincidence rate (CR%) greater than 80% has been recommended as a proper collection for breeding purposes [30]. In our analysis, the CR was found to be 93.96%, indicating that the mini-core collection selected from sugarcane germplasm is a good representative subset of the phenotypic diversity in the primary collection. Also, *t*- and Wilcoxon test results showed no significant difference in means and medians between the mini-core and primary sugarcane collections (Table 7). Furthermore, high values for VR% and VD% [31], which are the evaluating factors of core collections, reconfirm that the mini-core collection in this study incorporates the main portion of diversity of the primary sugarcane collection. Additionally, the *F* and Ansari–Bradley test results also revealed no significant loss of diversity in mini-core collection compared with the primary collection. Interestingly, for normally distributed traits, the mini-core collection was more diverse than the whole collection. However, for non-normally distributed traits, Ansari–Bradley test results showed non-significant difference between two collections, except for juice weight.

For comparing the Shannon–Weaver diversity indices between the two collections pairwise *t*-test was carried out. Accordingly, significant differences were found between the two collections, so that Shannon–Weaver index scores in the mini-core were significantly higher than in the primary collection. Higher variance in the core/mini collection relative to the entire collection has been reported frequently in previous researches [2,31,32].

5. Conclusions

The present study was designed to evaluate the primary germplasm of sugarcane (as a core collection) based on phenotypic traits and finally to construct a core set of accessions called the mini-core collection. This experiment led to the selection of a mini-core of 21 accessions (12.05% of the core collection of sugarcane) that retains the maximum diversity in the primary collection, as validated by comparing the Shannon–Weaver diversity index scores between the two collections. The average of coincidence rate of range in the 13 traits was 93.96%, indicating that the mentioned mini-core is a real representation of the core collection. Additionally, the findings of this study about both collections (primary/core and mini-core) have significant implications for the effective use of sugarcane germplasm, thereby enhancing our knowledge about important accessions with favourable traits, eventually, for sugarcane and energy cane improvement. In a sugarcane collection with a large germplasm, creating a mini-core or breeding collection is an efficient approach to identify the suitable parents for future breeding programs.

Acknowledgments: The authors are grateful to the Cane Development and Sidelong Industrial Research and Education Institute for providing helpful information about the sugarcane genotypes as well as giving facilities for the field trials. This study was funded by Agricultural Biotechnology Research Institute of Iran (grant number: 4/05/05/004/940008).

Author Contributions: This manuscript has derived from the PhD thesis of the first author and there are two supervisors, who are the two co-corresponding authors.

Conflicts of Interest: The authors declare no conflict of interest.

References

1. Henry, R.J.; Kole, C. *Genetics, Genomics and Breeding of Sugarcane*; CRC Press, Taylor & Francis Group: New York, NY, USA, 2010.

2. Todd, J.; Wang, J.; Glaz, B.; Sood, S.; Nayak, S.N.; Glynn, N.C.; Gutierrez, O.A.; Kuhn, D.N. Phenotypic characterization of the Miami World Collection of sugarcane (*Saccharum* spp.) and related grasses for selecting a representative core. *Genet. Res. Crop Evol.* **2014**, *61*, 1581–1596. [CrossRef]

3. Govindaraj, M.; Vetriventhan, M.; Srinivasan, M. Importance of genetic diversity assessment in crop plants and its recent advances: An overview of its analytical perspectives. *Genet. Res. Int.* **2015**, *2015*, 1–14. [CrossRef] [PubMed]

4. Zhang, H.; Zhang, D.; Wang, M.; Sun, J.; Qi, Y.; Li, J.; Wei, X.; Han, L.; Qiu, Z.; Tang, S.; et al. A core collection and mini core collection of *Oryza Sativa*. *Theor. Appl. Genet.* **2011**, *122*, 49–61. [CrossRef] [PubMed]

5. Odong, T.L.; Jansen, J.; Van Eeuwijk, F.A.; Van Hintum, T.J. Quality of core collections for effective utilisation of genetic resources review, discussion and interpretation. *Theor. Appl. Genet.* **2013**, *126*, 289–305. [CrossRef] [PubMed]

6. Frankel, O. Genetic perspective of germplasm collection. In *Genetic Manipulations: Impact on Man and Society*; Arber, W., Limensee, K., Peacock, W.J., Stralinger, P., Eds.; Cambridge University Press: Cambridge, UK, 1984; pp. 61–170.

7. Upadhyaya, H.D.; Ortiz, R. A mini core subset for capturing diversity and promoting utilization of chickpea genetic resources in crop improvement. *Theor. Appl. Genet.* **2001**, *102*, 1292–1298. [CrossRef]

8. Upadhyaya, H.D.; Gowda, C.L.L.; Sastry, D. Management of Germplasm Collections and Enhancing Their Use by Mini Core and Molecular Approaches. In Proceedings of the APEC-ATCWG Workshop, Taichung, Taiwan, 14–17 October 2008.

9. Upadhyaya, H.D. Establishing core collections for enhanced use of germplasm in crop improvement. *Ekin J. Crop Breed. Genet.* **2015**, *1*, 1–12.

10. Bhattacharjee, R.; Khairwal, I.S.; Bramel, P.J.; Reddy, K.N. Establishment of a pearl millet [*Pennisetum glaucum* (L.) R. Br.] core collection based on geographical distribution and quantitative traits. *Euphytica* **2007**, *155*, 35–45. [CrossRef]

11. Li, X.L.; Lu, Y.G.; Li, J.Q.; Xu, H.M.; Shahid, M.Q. Strategies on sample size determination and qualitative and quantitative traits integration to construct core collection of rice (*Oryza sativa*). *Rice Sci.* **2011**, *18*, 46–55. [CrossRef]

12. Studnicki, M.; Mądry, W.; Schmidt, J. Comparing the efficiency of sampling strategies to establish a representative in the phenotypic-based genetic diversity core collection of orchardgrass (*Dactylis glomerata* L.). *Genet. Plant Breed.* **2013**, *49*, 36–47.

13. Balakrishnan, R.; Nair, N.V.; Sreenivasan, T.V. A method for establishing a core collection of *Saccharum officinarum* L. germplasm based on quantitative-morphological data. *Genet. Res. Crop Evol.* **2000**, *47*, 1–9. [CrossRef]

14. Tai, P.Y.P.; Miller, J.D. A core collection for *Saccharum spontaneum* L. from the world collection of sugarcane. *Crop Sci.* **2001**, *41*, 879–885. [CrossRef]

15. Amalraj, V.A.; Balakrishnan, R.; Jebadhas, A.W.; Balasundaram, N. Constituting a core collection of *Saccharum spontaneum* L. and comparison of three stratified random sampling procedures. *Genet. Res. Crop Evol.* **2006**, *53*, 1563–1572. [CrossRef]

16. Dillewijn, C. *Botany of Sugarcane*; Chronica Botanica: Waltham, MA, USA, 1952.

17. Legendre, B.L. The core/press method for predicting the sugar yield from cane for use in cane payment. *Sugar J.* **1992**, *54*, 2–7.

18. SAS Institute Inc. *SAS OnlineDoc 9.3*; SAS Institute Inc.: Cary, NC, USA, 2011.

19. Hair, J.F.; Tatham, R.L.; Anderson, R.E. *Multivariate Data Analysis*, 5th ed.; Prentice Hall: Maryland, NY, USA, 1998.

20. Rasmussen, M.; Karypis, G. *gCLUTO—An Interactive Clustering, Visualization, and Analysis System*; CSE/UMN Technical Report TR04-021; University of Minnesota: Twin Cities, MI, USA, 2004.

21. Kim, K.; Chung, H.; Cho, G.; Ma, K.; Chandrabalan, D.; Gwag, J.; Kim, T.; Cho, E.; Park, Y. Power Core: A program applying the advanced M strategy with a heuristic search for establishing core sets. *Bioinformatics* **2007**, *23*, 2155–2162. [CrossRef] [PubMed]

22. Schoen, D.J.; Brown, A.H. Conservation of allelic richness in wild crop relatives is aided by assessment of genetic markers. *Proc. Natl. Acad. Sci. USA* **1993**, *90*, 10623–10627. [CrossRef] [PubMed]

23. Shannon, C.E.; Weaver, W. A mathematical theory of communication. *Bell Syst. Tech. J.* **1948**, *27*, 1–54. [CrossRef]

24. Raza, S.; Qamarunnisa, S.; Jamil, I.; Naqvi, B.; Azhar, A.; Qureshi, J.A. Screening of sugarcane somaclones of variety BL4 for agronomic characteristics. *Pak. J. Bot.* **2014**, *46*, 1531–1535.

25. Khan, I.A.; Raza, G.; Ismail, M.; Raza, S.; Ahmed, I.; Et, K. Assessment of contender sugarcane clones for morphological traits and biotic tolerance under agro-climatic conditions of Tando Jam. *Pak. J. Bot.* **2015**, *47*, 43–48.

26. Zhang, P.; Li, J.; Li, X.; Liu, X.; Zhao, X.; Lu, Y. Population structure and genetic diversity in a rice core collection (*Oryza sativa* L.) investigated with SSR markers. *PLoS ONE* **2011**, *6*, e27565. [CrossRef] [PubMed]

27. Agrama, H.A.; Yan, W.; Lee, F.; Fjellstrom, R.; Chen, M.; Jia, M.; Mcclung, A. Genetic assessment of a mini-core subset developed from the USDA rice genebank. *Crop Sci.* **2009**, *49*, 1336–1346. [CrossRef]

28. Zhang, Y.; Zhang, X.; Che, Z.; Wang, L.; Wei, W.; Li, D. Genetic diversity assessment of sesame core collection in China by phenotype and molecular markers and extraction of a mini-core collection. *BMC Genet.* **2012**, *13*, 102. [CrossRef] [PubMed]

29. Archak, S.; Tyagi, R.K.; Harer, P.N.; Mahase, L.B.; Singh, N.; Dahiya, O.P.; Nizar, M.A.; Singh, M.; Tilekar, V.; Kumar, V.; et al. Characterization of chickpea germplasm conserved in the Indian National Genebank and development of a core set using qualitative and quantitative trait data. *Crop J.* **2016**, *4*, 417–424. [CrossRef]

30. Hu, J.; Zhu, J.; Xu, H. Methods of constructing core collections by stepwise clustering with three sampling strategies based on the genotypic values of crops. *Theor. Appl. Genet.* **2000**, *101*, 264–268. [CrossRef]

31. Reddy, L.J.; Upadhyaya, H.D.; Gowda, C.L.L.; Singh, S. Development of core collection in pigeonpea [*Cajanus cajan* (L.) Millspaugh] using geographic and qualitative morphological descriptors. *Genet. Res. Crop Evol.* **2005**, *52*, 1049–1056. [CrossRef]

32. Mei, Y.; Zhou, J.; Xu, H.; Zhu, S. Development of Sea Island cotton (*Gossypium barbadense* L.) core collection using genotypic values. *Aust. J. Crop Sci.* **2012**, *6*, 673–680.

8

Identification of Optimal Mechanization Processes for Harvesting Hazelnuts Based on Geospatial Technologies in Sicily (Southern Italy)

Ilaria Zambon * (iD), Lavinia Delfanti, Alvaro Marucci, Roberto Bedini, Walter Bessone (iD), Massimo Cecchini and Danilo Monarca

Department of Agricultural and Forestry Sciences, DAFNE Tuscia University, Via San Camillo de Lellis snc, 01100 Viterbo, Italy; laviniadelfanti@unitus.it (L.D.); marucci@unitus.it (A.M.); r.bedini@unitus.it (R.B.); walter.bessone@regione.piemonte.it (W.B.); cecchini@unitus.it (M.C.); monarca@unitus.it (D.M.)
* Correspondence: ilaria.zambon@unitus.it

Academic Editor: Ole Wendroth

Abstract: Sicily is a region located in the southern Italy. Its typical Mediterranean landscape is appreciated due to its high biodiversity. Specifically, hazelnut plantations have adapted in a definite area in Sicily (the Nebroidi park) due to specific morphological and climatic characteristics. However, many of these plantations are not used today due to adverse conditions, both to collect hazelnuts and to reach hazel groves. Though a geospatial analysis, the present paper aims to identify which hazelnut contexts can be actively used for agricultural, economic (e.g., introduction of a circular economy) and energetic purposes (to establish a potential agro-energetic district). The examination revealed the most suitable areas giving several criteria (e.g., slope, road system), ensuring an effective cultivation and consequent harvesting of hazelnuts and (ii) providing security for the operators since many of hazelnut plants are placed in very sloped contexts that are difficult to reach by traditional machines. In this sense, this paper also suggests optimal mechanization processes for harvesting hazelnuts in this part of Sicily.

Keywords: hazelnuts; spatial analysis; mechanization processes; precision farming; rural landscape; Sicily

---

1. Introduction

The rural landscapes of Mediterranean Europe are characterized by their peculiar crops, whose agricultural practices have led to different land use changes [1]. In recent years, there has been a strong abandonment of agricultural areas [2,3], supporting a consequent reforestation development [1,4].

Hazelnuts represent ones of most produced nut crops in the Mediterranean contexts, as in Italy [5], since as agricultural products have relevant nutritional and economic value [6]. Given their profitability, they are also grown on unsuitable ground, due to the absence of land use policies (as in Langhe region in Italy) [2,7]. For example, Turkey imposed specific regulations for cultivating hazelnuts in given areas, where the maximum elevation is 750 m, the slope is more than 6% and IV or upper class of LCC [8]. According to such government regulations, potential hazelnut areas can be mapped with specific criteria (e.g., slope, elevation, and land use–land cover) using GIS technology [9]. Consequently, their detection may be useful to observe landscape changes, providing greater support to national and international institutions in the assessment of rural agriculture policies [10] and their latent consequences on local society, landscape, and production [11–13].

Defining hazelnut areas is possible through maps and satellite images by advanced computer programs such as Geographical Information Systems (GIS) and Remote Sensing (RS) technologies, which offer benefits in data management and acquisition [6,14]. In recent decades, GIS and RS have been appreciated within rural applications linked to resources at several spatial scales [9,15]. GIS presents a suitable tool for processing, analyzing, and collecting spatial information [7,16,17]. Spatial analysis reveals elevation, aspect, slope, and soil data using GIS methods and even investigates environmental situations, soil attributes, and topographic changes [6,9,18]. From RS technologies, land cover classification is regularly achieved by a multi-class scenery and supervised arrangement of textural or spectral characteristics at pixel level [19,20]. Remote sensing imagery permits to provide data about hazelnuts from satellite images [21], which can be then integrated to other database in GIS with the aim of securing sustainable development of rural areas [6,22–24]. Therefore, through Geographical Information Systems (GIS) and remote sensing with multi-temporal high-resolution satellite data, land use changes, vegetation cover, soil degradation, and further issues can be monitored integrally [25,26].

Remote identification of hazelnuts is not reasonably straightforward [2,7,27,28]. However, it is necessary (i) to optimize harvesting methods and (ii) to distinguish rural landscape dynamics and socio-economic and land use changes to achieve sustainable development [29,30]. Their detection usually takes place through a visual interpretation of very high resolution remote sensing imagery to exploit spectral and textural features, due to the absence of an automated method [20]. However, few studies have focused on mapping hazel groves with high resolution imagery [7,20,31–33]. Vegetation variables appear continuous and difficult to distinguish, e.g., biomass, fraction of vegetation cover, or leaf area index [28,34]. For instance, NDVI values appear very close for hazel groves and further woody vegetation [20]. In fact, it is difficult distinguishing hazelnuts from forest areas and other similar crops (such as olives) that are also typical of the Mediterranean landscape [35]. Their identification from other areas can decrease the inventory expenses by saving money and time [35]. The existence of vegetation maps, performed through Geographic Information Systems (GIS), can be useful for both qualitative and quantitative assessments of natural resources in a definite context [36–40].

The importance of having analytical parameters is essential to find hazelnut plants. The latter are usually located at an altitude of 500 and 1000 m [41]. Their typical altimetry is motivated by the degree of humidity and climate, with a slope between 6% and 30% [6]. Furthermore, the cultivation of hazelnuts is not recommended on steep slopes, since they are not able to prevent and hinder potential soil erosion processes [42,43].

Hazelnut production is frequently characterized by irregular plantations and inconstant density, from steep slopes and rough terrain environments [44]. There are several mechanization methods for collecting hazelnuts, aiming to rationalize costs and harvest production using appropriate existing technologies [45,46]. Several research activities have been launched to assess the collection of hazelnuts, minimizing the risks for the operators in the field (e.g., risk of overturning) [44]. Hazelnuts are usually planted in rows along which herbicides are distributed during the year on the herbaceous vegetation for improving mechanical operation during the harvest [47]. The major problem during the hazelnut collection concerns the situations of high slopes and terraces in addition to the risk of roll-over problems [44]. Furthermore, the intense hazelnut harvesting can lead to negative consequences (e.g., soil erosion) [47] and it is therefore necessary to evaluate how to optimize the collection depending on the soil characteristics.

The purpose of this paper is to identify hazelnuts with the aim of proposing strategies and optimizing mechanization systems through geo-spatial technologies. The case study focuses on 10 municipalities in the Sicily region, which are part of the National association of hazelnuts. In these contexts, many hazelnut plantations appear to be woods. Hazelnuts have well-adapted in the Nebrodi mountains [38], but very often are in problematic areas to reach and work in safety. The present paper aims to recognize the areas that really can contribute to the primary sector in economic terms, estimating the potential hazelnut cultivation, ensuring opportunities for cultivation and the security for operators during the harvesting according to the intrinsic characteristics of such context. In this framework, an optimization of collection and mechanization processes, depending on geo-morphological and territorial characteristics and avoiding possible pollution, was reached. This first examination estimates the biomass obtained from suitable hazelnut plants pruning, as a real solution to produce energy through thermo-chemical processes, i.e., combustion, gasification, and pyrolysis [48]. Finally, the work aims to suggest the consolidation of an agro-energetic district in this context. The latter provides several benefits, as it strengthens the local economy linked to the cultivation of hazelnuts and can start a reality based on the circular economy with the purpose of re-using agricultural residues for energy purposes.

2. Materials and Methods

2.1. Context of Study

Sicily is a Southern Italian region with many forests and fields designed for agricultural activities. Among these, hazelnuts have settled as one of the most visible crops in the north-eastern part of Sicily (along the Nebrodi mountains), given the confident morphological and climatic features [38]. The cultivation of hazelnuts in Sicily covers a surface area of 16,482 hectares, producing each year around 204,306 quintals. The diffusion of hazelnut trees in this context took place in 1890, after the crisis of gelsiculture. Today, thanks to their ease of adaptation, dense root system and profitable productivity, hazelnuts are the predominant yield of the Nebrodi agrarian landscape [38]. In this regard, the municipalities of the province of Messina of Castell'Umberto, Montalbano Elicona, Sant'Angelo di Brolo, Raccuja, Santa Domenica Vittoria, San Piero Patti, San Salvatore Fitalia, Sinagra, Tortorici and Ucria are part of the National Association of Hazel Towns ('Associazione Nazionale Città della Nocciola'), representing the region of Sicily.

2.2. Data Analysis and Materials

ESRI ArcGIS software was used to integrate data and accomplish spatial analysis [6]. GIS technology is decisive to spatial surveys for examining the context of the study. As computer-based system, it allows to capture, storage, recovery, analyze and display geographic data [17]. In this study, GIS techniques were used to overlay maps (vegetation map of Sicily, Corine Land Cover (CLC), and other geospatial data, as well as road system), to make elaborations examining where the hazelnuts are located and to hypothesize mechanization processes focusing on some of their morphological characteristics: DTM, slope, aspect and curvature. The National Terrain Model (DTM) map is the representation of the interpolation of orographic data from the map of the Military Geographic Institute. The resulting product is a 20 m regular step matrix, whose elements (pixels) show the values of the quotas. The Slope identifies the maximum rate of change in value from that cell to its neighbors. Principally, the maximum change in elevation over the distance (among the cell and its eight neighbors) finds the steepest downhill descent from the cell. The Curvature displays the shape or curvature of the slope and is calculated by computing the second derivative of the surface. The curvature, parallel to the slope, indicates the direction of maximum slope. A part of a surface can be concave or convex, by looking at the curvature value. It affects the acceleration and deceleration of flow across the surface: (i) a negative value indicates that the surface is upwardly convex at that cell, and flow will be decelerated, (ii) a positive profile indicates that the surface is upwardly concave at that cell, and the flow will be accelerated, and (iii) a value of zero indicates that the surface is linear. As the slope direction, aspect displays the downslope direction of the maximum rate of change in value from each cell to its neighbors. The values of each cell in the output raster designate the compass direction that the surface faces at such location, measured in degrees from 0 (north) to 360 (again north). Having no downslope direction, flat areas assume a value of -1.

The vegetation map was used as the base for the land use. It represents a convenient combination of the vegetal landscape, whose complex diversity reproduces the greatest physiographic, geomorphological, lithological, and bioclimatic variability of this region. In fact, the vegetation map is characterized by 36 phytocoenotic categories. As a result of years of research, it gives a summary of the widespread phytosociological and cartographic literature in Sicily. It was performed through Geographic Information Systems (GIS) at several scales (1:50,000, 1:25,000, and 1:10,000) and provides both qualitative and quantitative assessments of natural resources [36–40].

The vegetation map of Sicily was prepared at a 1:250,000 scale according several stages: (i) preparation of a GIS project (1:10,000 scale) with an inclusive database and thematic layers with georeferenced materials, (ii) photo-interpretation of the vegetation with satellite images (e.g., Landsat TM), orthophotos and digital data on the Technical Map of Sicily; (iii) validations with other maps, such as land use, vegetation or geology, (iv) validation of the photo-interpretation through field survey and verification, (v) digitization of the outcomes and further data, and (vi) phytosociological classification of the mapped types, categorized by 36 phytocoenotic classes [38]. Therefore, the vegetation map identifies all the existing crops in Sicily in a precise and detailed way. For instance, hazelnuts (identified with the code 202) occupy a surface area of about 9500 hectares.

2.3. Mechanization Framework

The cultivation of hazelnuts is characterized by several factors that make it difficult and dangerous to use mechanization systems for operators, at all stages of cultivation, especially in the harvesting phase. Some of these factors are predominantly irregular plantations, a high degree of acclimatization of the slopes (which also reach 35 degrees), uneven ground conditions, a lack or absence of business and interpersonal viability, presence of obstacles to the passage of machines, and unusual soil management with the abandonment of pruning residues on the ground.

The north-eastern part of Sicily along the Nebrodi mountains has seen the spontaneous diffusion of hazelnuts, which have easily adapted [38]. Despite their potential productivity, hazelnut plants are placed in very problematic environments, especially for the harvest phase, and therefore most of them are abandoned. Traditional vehicles have difficulty reaching these contexts (e.g., steep slopes that make it unsafe for operator intervention). Therefore, the currently-mechanization methodologies are equal to zero. In fact, harvesting is still by hand-picking in the few cultivated areas.

Focusing the prototype tested by [44], the present work suggests using a similar device that is self-propelled and easily transportable for harvesting in areas with poor or absent roads between farms (Figure 1). The device can move even under critical slope circumstances (even up to 30–35%) and overcoming substantial difficulties (e.g., terraces, where can be assemble the harvester to a mini crawler with hydraulic or hydrostatic transmission). In this manner, mechanization can be introduced in principally disadvantaged areas, with consequences in terms of safety for operators and a cost-benefit decrease. In operational stages, however, their prototype collects in a stationary position with the assistance of a suction line, permitting operation on highly sloped surfaces (more than 20%). The prototype tested by [44] is ideal for this Sicilian context, avoiding problems linked to steep slopes and movement among hazelnut groves.

Figure 1. Photo of the small-scale machine for nuts harvesting proposed in the study of [44].

3. Results

By means of a first GIS processing, the municipalities chosen are counted on a surface area of almost 4970 hectares (representing 52% of the hazels in the Sicily region) (Table 1). Comparing the vegetation map with the CLC, the accuracy of the first map was confirmed (Figure 2). While the CLC considered hazelnuts as forests, the vegetation map of the Sicily region highlighted their presence as hazelnuts (code 202). By comparing the two maps, 63% of the hazelnuts identified as "orchards" in CLC, while 22% are categorized as "deciduous forests" in CLC. As a first clarification, the CLC tends to aggregate hazelnuts in the category "orchards". Nonetheless, many fields of hazelnuts (22%) visually appeared as forests.

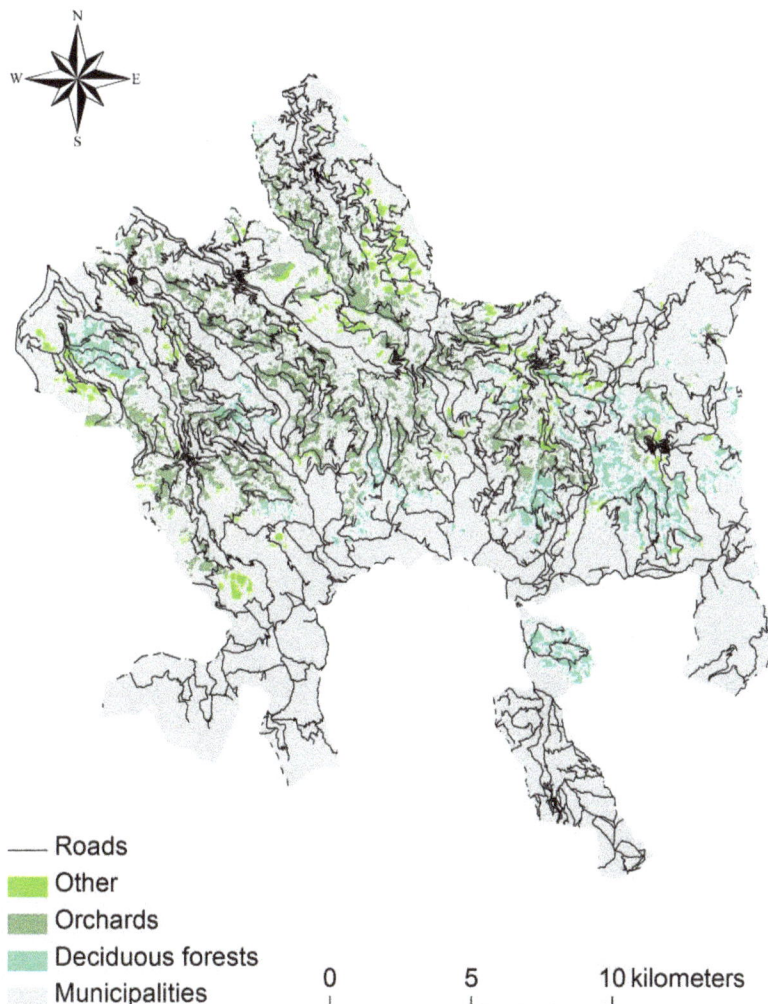

Figure 2. Hazelnut areas identified by the vegetation map of the region of Sicily. Each plot corresponds to the land use observed in CLC. Source: own elaboration.

Table 1. Hectares of hazelnuts belonging to vegetation map overlapped to CLC classes for each municipality.

Hectares of Hazelnuts Belonging to CLC Classes	Continuous Urban Fabric	Discontinuous Urban Fabric	Arable Crops in Non-Irrigated Areas	Orchards and Minor Fruits	Olive Grove	Annual Crops Associated with Permanent Crops	Complex Crop and Systems	Mainly Occupied Areas	Deciduous Forests	Areas with Natural Pasture at High Altitude	Areas Affected by Fires	Total of Hazelnuts
Code of CLC	111	112	211	222	223	241	242	243	311	321	334	
Castell'Umberto		20.88		85.16	17.95				0.24	0.00		124.23
Montalbano Elicona		5.16	4.07	91.99	1.47			35.49	457.65	16.82	0.20	612.85
Raccuja		8.46		315.92	15.40			21.75	25.37	4.64		391.53
San Piero Patti	3.32		7.21	423.81	66.77			91.16	205.15	17.86		815.28
San Salvatore di Fitalia		14.49		65.52	5.07		19.72	2.50	138.41	43.77		289.47
Santa Domenica Vittoria									118.95	0.00		118.95
Sant'Angelo di Brolo	0.22		0.71	471.31	26.53	23.51	2.18	38.21	19.68	82.73		659.21
Sinagra		1.87		455.59	44.39			3.12	1.26	9.75		515.98
Tortorici		11.97	0.47	669.69				0.56	92.44	63.45		838.59
Ucria		7.17		535.19	0.19			1.63	54.85	4.32		603.35
Total	3.54	70.00	12.45	3114.18	177.78	23.51	21.90	194.42	1114.00	243.33	0.20	4969.43
	0%	1%	0%	63%	4%	0%	0%	4%	22%	5%	0%	100%

Source: own elaboration.

Representing 43% of the hazelnuts considered, San Piero Patti and Tortorici are the municipalities that recorded the highest presence of hazel trees within their administrative boundaries. Hazelnuts plants prefer high altitudes (Table 2). In fact, 84% of them can be observed between 500 and 1000 m above sea level (with an average of 755 m by examining municipalities in analysis). Castell'Umberto, San Salvatore di Fitalia, Sant'Angelo di Brolo, and Sinagra are the municipalities that identified a larger percentage of hazelnuts at a moderate altimeter than the others, i.e., between 500 and 750 m. The municipality of Santa Domenica Vittoria demonstrated that 86% of its hazelnuts are above 1000 m, although it recorded, in quantitative terms, a reduced surface area for hazelnuts compared to other study contexts.

Table 2. Surface area (hectares) (top) and percentage (bottom) of hazelnuts for each municipality depending on DTM classes (meters).

	DTM classes (meters)					
	<250	250–500	500–750	750–1000	>1000	% Area Compared to Total
Castell'Umberto	0.0	10.7	85.5	12.9	0.0	2%
Montalbano Elicona	0.0	4.8	158.8	353.1	42.5	12%
Raccuja	0.0	0.5	143.1	195.0	18.3	8%
San Piero Patti	0.0	56.9	309.3	340.5	42.6	16%
San Salvatore di Fitalia	3.1	71.1	149.1	45.3	0.0	6%
Santa Domenica Vittoria	0.0	0.0	0.0	15.5	93.4	2%
Sant'Angelo di Brolo	0.8	70.5	346.9	191.8	0.0	13%
Sinagra	5.8	133.9	214.0	117.9	6.6	10%
Tortorici	0.0	31.9	313.1	368.4	58.2	17%
Ucria	0.0	24.9	173.0	317.6	43.3	12%
	<250 (%)	250–500 (%)	500–750 (%)	750–1000 (%)	>1000 (%)	Average DTM
Castell'Umberto	0%	10%	78%	12%	0%	627
Montalbano Elicona	0%	1%	28%	63%	8%	825
Raccuja	0%	0%	40%	55%	5%	802
San Piero Patti	0%	8%	41%	45%	6%	754
San Salvatore di Fitalia	1%	26%	56%	17%	0%	595
Santa Domenica Vittoria	0%	0%	0%	14%	86%	1088
Sant'Angelo di Brolo	0%	12%	57%	31%	0%	671
Sinagra	1%	28%	45%	25%	1%	614
Tortorici	0%	4%	41%	48%	8%	779
Ucria	0%	4%	31%	57%	8%	797
						755

Through the GIS program, maps concerning DTM, slope (classified in percentage terms), aspect, and curvature were produced (Figure 3). Starting from a DTM map the slope of the roads, which must be driven by the vehicles, and the specific slope of each hazelnut areas were calculated. Ambiguous contexts (such as a high degree of slope or altitude) have been assessed in a parallel analysis through available orthophoto investigation, even if the vegetation map detected hazelnut plants. Within this operation, the elevate degree of correctness of the vegetation map of Sicily can be confirmed.

Aspect   Slope   Curvature

DTM (in meters)

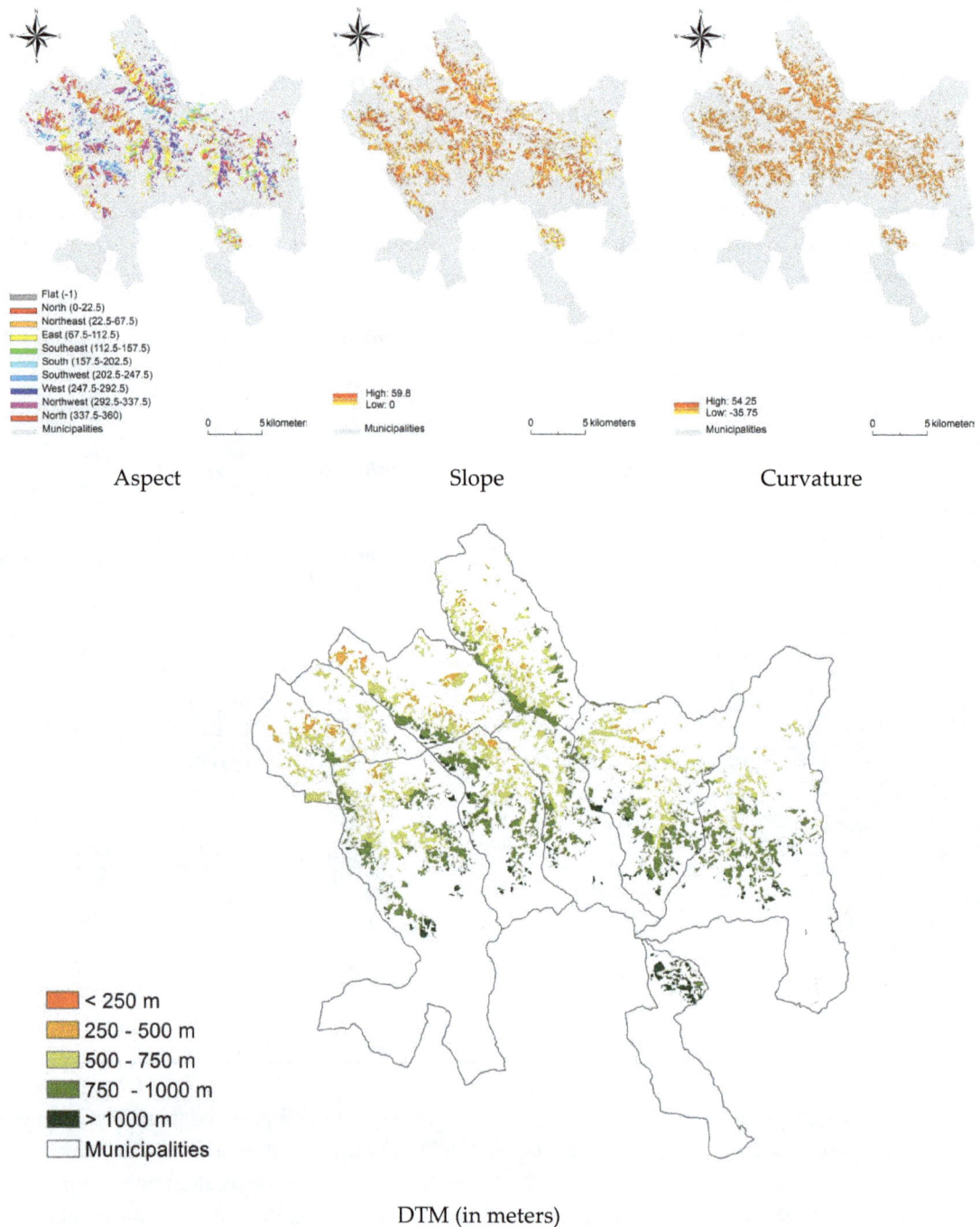

Figure 3. Morphological structures of the territory using GIS program. Source: own elaboration.

The slope of hazelnut areas was classified into seven classes: '1': 0%; '2': 1–10%; '3': 11–20%; '4': 21–30%; '5': 31–40%; '6': 41–50%; '7': >50%. The slope of the streets in the ten municipalities was classified into 17 classes: '1': 0%; '2': 1–2%; '3': 3–4%; '4': 5–6%; '5': 7–8%; '6': 9–10%; '7': 11–12%; '8': 13–14%; '9': 15–16%; '10': 17–18%; '11': 19–20%; '12': 21–22%; '13': 23–24%; '14': 25–26%; '15': 27–28%; '16': 29–30%; '17': >30%. Zones with a steep slope (>30%) and high altitude (>1000 m) are the ones to avoid for mechanized harvesting as it results in increased risk for operators when they should collect hazelnuts.

Through the raster calculator tool using GIS program, the territory was analyzed observing the most suitable places to introduce mechanization processes. Figure 4 displays the optimal contexts for hazelnuts (in legend with the label "0"), with minimal risk for operators, where the slope is minimal, with optimum altitudes to hazelnuts and ease in terms of mobility for the machines that need to reach such areas. There are also further favorable contexts for hazelnuts with good altitude and slopes. Finally, areas that should be avoided for greater risk for operators, due to their high altitude and slopes, discontinuous road system with strong slopes.

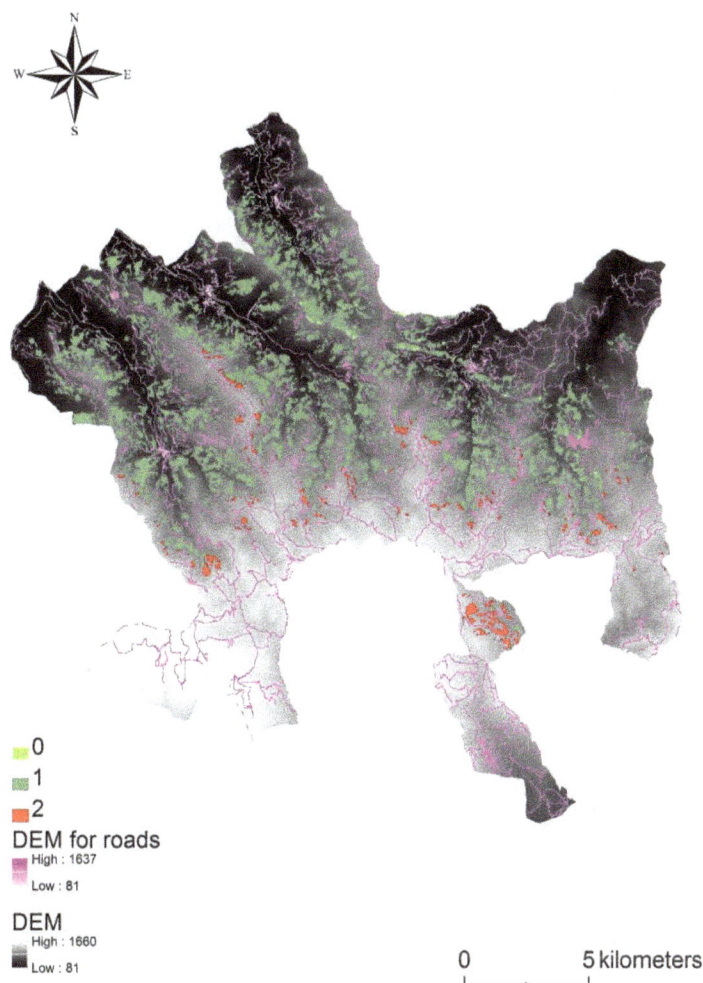

Figure 4. Possibility of mechanization. Legend 0: optimal areas for hazelnut, with minimal risk for operators; 1: favorable areas for hazelnut with good altitude and slopes. 2: areas to be avoided for greater risk for operators, including high altitude and slopes and road systems with strong slopes. Source: own elaboration.

Checking the results obtained, a region group elaboration was run using the GIS program. It identifies the degree of feasibility of cultivation and collection of hazelnuts depending on the morphological characteristics (Figure 5). Four groups of hazelnut areas can be observed. In this elaboration, the most optimal contexts emerge both to grow and manage the cultivation of hazelnuts and to provide the right security measures for the operators who must collect the hazelnuts (class "1"). In fact, in Figure 5, it is possible to clearly distinguish the southern zones, which are the ones that are higher in altitude (>1000 m), sloping (>30%) and mostly affect the safety of workers (class "4"). However, the best areas ("1") occupy only 430 hectares (about 9% compared to the total surface area of hazelnuts in the ten municipalities). Unsuitable contexts have a surface of 370 hectares (about 7% of the total surface area in analysis). The intermediate areas (classes "2" and "3" for the region group elaboration) are those that occupy the largest surface areas (almost 4200 hectares). Finally, 4600 hectares can be used as agro-energetic districts.

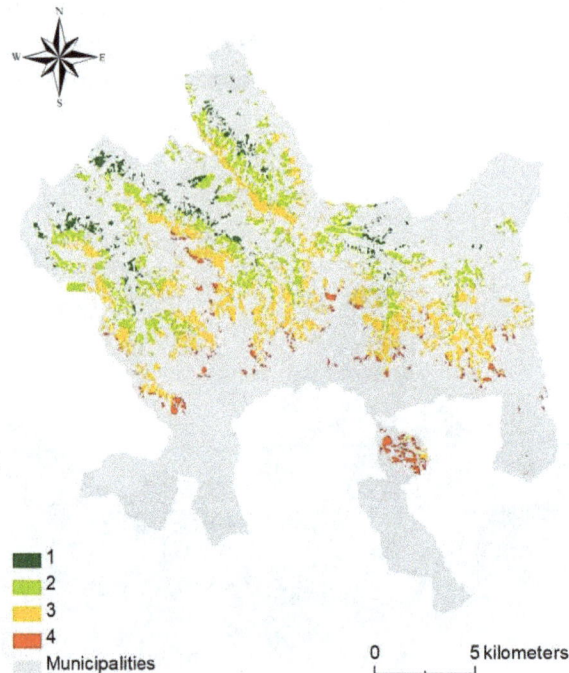

Figure 5. Region group elaboration using GIS program to identify the degree of feasibility of cultivation and collection of hazelnuts depending on the morphological and spatial characteristics. Source: own elaboration.

4. Discussion

Hazelnuts represents one of the major economic realities that constitute the primary sector of the Sicily region [49]. The latter is a unique Mediterranean context, given its climate, landscape, and peculiar characteristics [50,51]. Particularly, the Nebroidi park allows for the easy adaptability of hazelnuts [38]. However, high altitudes and the acclivity of slopes make it difficult to cultivate and harvest hazelnuts [42,43]. This study aims to identify the most favorable contexts to increase the growing and harvesting of hazelnuts using appropriate vehicles. First, the territorial characteristics should be considered, such as slope or the road system necessary to reach these contexts. Using and processing data through GIS technologies and databases obtained by remote sensing processes at local level was decisive.

Spatial data collection permitted the comparison of different databases. The vegetation map of Sicily has highlighted how a deep knowledge of the local contexts and the use of remote sensing and GIS technologies, in addition to a large bibliographic collection, allows for a detailed analysis, identifying several kinds of crops. In fact, limiting to a CLC map could causes an actual error in calculating surface areas destined for hazelnuts: only 63% of hazelnuts fall into the category of "orchards" in the CLC map. Data processing has confirmed the adequacy of the vegetation map of Sicily: most of the hazelnuts are found at slopes that are not too high (between 6% and 30%) [6] and at altitudes between 500 and 1000 m [41]. GIS processing has thus let to recognize the most appropriate areas for the hazelnuts, since their cultivation is not recommended on the steep slopes, since they cannot prevent and hinder environmental matters, as soil erosion processes [42,43]. Furthermore, when some contexts appeared uncertain (e.g., when the Vegetation map of Sicily detected hazelnut plantations along high degree of slope or altitude), a parallel analysis (orthophoto investigation of specific areas) assessed such outcomes, confirming the high correctness of the Vegetation map of Sicily.

Another issue that must be addressed in this paper concerns the collection of hazelnuts. In these contexts, traditional methods are still used, such as hand-picking. This makes the collection of hazelnuts expensive, wasteful, with high labor costs and long working hours. As a possible solution to the mechanization of harvesting the nuts in soils with planting distances and irregular with steep slopes (from 24 to 35%), the prototype, proposed by [44] can adapt to the most difficult conditions. It is not necessary to use other harvester machines. Very often, hazelnuts are located along inclined slopes or unconnected areas, where traditional means of mechanization fail to work optimally. The prototype (i) is smaller than existing machines, (ii) ensures agility in maneuvering and high stability in steep slopes, (iii) is easy to use and versatile, (iv) reduces capital amortization times, (v) is easy to be transported by simple means such as small trolleys or pickups, (vi) increases the capacity to collect hazelnuts, and (vii) improves working conditions (e.g., substantial reduction in the risk of biomechanical overload compared to manual harvesting). The prototype of [44] allows for simple collection, safeguarding the health and safety of workers, and reduces the time necessary for the hazelnut harvest (e.g., it separates hazelnuts from other elements such as weeds or leaves). Furthermore, as a work accessory, the prototype proposed by [44] fits to other machines depending on the working context.

Besides identifying the most suitable areas, the present paper also aims to offer a chance of sustainable development, such as increasing cultivation of hazelnuts, protecting the workers' safety and optimizing work times concerning picking hazelnuts given the intrinsic territorial adversity. The concepts of circular economy and agro-energy districts could be effectively applied in these territories [52,53]. From the point of view of agro-energetic districts, it is assumed that the former depends on several parameters, i.e., the cultivation type and site and the planting distance, defining the most appropriate use of residual biomass [48]. For intensive farming of hazelnut, the pruned biomass can reach about 1848 kg/ha [54]. Obtainable residual biomass from hazelnut trees pruning can be positively considered as an actual economic chance for this area. From our study, it is possible to estimate to get a biomass of 8500 kg (4600 hectares).

In conclusion, from the economic point of view, a greater cultivation of hazelnuts would also give more employment alternatives, increasing the employment status and leading to a valorization of local agriculture [55]. As in other region (e.g., Latium and Piedmont in Italy or in Turkey) where hazelnuts are important for the primary sector [7,8,56–58], they can be defined as an economic resource in Sicily since they could provide income opportunities in hilly and mountainous areas where other agricultural activities are limited by the hostile environment [55]. Potential revenue deriving from this kind of cultivation can be estimated depending on how many hectares are put back into culture [55]. Finally, hazelnuts are defined as one of the most profitable fruit, demonstrating a high degree of sustainability, mostly owing to the low input necessities for orchard management and the opportunity of using agricultural waste as potential biomass [48,52,54,55,58,59].

5. Conclusions

The present paper started from the collection and comparison of available materials. The GIS elaboration is decisive for analyzing the Sicilian context and discriminate the spatial database by choosing the most appropriate one. Using this method, the most suitable area for cultivation hazelnuts can be detected. Also, innovative mechanization processes should be employed since they are still undeveloped and can mitigate the physical obstacles to hazelnut production (e.g., discontinuous road system, high slope). Finally, a sustainable vision is offered with the aim to promote a circular economy and agro-energetic district in this Sicilian context based on hazelnut cultivation.

Acknowledgments: This study was supported by the SICILNUT Project founded by MIPAAF.

Author Contributions: Ilaria Zambon analyzed the data and wrote the paper; Lavinia Delfanti collected the materials (e.g., shapefile data) concerning Sicily; Roberto Bedini, Massimo Cecchini, and Danilo Monarca collected the materials concerning mechanization processes; Alvaro Marucci was involved in the critical review of the results obtained; and Walter Bessone revised the manuscript.

Conflicts of Interest: The authors declare no conflict of interest.

References

1. Bonet, A. Secondary succession of semi-arid Mediterranean old-fields in south-eastern Spain: Insights for conservation and restoration of degraded lands. *J. Arid Environ.* **2004**, *56*, 213–233. [CrossRef]

2. Godone, D.; Garbarino, M.; Sibona, E.; Garnero, G.; Godone, F. Progressive fragmentation of a traditional Mediterranean landscape by hazelnut plantations: The impact of CAP over time in the Langhe region (NW Italy). *Land Use Policy* **2014**, *36*, 259–266. [CrossRef]

3. Sitzia, T.; Semenzato, P.; Trentanovi, G. Natural reforestation is changing spatial patterns of rural mountain and hill landscapes: A global overview. *For. Ecol. Manag.* **2010**, *259*, 1354–1362. [CrossRef]

4. Sluiter, R.; de Jong, S.M. Spatial patterns of Mediterranean land abandonment and related land cover transitions. *Landsc. Ecol.* **2007**, *22*, 559–576. [CrossRef]

5. FAO. Food and Agricultural Commodities Production. Available online: http://faostat.fao.org/site/339/default.aspx (accessed on 1 July 2010).

6. Aydinoglu, A.C. Examining environmental condition on the growth areas of Turkish hazelnut (*Corylus colurna* L.). *Afr. J. Biotechnol.* **2010**, *9*, 6492–6502.

7. Reis, S.; Yomralioglu, T. Detection of current and potential hazelnut (*Corylus*) plantation areas in trabzon, north east Turkey using GIS and RS. *J. Environ. Biol.* **2006**, *27*, 653–659. [PubMed]

8. TURKSTAT. Turkish Statistical Institute. Available online: www.turkstat.gov.tr (accessed on 12 May 2001).

9. Sarıoğlu, F.E.; Saygın, F.; Balcı, G.; Dengiz, O.; Demirsoy, H. Determination of potential hazelnut plantation areas based GIS model case study: Samsun city of central Black Sea region. *Eurasian J. Soil Sci.* **2013**, *2*, 12–18.

10. London Economics. *Evaluation of the CAP Policy on Protected Designations of Origin (PDO) and Protected Geographical Indications (PGI)*; European Commission—Agriculture and Rural Development: Bruxelles, Belgium, 2008; p. 275.

11. Martinez-Casasnovas, J.A.; Ramos, M.C.; Cots-Folch, R. Influence of the EU CAP on terrain morphology and vineyard cultivation in the Priorat region of NE Spain. *Land Use Policy* **2010**, *27*, 11–21. [CrossRef]

12. Van Berkel, D.B.; Verburg, P.H. Sensitising rural policy: Assessing spatial variation in rural development options for Europe. *Land Use Policy* **2011**, *28*, 447–459. [CrossRef]

13. Westhoek, H.J.; van den Berg, M.; Bakkes, J.A. Scenario development to explore the future of Europe's rural areas. *Agric. Ecosyst. Environ.* **2006**, *114*, 7–20. [CrossRef]

14. Official Gazette. *The Regulation of the Law Planning Hazelnut Production and Determining Hazelnut Plantation Areas*; Official Gazette: Ankara, Turkey, 2009; pp. 27289.14.

15. Dengiz, O.; Ozcan, H.; Köksal, E.S.; Kosker, Y. Sustainable Natural Resource Management and Environmental Assessment in The Salt Lake (Tuz Golu) Specially Protected Area. *J. Environ. Monit. Assess.* **2010**, *161*, 327–342. [CrossRef] [PubMed]

16. Lioubimtseva, E.; Defourny, P. GIS based landscape classification and mapping of European Russia. *Landsc. Urban Plan.* **1999**, *44*, 63–75. [CrossRef]

17. Longley, P.A.; Goodchild, M.F.; Maguire, D.J.; Rhind, D.W. *Geographic Information Systems and Science*; Bath Press: London, UK, 2001.

18. Bolca, M.; Kurucu, Y.; Dengiz, O.; Nahry, A.D.H. Terrain characterization for soils survey of Kucuk Menderes plain, South of Izmir, Turkey, using remote sensing and GIS techniques. *Zemdirb. Agric.* **2011**, *98*, 93–104.

19. Wilkinson, G. Results and implications of a study of fifteen years of satellite image classification experiments. *IEEE Trans. Geosci. Remote Sens.* **2005**, *43*, 433–440. [CrossRef]

20. Reis, S.; Taşdemir, K. Identification of hazelnut fields using spectral and Gabor textural features. *ISPRS J. Photogramm. Remote Sens.* **2011**, *66*, 652–661. [CrossRef]

21. Lillesand, T.M.; Kiefer, R.W. *Remote Sensing and Image Interpratation*; The Lehigh Press: New York, NY, USA, 2000.

22. Cohen, Y.; Shoshany, M. A national knowledge-based crop recognition in Mediterranean environment. *Int. J. Appl. Earth Observ. Geoinf.* **2002**, *4*, 75–87. [CrossRef]

23. Grauke, L.J.; Thompson, T.E. Rootstock development in temperate nut crops. Genetics and breeding of tree fruits and nuts. *Acta Horticult.* **2003**, *622*, 553–566. [CrossRef]

24. Yomralioglu, T.; Inan, H.I.; Aydinoglu, A.C.; Uzun, B. Evaluation of initiatives for spatial information system to support Turkish agriculture policy. *Sci. Res. Essay* **2009**, *4*, 1523–1530.

25. Mundia, C.N.; Aniya, M. Analysis of land use/cover changes and urban expansion of Nairobi city using remote sensing and GIS. *Int. J. Remote Sens.* **2005**, *26*, 2831–2849. [CrossRef]
26. Yuan, F.; Sawaya, K.E.; Loeffelholz, B.; Bauer, M.E. Land cover classification and change analysis of the Twin Cities (Minnesota) metropolitan area by multi temporal Landsat remote sensing. *Remote Sens. Environ.* **2005**, *98*, 317–328. [CrossRef]
27. Franco, S. Use of remote sensing to evaluate the spatial distribution of hazelnut cultivation: Results of a study performed in an Italian production area. *Acta Horticult.* **1997**, *445*, 381–388. [CrossRef]
28. Kavzoglu, T. Increasing the accuracy of neural network classification using refined training data. *Environ. Model. Softw.* **2009**, *24*, 850–858. [CrossRef]
29. De Aranzabal, I.; Schmitz, M.F.; Aguilera, P.; Pineda, F.D. Modelling of landscape changes derived from the dynamics of socio-ecological systems: A case of study in a semiarid Mediterranean landscape. *Ecol. Indic.* **2008**, *8*, 672–685. [CrossRef]
30. Tzanopoulos, J.; Jones, P.J.; Mortimer, S.R. The implications of the 2003 Common Agricultural Policy reforms for land-use and landscape quality in England. *Landsc. Urban Plan.* **2012**, *108*, 39–48. [CrossRef]
31. Fabi, A.; Varvaro, L. Remote sensing in monitoring the dieback of hazelnut on the 'Monti Cimini' district (Central Italy). *Acta Horticult.* **2009**, *845*, 521–526. [CrossRef]
32. Taşdemir, K. Exploiting spectral and spatial information for the identification of hazelnut fields using self-organizing maps. *Int. J. Remote Sens.* **2012**, *33*, 6239–6253. [CrossRef]
33. Yalniz, I.; Aksoy, S. Detecting regular plantation areas in satellite images. In Proceedings of the IEEE 17th Signal Processing and Communications Applications Conference, Antalya, Turkey, 9–11 April 2009.
34. Kimes, D.S.; Nelson, R.F.; Manry, M.T.; Fung, A.K. Attributes of neural networks for extracting continuous vegetation variables from optical and radar measurements. *Int. J. Remote Sens.* **1998**, *19*, 2639–2663. [CrossRef]
35. Aslan, Ü.; Özdemir, İ. Separation of Agricultural Aimed Plantations from the Forest Cover by Using the LANDSAT-5TM and SPOT-4 HRVIR Data in Turkey. International Archives of Photogrammetry. *Remote Sens. Spat. Inf. Sci.* **2004**, *36*, 324–327.
36. Biondi, E.; Calandra, R. La cartographie phytoécologique du paysage. *Écologie* **1998**, *29*, 145–148.
37. Biondi, E.; Catorci, A.; Pandolfi, M.; Casavecchia, S.; Pesaresi, S.; Galassi, S.; Pinzi, M.; Vitanzi, A.; Angelini, E.; Bianchelli, M.; et al. Il Progetto di "Rete Ecologica della Regione Marche" (REM), per il monitoraggio e la gestione dei siti Natura 2000 e l'organizzazione in rete delle aree di maggiore naturalità. *Fitosociologia* **2007**, *44*, 89–93.
38. Gianguzzi, L.; Papini, F.; Cusimano, D. Phytosociological survey vegetation map of Sicily (Mediterranean region). *J. Maps* **2016**, *12*, 845–851. [CrossRef]
39. Pedrotti, F. *Cartografia Geobotanica*; Pitagora Editrice: Bologna, Italy, 2004; p. 248.
40. Rivas-Martínez, S. Notions on dynamic-catenal phytosociology as a basis of landscape science. *Plant Biosyst.* **2005**, *139*, 135–144. [CrossRef]
41. Duran, C. Drought and vegetation analysis in Tarsus River Basin (Southern Turkey) using GIS and Remote Sensing data. *J. Hum. Sci.* **2015**, *12*, 1853–1866. [CrossRef]
42. Ozturk, I.; Tanik, A.; Seker, D.Z.; Levent, T.B.; Ovez, S.; Tavsan, C.; Ozabali, A.; Sezgin, E.; Ozdilek, O. *Technical Report on the Land-Use Methodology Being Tested and Draft Land-Use Plans, Testing of Methodology on Spatial Planning for ICZM*; Akçakoca District Pilot Project; ITU: Istanbul, Turkey, 2007.
43. Tanik, A.; Seker, D.Z.; Ozturk, I.; Tavsan, C. GIS based sectoral conflict analysis in a coastal district of Turkey. *Int. Arch. Photogramm. Remote Sens. Spat. Inf. Sci.* **2008**, *37*, 665–668.
44. Monarca, D.; Cecchini, M.; Colantoni, A.; Bedini, R.; Longo, L.; Bessone, W.; Caruso, L.; Schillaci, G. Evaluation of safety aspects for a small-scale machine for nuts harvesting. In Proceedings of the MECHTECH 2016 Conference—Mechanization and New Technologies for the Control and Sustainability of Agricultural and Forestry Systems, Alghero, Italy, 29 May–1 June 2016; pp. 32–35.
45. Monarca, D.; Cecchini, M.; Massantini, R.; Antonelli, D.; Salcini, M.C.; Mordacchini, M.L. Mechanical harvesting and quality of "marroni" chestnut. *Acta Horticulturae* **2005**, *682*, 1193–1198. [CrossRef]
46. Formato, A.; Scaglione, G.; Ianniello, D. Application of software for the optimization of the surface shape of nets for chestnut harvesting. *J. Agric. Eng.* **2013**, *44*. [CrossRef]
47. Recanatesi, F.; Ripa, M.N.; Leone, A.; Luigi, P.; Luca, S. Land use, climate and transport of nutrients: Evidence emerging from the Lake Vicocase study. *Environ. Manag.* **2013**, *52*, 503–513. [CrossRef] [PubMed]

48. Bilandzija, N.; Voca, N.; Kricka, T.; Matin, A.; Jurisic, V. Energy potential of fruit tree pruned biomass in Croatia. *Span. J. Agric. Res.* **2012**, *10*, 292–298.

49. Cotugno, L. Territory and Population—Demographic Dynamics in Sicily. *Rev. Hist. Geogr. Toponomast.* **2011**, *6*, 81–91.

50. Barbera, G.; Cullotta, S. An inventory approach to the assessment of main traditional landscapes in Sicily (Central Mediterranean Basin). *Landsc. Res.* **2012**, *37*, 539–569. [CrossRef]

51. Colantoni, A.; Ferrara, C.; Perini, L.; Salvati, L. Assessing trends in climate aridity and vulnerability to soil degradation in Italy. *Ecol. Indic.* **2015**, *48*, 599–604. [CrossRef]

52. Colantoni, A.; Delfanti, L.M.P.; Recanatesi, F.; Tolli, M.; Lord, R. Land use planning for utilizing biomass residues in Tuscia Romana (central Italy): Preliminary results of a multi criteria analysis to create an agro-energy district. *Land Use Policy* **2016**, *50*, 125–133. [CrossRef]

53. Colantoni, A.; Longo, L.; Gallucci, F.; Monarca, D. Pyro-Gasification of Hazelnut Pruning Using a Downdraft Gasifier for Concurrent Production of Syngas and Biochar. *Contemp. Eng. Sci.* **2016**, *9*, 1339–1348. [CrossRef]

54. Cecchini, M.; Monarca, D.; Colantoni, A.; Di Giacinto, S.; Longo, L.; Allegrini, E. Evaluation of biomass residuals by hazelnut and olive's pruning in Viterbo area. In Proceedings of the International Commission of Agricultural and Biological Engineers, Section V. CIOSTA XXXV Conference "From Effective to Intelligent Agriculture and Forestry", Billund, Denmark, 3–5 July 2013.

55. Cerutti, A.K.; Beccaro, G.L.; Bagliani, M.; Donno, D.; Bounous, G. Multifunctional ecological footprint analysis for assessing eco-efficiency: A case study of fruit production systems in Northern Italy. *J. Clean. Prod.* **2013**, *40*, 108–117. [CrossRef]

56. Gönenc, S.; Tanrıvermis, H.; Bülbül, M. Economic assessment of hazelnut production and the importance of supply management approaches in Turkey. *J. Agric. Rural Dev. Trop. Subtrop.* **2006**, *107*, 19–32.

57. Petriccione, M.; Ciarmiello, L.F.; Boccacci, P.; De Luca, A.; Piccirillo, P. Evaluation of 'Tonda di Giffoni' hazelnut (*Corylus avellana* L.) clones. *Sci. Horticult.* **2010**, *124*, 153–158. [CrossRef]

58. Di Giacinto, S.; Longo, L.; Menghini, G.; Delfanti, L.M.P.; Egidi, G.; De Benedictis, L.; Salvati, L. A model for estimating pruned biomass obtained from *Corylus avellana* L. *Appl. Math. Sci.* **2014**, *8*, 6555–6564. [CrossRef]

59. Zambon, I.; Colosimo, F.; Monarca, D.; Cecchini, M.; Gallucci, F.; Proto, A.R.; Colantoni, A. An innovative agro-forestry supply chain for residual biomass: Physicochemical characterisation of biochar from olive and hazelnut pellets. *Energies* **2016**, *9*, 526. [CrossRef]

9

Analysis and Diagnosis of the Agrarian System in the Niayes Region, Northwest Senegal (West Africa)

Yohann Fare [1,*], Marc Dufumier [1], Myriam Loloum [1], Fanny Miss [2], Alassane Pouye [3], Ahmat Khastalani [3] and Adama Fall [4]

[1] Unité d'Enseignement et de Recherche Agriculture Comparée et Développement Agricole, AgroParisTech. 16, rue Claude Bernard, F-75231 Paris CEDEX 05, France; Marc.Dufumier@agroparistech.fr (M.D.); myriam.loloum@gmail.com (M.L.)

[2] École Nationale du Génie Rural, des Eaux et des Forêts, AgroParisTech, 19 Avenue du Maine, 75732 Paris CEDEX 15, France; fanny_miss@yahoo.fr

[3] Ecole Nationale Supérieure d´Agriculture (ENSA) de Thiès, B.P A 296-Thiès, Sénégal; alassanepy@hotmail.com (A.P.); khastalani@yahoo.fr (A.K.)

[4] SOS SAHEL International, 21001 Thiès, Senegal; faladama@hotmail.com

* Correspondence: yohannfare@gmail.com

Academic Editor: Les Copeland

Abstract: The agrarian system Analysis and Diagnosis is used for this study, the goal of which was to provide a corpus of basic knowledge and elements of reflection necessary for the understanding the *Niayes* farming systems dynamics in Senegal, West Africa. Such holistic work has never been done before for this small region that provides the majority of vegetables in the area, thanks to its microclimate and access to fresh water in an arid country. Reading of the landscape and historical interviews coupled with fine-tuned household surveys were used to build a typology of agricultural production units (each type being represented by a production system). The main phases within the region's history were distinguished. Before colonization, agriculture was based on gathering and shifting agriculture (millet and peanut) in the southern region and transhumant stockbreeding in the North. During colonization, market gardening became a source of income as a response to cities' increasing demand. Two major droughts (in the 1970s and 1980s) have accelerated this movement. Extension of market gardening areas and intensification of activities were made possible by Sahelian migrants' influx and the creation of *mbeye seddo*, a contract that allows for sharing added value between the employer and seasonal workers, named *sourghas*. Over the past 20 years, the "race for motorization" has created important social gaps (added value sharing deserves review) and a risk of overexploitation of groundwater.

Keywords: comparative agriculture; survey on farming; socioeconomic differentiation; Senegal

1. Introduction

From 3.3 million inhabitants in 1961, Senegal will count its population at about 17 million in 2020. Feeding an unceasingly growing population and supplying towns with fresh and quality vegetables as well as fruits are challenges for the country, which has mainly Sahelian conditions. Due to urban growth, competition for foreign products and devaluation of West African FCFA (Franc de la Communauté financière en Afrique, the currency of eight independent states in West Africa), Senegal saw an increase in imports of onions, potatoes, rutabagas and carrots. Production from the Niayes agricultural region seems to play a non-negligible role in importation to satisfy national needs. In fact, Niayes farmers provide anywhere from half to two-thirds of the national production of fresh vegetables (tomatoes, onions, rutabagas, cabbages, carrots, etc.) [1]. Despite the dynamism of the

Niayes production systems, there has never been any holistic work to prepare Senegalese policy makers to face the challenges of feeding the cities that are growing with Niayes production. Our research was conducted within agricultural development projects, led by a non-governmental organization (NGO) named SOS SAHEL, and aims at:

- Understanding conditions of market gardening development in the Niayes (environmental parameters: Soil fertility, water resources, etc.) and the diversity of production units in this region to fine-tune forthcoming projects.
- Determining the conditions necessary for the Niayes to better contribute to the country's agricultural production. Beyond technological choices, we had to understand the economic rationality of Niayes' production units. Understanding their rationality will help this project to better suggest social, economic and political actions that would be able to boost production and create jobs while respecting the environment.

To address these objectives, this study has adopted as a theoretical reference the concept of the "Agrarian System" developed at the French Agricultural Research National Institute, Agrarians Systems and Development Department (INRA-SAD) [2–4]. This holistic methodology has been used to describe many situations in world agriculture but not in the Niayes [5]. The extension of this approach to Niayes would enrich our knowledge of the world's agriculture, in addition to providing answers to the abovementioned questions.

Originating from comparative agriculture studies and systems research applied to agriculture, the concept is defined as "a mode of exploitation of a given agro-system, historically constituted and long-lasting, adapted to the bioclimatic conditions of a given space, and meeting the requirements and social needs of the time" [2]. "Agrarian System analysis and diagnosis" is the survey method related to the concept of an agrarian system [3,4]. It is an all-encompassing methodology, combining different levels of analysis and therefore capable of making sense of agricultural activities operated by farmers within a given agricultural district, in a way that accounts for both ecological and socioeconomic dimensions. This holistic approach aims at describing farmers' social and economic practices and techniques and at understanding the phenomena that influence them [4,6]. It also allows for evaluating the sustainability of a region's agriculture and collecting all elements for future transformations and forecasting, with or without any project-like intervention.

For this research project, data have been collected through observation as well as questionnaires and a literature review. The data collection has been organized at different scales from the general level (national situation) to the particular (region, then farms, plots and/or herds) following an iterative analysis–synthesis process. This process obeys the methodology of comparative agriculture, which "favours the usage of a telescopic change of scale, particularly between the three levels of analysis privileged by us, i.e., that of the plot or herd for examining practices, that of the production unit or farm for integrating different cropping and livestock systems, and that of the (more or less vast) region or country for the pertinent application of the agrarian system concept" [3,4]. Understanding the differentiation of previous agrarian systems is an essential step to understanding the dynamics of the production system currently under survey. Within this current agricultural system, farms have either always used the same techniques or implemented similar "strategies" due to different access to production means (land, work, capital) and the heterogeneity of regional conditions [3]. The technical–economic evaluation of the farms surveyed allows for a better understanding of the differences between them [7]. Finally, an analysis of current projects and agricultural policies helps with developing future perspectives and possibly generates some corrective measures to implement from the general interest point of view.

2. Materials and Methods

The survey area is the Niayes, a small agricultural region of Senegal specialising in market gardening. The boundaries of our study area are set based on geomorphology, climate and main agricultural trends as an important market gardening area. The main boundaries of our study area were set as follows: in the West, the Atlantic Ocean; in the East, the main rain-fed agricultural areas (millet–peanuts system); in the North, the Senegal River; and in the South, the Cape Verde Peninsula (Figure 1). This general delimitation was enhanced on site by observations that allowed the project to determine zones, as relatively homogenous sub-units of our survey area.

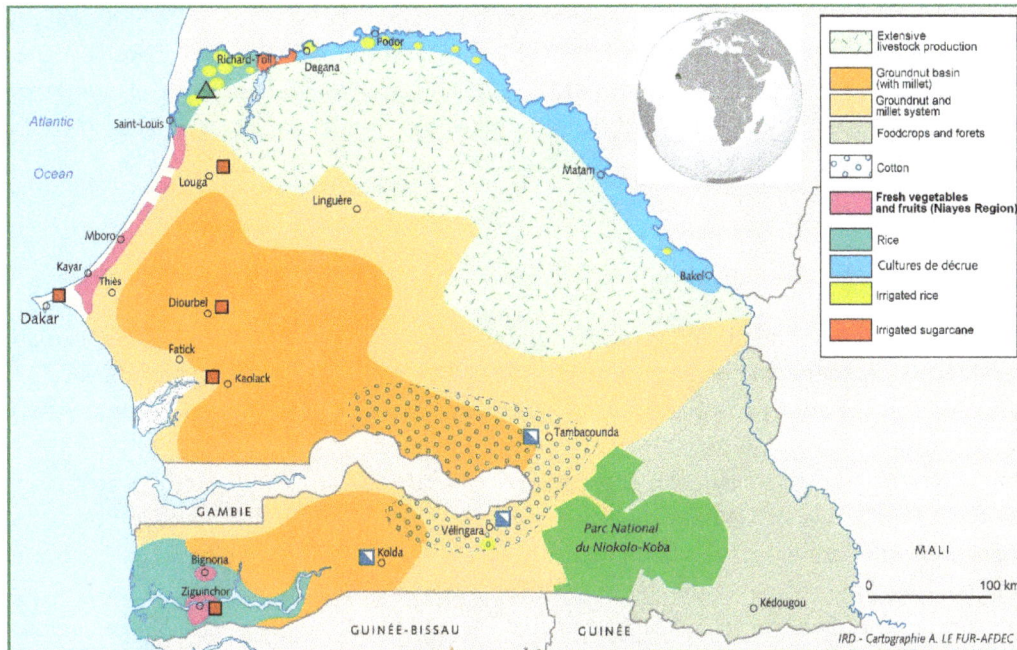

Figure 1. Location of the Niayes within the main agricultural regions of Senegal.

On-site data collection was conducted from 2010 to 2013 with the support of SOS Sahel International teams, a non-governmental organization (NGO). These surveys were completed by five of the authors, sometimes accompanied by local interpreters. Intermediate results have been presented in four Master's theses, supervised by the main author [1,8–10]. Citizens' and farmer organizations' acceptance was facilitated by one of the authors' work with the market gardening union association of Niayes.

First, the study aimed at interviewing groups of elderly farmers who were witnesses to the recent evolution of Niayes agriculture. The topic chosen for discussion during the interviews was Niayes' agricultural history. Souvenirs that we collected directly dated from the 1940s. The participant groups were sometimes associated with people in charge of the farmers' organizations, retired people from technical services who stayed and lived in their former working place. Such information was required in order to understand the rationale for their installation. With the purpose of dividing the long historical period into sequences and locating the souvenirs in time, we used some outstanding events as references: the Second World War; the Leopold Sedar Senghor, Abdou Diouf and Abdoulaye Wade presidential terms; the 1995 CFA franc devaluation; and the 2008 price surges. Thirty historical interviews coupled with 100 related to farms' life cycles were coordinated by the main author, who also took part in many of them. About 80 fine-tuned surveys were used to help establish economic results at the household level. The objective was to obtain the number of workers in each household and those external to the households, the number of people to feed, the technical details for each cultural and production system, locating them in time in landscapes, yearly and inter-yearly rotations, crop shifts, income, inputs and manpower, produce destinations (for self-consumption, sales, transformation), and access to credit. The choice of farms type for the surveys was guided by the farmers' organizations, based on rather general typology standards, but the list of farms was validated based on their gardening techniques and practices related to their means of irrigation and their location in the landscape, and therefore gardening soil types. Each investigation was led by an investigator accompanied by an interpreter, as open or half-open survey type using a survey guide. During the study, we found that sometimes it was necessary to come back several times to have more accurate information and be more precise about different periods of the year. Investigations at the market and within the Senegalese National Agency on statistics and demography provided price evolution related to different productions in the survey area. Finally, we met some managers of NGO projects, national projects, and microfinance institutions, as well as officers of technical services in the area. Those people know more about the past and current projects operating on the agrarian system. They also offered some assistance during surveys; they helped during the selection of relevant villages, shared contacts of rural leaders and sometimes introduced us to villagers.

2.1. The Survey Area

2.1.1. Boundaries

The Niayes area, a small and unique region in Western Senegal. covers approximately 2759 km^2 [11,12],less than two percent of the whole country's surface. Niayes covers the area situated between Dakar in the south and Saint-Louis in the north, along the country's northern coast and 5–30 km in width.

The Niayes belongs to a particular ecosystem of the Sahelian strip. Although located at latitude 14°30' and 16° North, this area is nevertheless characterized by a tropical climate of Sub-Canarian type and an original Sub-Guinean vegetation [13]. Such an exception is produced by a marine tradewind, relatively fresh and humid, generated by the Azores anticyclone (Northern Atlantic), which protects the northern coast of Senegal from warm and dry harmattan winds [14]. The Niayes agricultural region's relief is a succession of sand dunes and inter-dune basins. These quaternary-aged dunes are established in a marly and limy substratum dating from the Eocene and Paleocene [15,16]. Rainfall is relatively low, between 350 and 450 mm rain per year, but water availability is important in inter-dune basins. This profusion is tied to the "Nappe des Sables Quaternaires"—NSQ—groundwater, circulating in sandy and sandy-clayey deposits, relatively permeable, on a marly and marly–limy substratum of the Eocene, which is impermeable [17,18]. In current landscapes, from the continent to coastal area, four types of dunes succeed each other:

- Long red and levelled dunes, oriented from northeast to southwest, running parallel with the shore line, witnessing the regressive and dry phase (Ogolien) of the last ice age (18,000 years ago). These dunes are steady and indicate the borders of the Niayes and the Dieri (the local name for the ancient peanut basin).
- Shorter and higher red dunes, sheltering much more than anywhere else humus-bearing soils and peat bogs. These dunes would be resulting from Ogolian dunes alteration by a marine tradewind [16]. They are in general permanent but revive in some places.
- Semi-permanent yellow dunes, forming a coastal dune stretch with variable width from one to four kilometres. Of 20 to 30 m high, they end with an abrupt front in the windy part. They would have been settled in the regressive phase after 5500 BCE. The depressions between dunes bear less humus [16].
- White coastal dunes, sharp, forming a large band ranging from some dozens of meters to 300 m from the beach. The sand was brought by recent coastal accumulation (by 1800 years ago). They are partly maintained of filao trees (*Casuarina equisetifolia*) [14,19,20].

The Niayes area is highly populated and characterized by a dynamic agriculture dominated by market gardening activities (more than half of the national production). The Niayes make an important contribution to the fresh vegetable supply in Dakar. More than two million people lived in Dakar in

2006, more than 20% of Senegal's whole population, and more than two-thirds of Senegal's population live less than 60 km from coastal areas [21].

2.1.2. Zoning Elements

Tied in with geomorphology, proximity with NSQ groundwater and soil types, and dune and basin succession can be divided into different zones (Figures 2–4 and Table 1):

- A coastal dune string (A), composed of raw mineral soils. On these dunes, which can be potentially mobilized by wind, colonial forestry administration by the Senegalese government established in successive stages a filao trees strip (*Casuarina equisetifolia* L.), a vegetal barrier protecting against the sand-silting risk to the basins between dunes.

- Small-sized depressions, between levelled and scarcely collected dunes (B). On this area, which stretches a few hundreds of meters behind the filao trees strip, the groundwater level is not very deep (two to five meters). For that reason, undoubtedly, depressions are named "Ndioukis," meaning "drawing water with a bucket using a pulley." Soils are of a siliceous type, poor in organic materials and sometimes salinated. Ndioukis are much more numerous in the North of the Niayes (Northern zone: from the South of Saint Louis to the Lompoul s/mer-Kebemer axis; Central zone: between Lompoul-Kebemer axis and Kayar-Thies axis (corresponding to the Eastern boundary and Cape Verde Peninsula: Southern zone: East of the Cape Verte Peninsula, Keur Moussa, Sébikotane, Bayakh, Diender).

- A dune area partly fixed by scarce shrubby and wooded vegetation, with sandy-soiled inter-dune depressions, slightly moist and of ochre colour. This area (C) covers the intermediate part of Niayes (from 300 to 2000 m towards lands, to the west). Depressions are of varied sizes and forms.

- The ancient riverbed or lake zones (D and E) resulting in large depressions; the NSQ is at grounds' level. Depressions are embanked, clayey or even peaty (local name: *xour*; we classify them under subzone D) or clayey and muddy soils (called *ban* in Wolof) (subzone E). The relics of Guinean vegetation can be seen in this area. As we can see in the literature, the term "Niayes" was first associated with this well knows and early populated area. We call this part the "peaty Niayes zone." The width of that area is much more important in the south and centre than it is in the north (Figure 3).

- The last dunes area, where access to water allows market gardening with adapted drainage means. Soils are of dior type in height and of deck dior in the bottom (F). This constitutes the boundary with the Dieri.

- In the Dieri eastern boundary (G and H), which marks the start of the Senegalese peanut basin, the first subzone (G) is not cultivated but reserved as a livestock way and for forestry produce collection. The 'H' area in the diagram (Figure 2) is cultivated with a millet/peanut system under cover of *Acacia albida* Del.

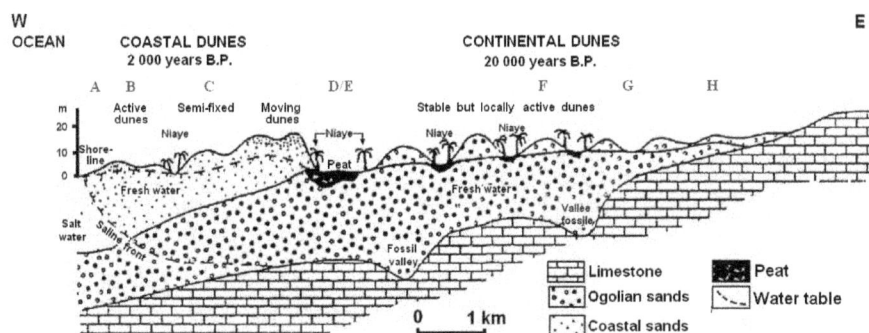

Figure 2. Geological section of the Niayes and location of agro-ecological zones (adapted from Pezeril et al. [22]).

Source :
Topographic map
n°ND-28-XX from
the Direction des
Travaux
Géographiques et
Cartographiques du
Sénégal. In
addition, we have
based our zoning
on remote sensing
maps (Google
Earth) and field
observations.

Legend:

■ A : Forests (*Casuarina equisetifolia* forests in the coastal zone are on sandy soils)
■ B : *Ndioukis* (basins with *dior* soils)
■ D et E : Peaty basins (*xour* or *ban* soils)
▨ E : Clayey but without peat basins (*ban* or *deck* soils)
■ C and F : Dunes semi-fixées (sur sols *dior* avec deck *dior* / *mbamb* en interdune)
■ Active dunes (*dior* soils)
■ G and H : Dieri's ogolian dunes (*dior*)
■ Body of water
■ Ancient wetlands (salty soils)

0 ———————— 30 km

☐ Transect in figure 4

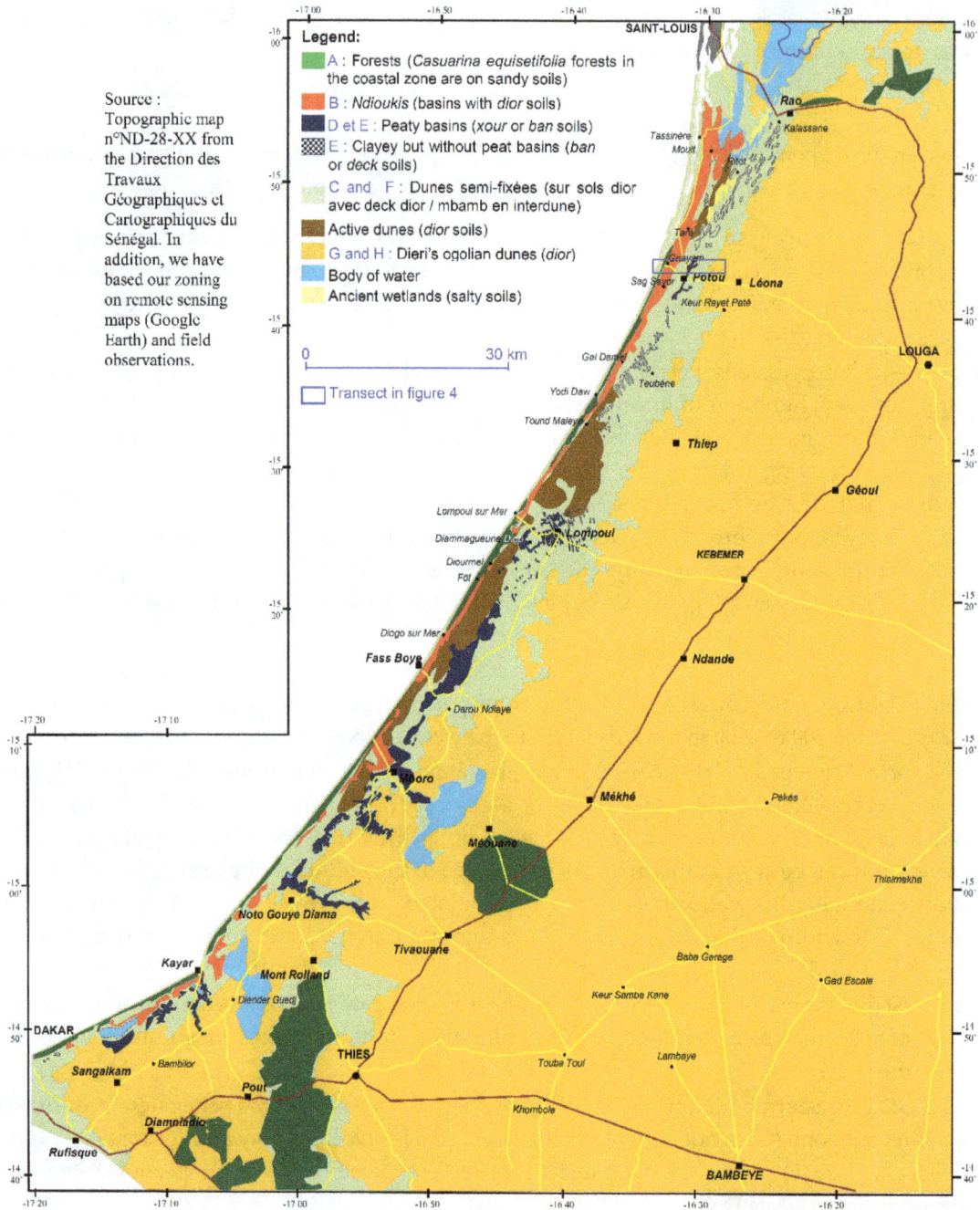

Figure 3. Zoning of the survey region.

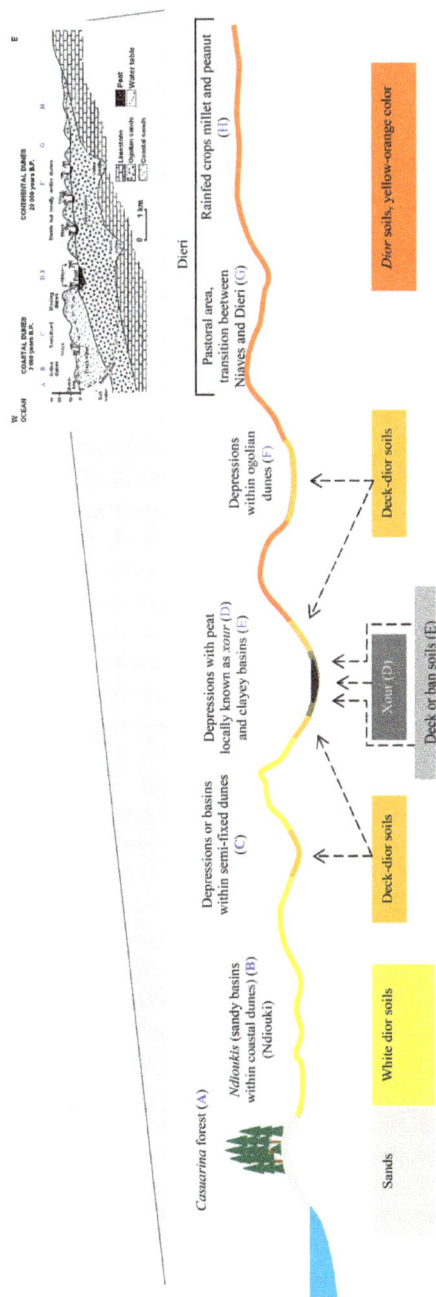

Figure 4. Diversity of depressions and soils, according to topography (and therefore proximity of the water table).

Table 1. Diversity of depressions and soils, according to topography (and therefore proximity of the water table).

Zones	Ndioukis (B)	Semi-Fixed Dunes (C)	Peaty Niayes (D and E)	High Basins of Continental Dunes (F)	Dieri	
					Pastoral Transition Area and Dieri Area (G)	Rainy Cultivating Area (H)
Altitude (≈level difference between thalweg bottom and dune top)	Five to ten meters (m)	10–20 m	10–15 m (more important level difference towards Fass Boye with about 20 m summits)	15–25 m	25 to 30 m	30 to 45 m
Access to water	Water ground is in average two to five meters.	Relatively easy access (water ground is at seven to fifteen meters)	Water is almost flushing in the basin bottom.	Difficult access (15–20 m)	Very deep water ground (>25 m), for annual plants, available water is rain).	
Soils	Sandy and very poor in organic material.	Sandy (*dior*) in height and sandy and muddy or *deck-dior* in basin bottom.	Clayey and peaty (*xour*) in basin bottom, clayey and sandy in mid-slope (*bam*)	Sandy (*Dior*)	Sandy *dior* type, with very slight textural variation on the top of dunes (weak textural stability) in slope bottom (with few silts accumulated by colluviation).	
Spontaneous and woody vegetation	Steppe, with scarce *Casuarina equisetifolia*, due to close presence of filao trees strip.	Various acacias and combretacea. Cactacea and euphorbiacea disseminated by humans.	Guinean-type vegetation with palm trees (*Elaeis guineensis* and *Cocos nucifera*) With aquaphyle plants (*Nymphaea lotus, Phragmites vulgaris, Typha australis*)	Various acacias and combretacea. Cactacea and euphorbiacea disseminated by humans.	Shrubby savanna with *Detarium senegalensis, Cassia sieberiana, Celtis integrifolia, Prosopis africana* and *Securidaca longipediculata*	Wooded parks with *Acacia albida* and *Adansonia digitata*.

2.2. Definition of the Agricultural Surface

Farmers were generally able to provide us with information on the surface of plots they cultivated in hectares. When that was not the case, we sometimes had to measure the surface or estimate the total sown surface (e.g., sum of all disseminated plots owned by the same farmer). The land reserve of each farmer has also been estimated.

2.3. Economic Analysis

2.3.1. Evaluation of Performances of Cropping Systems and Livestock Farming System

Gross average product (GP) of a cropping system or livestock:

$$GP = Q \times P, \tag{1}$$

where Q is the quantity harvested and P is the price/unit.

Food prices are subject to change year-round. For market gardening produce, we used the average price at the peak period of sales. This was easy enough to set up due to the perishable quality of market gardening produce. The price of livestock produce is more stable, except for sheep sold for yearly festivities (end-of-year festivities, "Tabaski" and "Eid," which is the name of the end of the Ramadan fast).

Net value added (NVA): Evaluation of value added allows for assessing the creation of wealth from the perspective of the community. In order to compare the performances of the different cropping systems, the economic results were modelled according to a linear model:

$$NVA/worker = a^*S/worker - b,$$
$$a = (GP - IC_p - Amt_p)/hectare \tag{2}$$
$$b = (IC_{np} + Amt_{np}),$$

where S is the area farmed (in hectares), GP is the gross average product per farmed hectare, IC is the cost of variable inputs, i.e., value of goods and services that have been transformed or fully consumed during the production process (fertilizers, manure, pesticides, etc.) proportional (p) or non-proportional (np) to the farmed area; Amt: is the average cost per hectare of the amortization and maintenance of equipment and fixed assets, proportional (p) or non-proportional (np) to the farmed area; and NVA is the net production of wealth per worker, i.e., the net productivity of labour.

2.3.2. Typology of Production Systems

To construct the typology, we used structural variables (length of the farm, type of irrigation equipment—manual, motorized, intermediate, number of family workers) and functional variables. In comparative agriculture studies, there are three main groups of farm types: family, family business, and capitalist. In Table 2, we give some criteria related to these three models of farms.

Table 2. Types of models of agricultural production units [23,24].

	Entrepreneurial Agricultures	⟹	Family Agricultures
Types of production units	Enterprise farm	Family business farm	Family farm
Labour	Specialize in supervision of hired labour	Work on farm and supervise hired labour	Family labour; no permanent hired labour

2.3.3. Evaluation of Performance of Each Type of Production System

To refine the interpretation of results of farms' typology, and really measure the net family's revenue (RA), it is recommended to withdraw the part of income redistributed to working persons outside of the family (salary or benefit sharing) [3]; we have built a second linear model as follows:

$$RA/\text{family worker} = A{*}S/\text{family worker} - B,$$
$$A = (GP - IC_p - Amt_p - \text{land property taxes} - \text{rents})/\text{hectare} \qquad (3)$$
$$B = (IC_{np} + Amt_{np} + \text{taxes} + \text{salaries} - \text{subsidies})/\text{family worker}.$$

Evaluation of the farm's income allowed the study to assess the profitability of the activity from the farmer's point of view (versus net value added, which represents the creation of wealth from the perspective of the community). While value added and productivity measure the intrinsic economic efficiency of the production system as a value creation process, it is the farm income that is in a position to express the share of value added (potentially increased by the subsidies received), enabling the farmer to support his family and, if possible, to invest so as to increase his capital and, ultimately, the productivity of his farm [1–3].

For each production system-type we have evaluated the average annual income. We then constructed a representation of economic performance per family worker according to the available area per family worker. Finally, we have compared this annual income/family worker to:

- A survival threshold (ST): such a comparison informs us about the future of the farm and its capacity to develop (Figure 5).

    o   If RA > ST, then the production unit is increasing in wealth, which enables the farmer to make some additional net investments.

    o   If RA < ST, then the production unit is even less able to make any additional net investments, and cannot even entirely renew its means of production and remunerate its labour power at the market price. Such a farm is in crisis, losing assets and facing basic needs; in the end, it may disappear, shift to another activity (labourers moving to wealthy neighbouring farms), or the owner may choose to migrate.

In the evaluation of this "survival threshold" (per working person and per person to feed within the family) we asked family members what their basic needs were for a given year.

- The regional opportunity cost of labour: informs about the economic interest farmers may have in dedicating their work to current production system or a possible shift to another competing activity (e.g., urban migration).

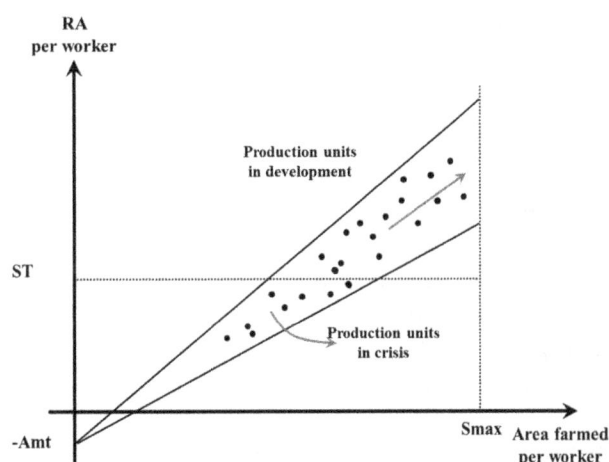

Figure 5. Production units in development and production units in crisis within the same agrarian system (adapted from [2]).

3. Results

3.1. History and Main Features of the Agrarian System

3.1.1. Original Ecosystem

Before agriculture, the vegetation of the coastline between Dakar and Saint-Louis was a relic of Guinean vegetation, also existing in the south of the country [13,25]. This distinguishes the Niayes vegetation from that of the same latitude in Senegal. Archaeological work, being rare, does not allow researchers to get precise data on the origins of agriculture in this area [26]. Using as a basis Arabic and Portuguese travellers' narratives, it can be said that the Niayes was unoccupied in the 15th century, but was visited by *Serer* populations extracting palm wine, oils and oinments from *Elaeis guineensis* Jacq. palm trees [27].

3.1.2. Settling Processes and Pre-Colonial Agriculture

The first most important settlements started in the 13th century. This settling process would be subject to successive migration [28–30]:

- In the Southern part, a temporary *Manding* occupation before the 13th century and undoubtedly following the Ghana Empire dislocation. In their raid to the south, they settled around Mont Rolland in the 13th century, where they practiced shifting agriculture and palm tree exploitation in coast. *Lebous* practiced coastal fishing.
- In the north and centre, towards 1680, *Fula* from Senegal River (*Waalowaalbes*) and from the edge of the current Louga (*Jeerinkkobes*). Hamlets were established along the coast by populations who already know the area due to transhumance: the Niayes provides pasture in the dry season.
- All along the area, *Wolofs*, after a temporary presence regulated by seasons, were installed in the 18th century to escape instability in the *Joloof* kingdom and slave raids in the *Walo*, *Cayor* and *Baol*. They occupied the region without established rules and practiced shifting agriculture (millet, peanut with 15 years of fallow lands).

3.1.3. Pre-Colonial Agriculture (Before 1885)

Oil palm exploitation in the xour basin (area D in the Figure 2) is undoubtedly the most ancient method of agricultural land use in the Niayes. The people practiced an economy of gathering, which provided in small quantities fruit, wine and produce for basket-making [31]. In the central and southern Niayes areas, the first emigration waves of *Lebou* and *Serer* brought sedentary agriculture to the area. This was an itinerant slash-and-burn agricultural system with exploitation of the dry Acacia seyal forest in the Dieri area (H) between Dakar and Thies [31,32]. With this self-subsistence agriculture, priority was given to food and textile fibres' supply. Until the early 19th century, the main crops were: the vouandzou (*Voandzeia subterranean* L.), rich in proteins; sesame (*Sesasum indicum* L.), which were grains rich in calcium and grilled for consumption; sweet melon (*Colocynthis citrullus* L.), rustic, of large size called *béref* in *Serer* and *Wolof*; cassava (*Manihot esculenta* Crantz), of an American species; and a few shrubby cotton plants (*Gossypium hirsutum* L.). It cannot be excluded that around humid basins flood-recession kitchen gardens have existed and planted with sweet potatoes (which is also American), African rice and gombo). *Fula* transhumant travelled regularly to the northern area, where some *Fula* hamlets were established.

3.1.4. Introduction of New Cultures by Colonial Administration

The creation, in the 17th century, of the Saint-Louis trading post and, in the 19th century of Richard Toll's experiment gardens marked a determination to develop colonial agriculture. Numerous fruits and vegetables were experimented with at Richard Toll's and Gorée. In Senegal, peanut cultivation organised by agricultural services started in the Cape Verde Peninsula in the 19th century. The Dieri production development is concomitant with the Cayor Kingdom annexation by France and railway establishment from Dakar to Saint-Louis in 1885.

3.1.5. Fruit and Vegetable Market Development during the Colonial Period (1885–1960)

The Niayes' potential to supply Dakar with fruits and vegetables was recognized in the early 20th century. Political and economic decision-making, accompanied by agricultural research and a real climatic advantage, were decisive motivators of the first vegetable producers. Three elements drove market gardening development. The first element was clearing the concerned area of malaria and trypanosomiasis. By 1906, experts were sent to have oversight of area conditions. Clearing and prophylaxis measures were taken afterwards. The second is cities' growth. For Dakar and its surroundings, French installation outside Gorée being negotiated in 1867, the population increased from 8937 inhabitants in 1891 to 18,447 in 1909, including 2000 Europeans [31]. In the North, there was Saint-Louis' influence; along the railway were established markets that helped sell off horticultural production. In the 1910s, various vegetables from Europe were subject to more advanced experimentation. For that, a training support document has been published. After the First World War, the need to produce more locally was evident: during the War, the need for fruits and vegetables from temperate countries and from the West Indies was hardly satisfied by importation. Military forts were also among the vegetable requesters. The third and last element regards the limitation on the peanut policy as much. The area was not adapted to peanut seeds envisaged for the Peanut Basin. Appropriate deforestation would threaten the Niayes area. In 1908, another agricultural orientation was considered: reforest and protect forest to restore ecosystem, and settle dunes and develop market gardening. In order to allow the growth of fruits and vegetables on the northern coast, the forestry administration initiated a dunes settlement operations with a dense population of filao trees (area A in Figure 2).

A proactive policy was developed. In 1920, Governor Ponzio, aiming to motivate market gardening, decided to reduce the related business taxes. This decision had an impact on market gardening in the Southern Niayes Area and, to a lesser extent, in the Central Area (Mboro). In 1937, a policy was elaborated to organize migration from the peanut basin (where demographic pressure was strong) to the Niayes. An agricultural station was also installed in Mboro in 1937. With regards to the northern area, the French introduced potatoes and some other vegetable species to facilitate supply to the forts (Saint-Louis, etc.). These new species are cultivated counter-season, which allowed for easy mobilization of manpower. Rain-fed crops (millet, peanut, niebe) were maintained on ledges to insure cereal and oleaginous ration for family and produce fodder for animals (dead leaves of peanuts, millet and cowpea). The introduction of vegetal species requiring regular watering (potatoes, chili peppers, cabbage, bitter eggplant) modified agricultural and food systems. New species were consequently planted near water points in *xour* and *ban* area (zones D and E). Cabbage was only produced in zones D and E. As its root is not deep enough, this plant is much more sensitive to drought and being planted in soils that are flooded in rainy seasons, next to rivers or lakes. So, in zones D and E, cabbage was cultivated in small quantities, as a single crop or associated with maize.

In the Ndioukis area (zone B), producers had to dig *céanes* in depressions, and basic wells were reinforced with straws and wood. Potatoes gradually replaced sweet potatoes, in shifts with cassava and maize. Women cultivated shrubby species (chili peppers, bitter eggplant) in small and fenced vegetable gardens not exceeding 300 m². Each year, producers in the northern Niayes area left half of their cultivation area fallow. During dry seasons, they cultivated cassava, maize and cowpea. Apart from potatoes, each cultivated species was entirely destined for family consumption. Surplus potatoes were transported by donkey to be sold in Saint-Louis. The revenues allowed producers to have new production tools (cast-iron buckets, hoes, iler which is a long-handled scuffle hoehilaire, cans, rakes, axes and machetes).

With the 1942 Plan bearing his name, Robert Sagot, an Agricultural General Inspector of the AOF and Director of the Agronomic Centre of Bambey from 1928 to 1942, designed the region's supply program on a local basis. Among his supporting measures was the objective to increase the production of Dakar's vegetable gardening belt from 12,000 tons in 1938 to 17,000 tons in 1944. This encouragement was furthermore seen as decisive for the development of market gardening, already existing in Cape Verde Peninsula and the Niayes territory. Soil saturation near Dakar encouraged improvement of the northern area, which was of poor quality. In fact, in 1945, to better rationalize production and improve quality, farmers in Dakar formed a syndicate named le Syndicat des Jardiniers et Maraîchers du Cap-Vert (SYNJARMAR). This syndicate initiative provided inspiration and market gardening extended little by little all along the Niayes. Training increased, coupled with campaign planning.

3.1.6. Impacts of the Drought in the 1970s–1980s

Drought in the 1970s caused considerable changes in the Niayes' agricultural activities and population density. For example, in the Southern Niayes, dune slopes with loamy sandy soils (*dior*), still less cultivated in cereals and peanuts, became unproductive and were finally abandoned. These fallow areas were subject to sale to civil servants on land close to Dakar's suburbs. These citizens had already invested in fruit arboriculture. This had been greatly developed, mainly along the "Niayes' route" (the Thiaroye–Malika–Rufisque axis) which was opened on the national highway situated next to the main market for horticultural produce selling in Thiaroye. Aviculture developed in Dakar's periphery, but in the 1980s drought persisted; NSQ groundwater carried on decreasing. In basins where ponds formerly existed, market gardening occupied the lowest lands where it was possible to access water because of farmers digging basic wells. Peaty soils (*xour*) on which rice was cultivated became dewatered (except in wintering). People cultivated potatoes on these soils henceforth and cabbage on the silty–clayey slopes where sweet potatoes were cultivated before. To combat drought and satisfy constantly increasing demand, some wealthy farmers started to buy motor-pumps, pre-built cemented wells, and basins interconnected by PVC (polyvinyl chloride) plastic piping. This allowed for cultivating its wide surface and multiplying campaigns as it was possible to cultivate even late in the dry season. The first established families (*Lebous* in the South and Centre, and the *Wolof* or *Fula* in the North), who had access to best soils and other sources of income (fishing, cattle breeding), could obtain such materials. Traders also had the means to buy land (without originating from the villages) and to invest in drainage materials and water distribution. Despite motorization, some activities required manual work (picking, weeding, harvesting); this slowly encouraged the family to seek manpower outside of their household.

The Niayes absorbed part of a rural exodus, with migrants seeking income-generating activities throughout the year. Much of the land had not been turned to profit; accordingly, the chiefs of the villages granted ownership rights to each newcomer. The Ndioukis were by preference saturated because of easy water access (manual watering was easier) and the availability of organic materials (litter from filao trees). This beneficial ownership was from time to time transformed into real ownership due to alliances (mainly marriages). After Ndioukis' saturation, the migration waves were oriented to semi-fixed dunes, but access to water was more difficult.

A 5–6-month seasonal contract was negotiated between the already established cultivators and job hunters. These seasonal workers, who are present mainly during dry seasons, are called *sourghas*. The worker is fed and accommodated on the farm. He works by himself on a dedicated plot (e.g., from 800 to 1200 m² in manual culture; 2500 m² in the connected basin system). The farm owner provides inputs (seeds, animal excrements, and eventually synthesis fertilizers) and production tools. The worker must manage all the production activities (nursery preparation, picking out, fertilization, watering, weeding, harvesting), and sell the produce at the market (the farmer provides the bags and the transportation). He gets half of the production's added value. A new type of agricultural venture, the family business, thereby emerged. A cultivator–employer, by getting 50% of the added value generated by *sourghas*, increases his income considerably. This sharing method, which is called *mbeye seddo*, has created a gap with regards to unequal access to drainage means, and has led to an unequal ability to make profit in the available basins. Remarkably, access to heavy equipment was not enough. It was also necessary, in order to secure production campaigns, to have a treasury: fuel to ensure motor-pumps operations 70 to 90 days per campaign based on cultivated species, fertilizers, organic manuring, daily costs related to *sourghas* (rice, meat, tea, soap, etc.).

3.1.7. Since 1995, an Increasing Level of Equipment and Production

Among the objectives of the 1995 CFA Franc devaluation was "the agricultural income enhancement by (1) encouraging exportations and internal demand for local agricultural produces, (2) increasing price paid to cultivators for these produces and creating new opportunities for activities upstream and downstream production. Analysts forecast a rise of market gardening channels due to the competitiveness and profits within European markets (French bean) as well as African markets for basic market gardening produce (onion, tomatoes)". For instance, in Senegal, the national production of onions has experienced intense growth, but this was mainly the case after the CFA Franc devaluation. This growth significantly increased prices among producers.

The drought between the 1970s and 1980s was conducive to the settlement of numerous rural migrants in the Niayes, and allowed ancient farmers of the peanut basins or ancient transhumant breeder of the Ferlo to find alternative income in this trading agriculture; moreover, farms could develop activities requiring manpower, by leaning on *sourghas*. The phenomenon intensified with the CFA Franc devaluation (in 1994), which encouraged cultivators to produce trading cultures (mostly onions, but also tomatoes, carrots, and cabbages), more intensely (two cycles per year in the northern area; three to four cycles in the central and southern areas), and in larger quantities. Family business farms, which were able to make much more savings than family farms, could reinvest in materials, recruit more *sourghas*, and extend their surface area. Some farmers have limited their responsibilities to supervision, entrusting almost all land to the *sourghas* by providing them a monthly salary instead of benefit-sharing. Capitalist arrangements became more and more numerous in the 2000s.

By the end of the 2000s, the following coexisted in the Niayes':

- Capitalist arrangements (all workers are employees), which little by little concentrated peat bog land in the southern and central zones. They were equipped with powerful motor-pumps and irrigation networks. Watering was done using a hose. Workers receive monthly wages.
- Family businesses (farms with both family workers and employees) in the entire region, of variable size depending on equipment (simple wells using pulley for drainage; small motor-pumps feeding a concrete basin network; big motor-pumps associated with spraying hoses). Workers, *sourghas*, are fed and accommodated and receive by the end of the year half of the added value.
- Family businesses (all workers are family members), mostly of humble size, with equipment varying from wells to basin networks. They are situated mostly in the central and northern Niayes zones.

3.2. Contemporary Agrarian System Analysis

3.2.1. Determining the Economic Thresholds

The survival threshold (meaning the minimal level of necessary resources, Table 3) was estimated for an average family at CFA 149,000 per working person and per year (227 Euros). The evaluation of the "survival threshold" (per working person and per person to feed within the family) was possible by asking family members (including working and non-working members) what were their basic needs for a given year (i.e., the goods needed to ensure maintenance and reproduction in decent conditions). This threshold includes food and non-food expenses as well as self-consumption. This indicator was set for a family with an average of 12 persons including seven working persons (three men and four women), two retired persons (a man and a woman of more than 65 years old) and three young children (Table 3). This average family was considered based on data collected during surveys. The survival threshold per working person shows what each of them should own at the minimum to support their family. In Senegal, the work situation is extremely unstable: Dakar hosts more and more migrants from the countryside, for which the unemployment rate is very important. In such a condition, using the minimum legal salary as the farm survival threshold would be irrational.

Table 3. Model of the survival threshold for a family composed of seven working people, two retired persons, and three children.

Item	Qty	Unit	Unit Price (FCFA)	Expense (FCFA)	Frequency	Ratio	Yearly Expenses (FCFA)
Food							
Rice	3	kg/day	300	328,500	every day	1.00	328,500
Millet	3.7	kg/day	150	202,575	every day	1.00	202,575
Fresh mill	3.7	kg/day	30	40,515	every day	1.00	40,515
Peanut paste	0.4	kg/day	240	23,360	2 days over 3	1.00	23,360
Crushed peanut	0.4	kg/day	480	23,360	1 day over 3	1.00	23,360
Charge in water	1	Daily rate	150	54,750	every day	1.00	54,750
Condiments	1	Daily rate	400	146,000	every day	1.00	146,000
Total							**819,060**
Clothes							
Uniform for children	9	set	2500	22,500	3 times per year (start of academic year, tabaski and other)	1.00	22,500
Sandals for women	6	pair	750	4500	2 times a year	1.00	4500
Uniforms for women	16	Set of 3 pagnes	4000	64,000	3 times	1.00	64,000
Sandals for men	8	pair	750	6000	2 times a year	1.00	6000
Uniforms for men	12	uniform	4000	48,000	2 times	1.00	48,000
Sandals	8	pair	750	6000	2 times a year	1.00	6000
Total							**151,000**
Consumable							
Soap	12	month	1500	18,000	pm	1.00	18,000
Battery or oil	1	Monthly rate	1500	1500	pm	1.00	1500
Health care	12	month	3000	36,000	pm	1.00	36,000
Total							**55,500**
Utensils							
Torch	8	Unit	1000	8000	1 year over 4	0.25	2000
Cooking pot	4	Unit	4000	16,000	1 year over 2	0.50	8000
Basin	2	Unit	3000	6000	1 year over 4	0.25	1500
Bucket	2	Unit	3000	6000	1 year over 4	0.25	1500
Spoon	4	Unit	500	2000	1 year over 5	0.20	400
Cup	10	Unit	300	3000	1 year over 5	0.20	600
Tray	4	Unit	2000	8000	1 year over 5	0.20	1600
Plate	10	Unit	300	3000	1 year over 5	0.20	600
Mortar	2	Unit	2500	5000	1 year over 5	0.20	1000
Pestle	2	Unit	250	500	1 year over 5	0.20	100
Total							**17,300**
Survival threshold							1,042,860 (≈1,590 €)
Per working person							148,980 (≈227 €)

Acronyms: Qty = quantity.

Another threshold, the opportunity cost of labour, was estimated based on the salary obtained by a non-qualified worker. In the survey area, the labour force is remunerated at 1000 FCFA a day (about one euro and fifty cents) for basic activities (as a farm labour for instance). We considered that a labourer works 300 days a year.

3.2.2. Typology of Production Systems

Three criteria have brought social differentiation among the Niayes' households: access to land, importance of livestock, and access to capital and/or treasury (with market gardening requiring an important pre-financing level). The typical trajectories were as follows:

- Main *Lebou* owners. Installed very close to southern peat bogs' surrounding, they had access to most fertile lands in Southern Niayes and could develop a more diversified culture and access to motorization by converting a part of land inheritance into money and/or making savings from fish sales.
- Wealthy families descended from *Fulani* or *Toucouleur's* stockbreeders. Established very early in the pasture area (G), they could start market gardening and were at once able to finance campaigns (treasury from stock sales), get materials (transportation of manure, crops, pumps, and hoses), and fertilize fields: these families have important land reserves, both in *Niayes* and *Dieri*; herds are parked in the Dieri in rainy season and dried manure from stockyards are transferred to the Niayes progressively with culture installation.
- Wealthy *Wolof* families also settled very early. They started off with combined rain-fed agriculture in the Dieri and market gardening in the Niayes during the dry season. Little by little, with the peanut channel crisis and drought, they became established in the Niayes. Those who saved capital in the form of herds could later get enough materials.

Nowadays, these three groups implement production systems that largely call for *sourghas* (seasonal workers).

Here are the two most frequent settlement processes for new farmers in the Niayes:

- They are cultivators arriving in a place with no other means than their ability to work. They started as *sourghas*, moving between the peanut basin and the Niayes. Then, they progressively made alliances (marriages) and became autonomous, but very slowly (free land lending).
- They are cultivators and tradespeople (*bana bana*). They were either in charge of peanut cooperatives, or peanut collectors in the Dieri, or hawkers of various goods. They knew the Niayes from their trading activity, or from associative movements (agricultural syndicates). Owning some capital, they could get land in the peat bogs, and became rapidly equipped with motor-pumps. Today, *bana bana* merchants still buy land to cultivate that is on the way to decapitalization; *bana bana* can also open new fields in area where watering is more difficult. Their heavy equipment allows them to resolve drainage issues. The *bana bana* group generally calls for paid manpower (Table 4).

In terms of farms' organization, there are three main methods:

- Direct control, in which all activities are accomplished by family members. During heavy periods, they manage the work with assistance from villagers. In comparative agriculture, this is called "a family farm".
- The *mbey seddo*, in which the gross added value is split fifty-fifty between the employer and the sourgha. The *sourghas* are foreigners—Guinean, Malian, Gambian—or Senegalese from poor villages of the Dieri. These groups are generally present in the Niayes for about six months a year. In comparative agriculture, this is called "a family business farm".

- The use of employees (with a monthly salary and not a share of added value). The head of the business hires labour on a monthly basis. In comparative agriculture, this is called "a capitalist farm". During heavy periods, the head also calls for workers, who are paid on a daily basis, with money or in kind. The labourer often comes from a nearby village or are manual farmers who devote a quarter of their time to such employment and the rest to his family's vegetable garden.

Also, there are three levels of equipment that give evidence of unequal productivity (Table 5 and Figure 6), regardless of the type of farm organization (family farm, family business or capitalist):

- Manual: water drawing and distribution are manual (with a bucket-pulley and a bucket, respectively);
- Combined: water drawing is motorized (motor-pump) but water is distributed manually (bucket);
- Almost entirely motorized: use of a motor-pump, then water is distributed via a network of vinyl tubing (PVC) and sprayed through a hose.

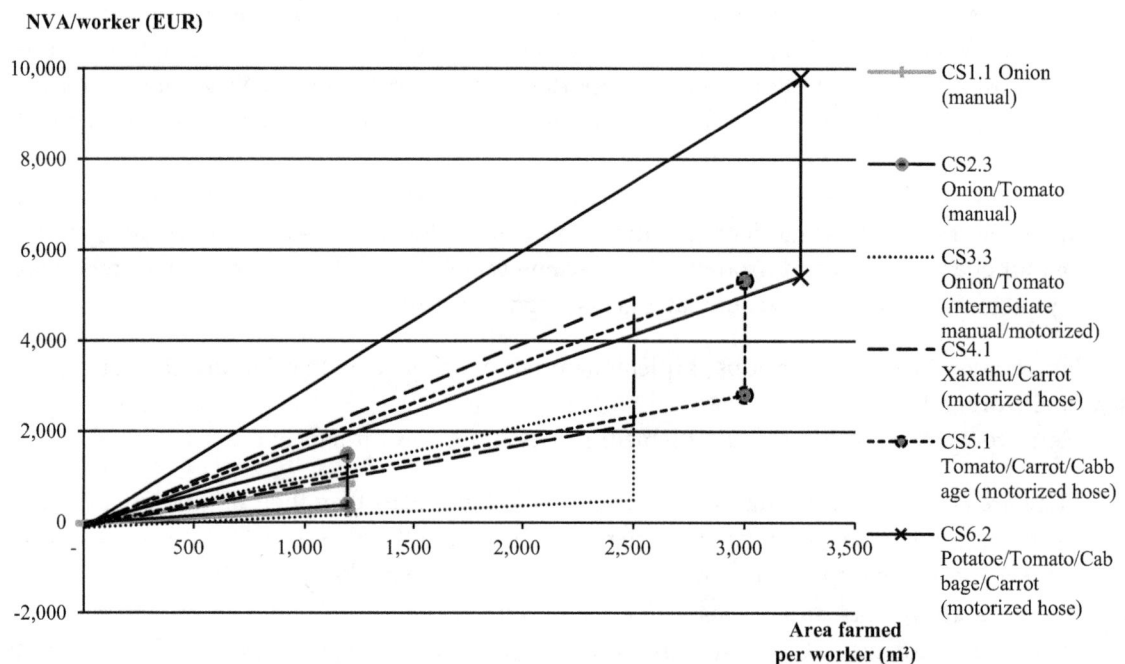

Figure 6. Comparative labour productivity of manual (CS1.1), semi-motorized (CS2.3), and motorized (CS 4.1, 5.1 and 6.1) cultivation systems. NVA: net value added; CS: cropping system.

3.2.3. Land Owners' Accumulation of Wealth

To describe this process of social distinction, let us consider the case of a family cultivator initially doing manual work (CS 1.1). Alone, doing manual work, the farmer cannot cultivate more than 1200 m^2 of onions. If the farmer gets associated with a *sourgha* (CS 2.1), he doubles the surface and increases his income by 50% for the surface unit (Figure 6). This "fruit-part or benefit-sharing contract" is named *mbeeye seddo*. As the employer is committed to his plot and does not provide more than a few days of supervision, the *mbeeye seddo* is advantageous for him. In fact, the income difference is more and more important if considering the man/day ratio (a ratio of one to four or one to five following obtained income, in Table 6 and Figure 7).

Table 4. Niayes agricultural ventures' historical trajectories and current production systems (PS).

Main Types Based on Social Organization of the Farm	Sub-Types Based on Irrigation Capacities of the Farm	Basin Types	Origins/Social Trajectories of the Farmers
PS1: Familial	PS1.1: Familial Manual	B	Recent migrants in zone B, coming in the 1980's–1990's from zone H, with no other means but their labour force. *Fulani* shepherds starting agriculture (treasury brought by cattle selling)
PS2: Familial with employees	PS2.1: Employer Manual	B, C	Idem, but with some amount as economy brought from the *Dieri*. Thanks to success in livestock farming, they were gradually able to employ *sourghas*.
PS3: Capitalist	PS3.1: Capitalist manual	B, C, D	*Bana-bana* traders from *Dieri* knowing the *Niayes* from their work.
PS1: Familial	PS1.2: Familial semi-motorized (pumps and basins)	C, F	Recent migrants with consistent means. *Fulani* shepherds from zones C, F and B, with few cattle in the beginning.
PS2: Familial with employees	PS2.2: Employer semi-motorized (motor-pumps and basins)	C, D	*Fulani* shepherds from C, F and B zones, with access to larger surfaces and integration agriculture-breeding.
PS1: Familial	PS1.3: Familial motorized (hoses, drop-by-drop)	D, E	Young people from large families. Recent comers, former employees in cities or traders having invested little by little in the area through harvesting campaigns.
PS2: Familial with employees	PS2.3: Employer motorized (hoses, drop-by-drop)	D, E	Descendants of wealthy families; marabout notables; *bana bana* traders; former employees in cities (early retired or in reorientation, former immigrant). Systems maximizing peat bog valuing.
PS3: Capitalist	PS3.3: Capitalist motorized (hoses, drop-by-drop)	D, E, F	Wealthy *bana bana* traders; *marabout* notables or their relatives.

PS: Production system.

Table 5. Comparative performance of the various irrigation methods used in the Niayes.

	Strictly Manual	Manual + Basins	Motor-Pumps + Basins	Motor-Pumps + Hoses	Drop-by-Drop
Volume of water (l/m²/j)	9	9	9	7	4
Maximal watered surface S_{max}/working person/campaign	800 m²/working person (cabbage) to 1200 m²/working person (onion) (1 ha p. 10 working persons)	1200–1500 m² (1 ha p. 8 working persons)	2000–2500 m²/working person	2 ha p. 2 working person + 2 trainees (sometimes 4 working persons + 4 trainees), so 5000 m²/working person (sometimes 2500)	1 ha p. 1 working person, even more
Cultivation ability ratio (basis1 = S_{max} in manual)	1	1.2–1.5	2–2.5 (+40% of basins)	3.5	10
Number of campaigns maximum/year/working person	2	2	2	3–4	2–3
Investment (annual depreciation per ha)	230 €	380 €	300 €	160 €	4000 €

S_{max}: maximum surface; ha: hectare.

Table 6. Added value partition between employer and *sourgha* within an onion manual cultivation system using a *mbeye seddo* (benefit sharing contract).

Plot of 1200 m²	GAV (€)	GAV Sourgha (€)	GAV Employer (€)	Workload (Man Days)				Labour Productivity GAV/md (before Sharing)	Income/md Sourgha	Income/md Employer and Family
				Total	Employee (Sourgha)	Employer	Punctual Labourers			
Low yields	301	151	151	183	138	26	20	1.6 €	1.1 €	6 €
Average yields	684	342	342	196	138	32	27	3.5 €	2.5 €	11 €
High yields	875 €	438	438	203	138	35	31	4.3 €	3.2 €	13 €

GAV: Gross Added Value = Gross Product – Intermediate consumptions. md: man-day.

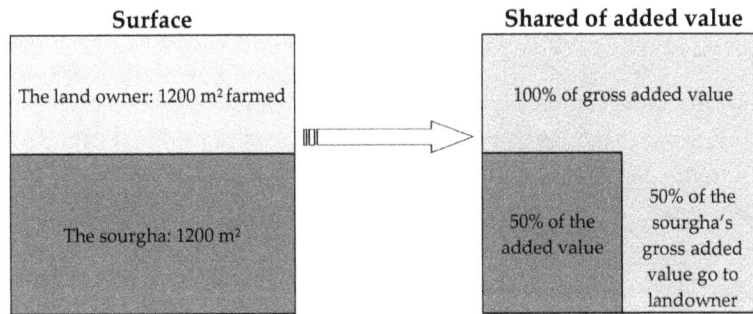

Figure 7. Representation of unequal development due to income partition between land owners and *sourghas*—Example of 1200 m², which is the maximum surface in manual agriculture—*The land owner accomplish his own farming on 1200 m² farmed (that is the maximum surface for one person with manual equipment) and keeps 100% of the added value he has generated. The sourgha who borrows 1200 m² to the landowner reassign 50% of the added value he has generated to the landowner. The two persons had the same level of effort but one (landowner) earns three times the income of the second (sourgha).*

In Figure 8, we present results from surveys over 57 manual farms. Family manual farms (PS1.1) and some of the family businesses that have low use of *sourghas* (PS2.1) can hardly manage to reproduce their business from year to year (cluster 1). A lot of them fall under the survival threshold and almost all are tempted to sell their labour force in cities (with little hope) or in capitalist-type farms that hire experimented workers at 30,000 FCFA (45 euros) a month. Hiring *sourghas* is a solution adopted by wealthier farms (with important herds and, therefore, able to provide campaign costs and *sourghas'* living expenses in advance), if they have available land reserves. Such collaboration allows the heads of employers' manual farms (PS2.1) to sensitively increase surfaces per family member by entrusting a part to *sourghas* and thereby make more profit (cluster 2). Access to land coupled with the ability to save money and then hiring employees is a major shift in the history of farming in the Niayes region.

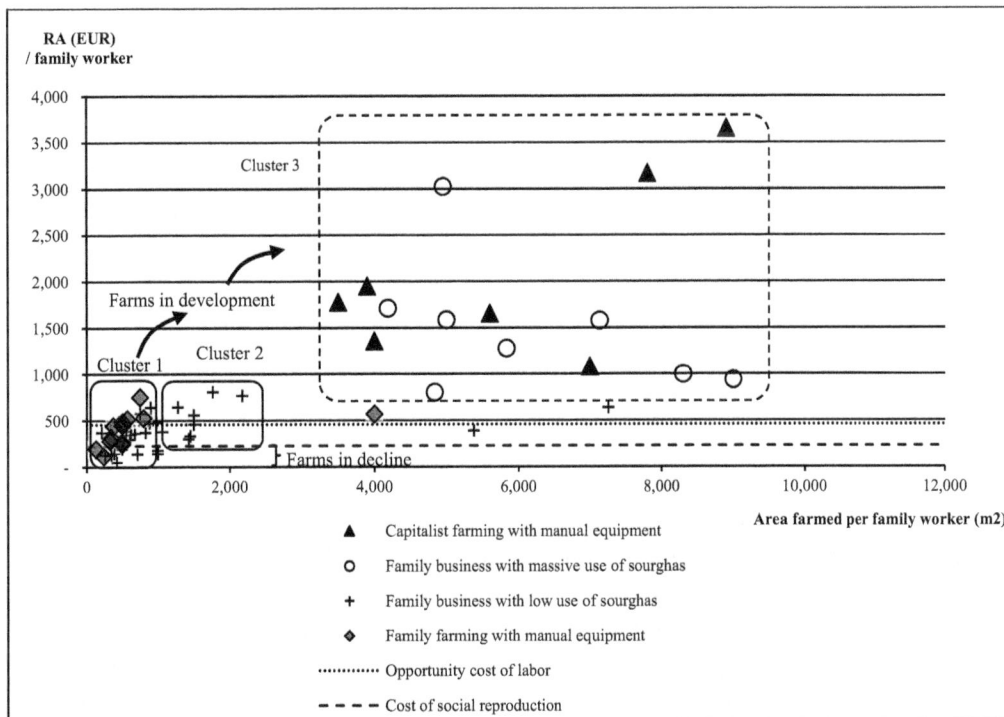

Figure 8. Comparative performance of manual production systems.

3.2.4. Motorization Allows an Acceleration of Enrichment

Intermediate motorization (motor-pump + basin network) or complete motorization (motor-pump + sprinkling) allows cultivators who are able to afford it to dramatically exceed the survival threshold. That applies to family, family business, or capitalist-type farms (Figure 9). Fruit-part contracts remain valid in combined systems. Yet, for many farms, purchasing a motor-pump would be an opportunity to "liberate themselves" from *sourghas* and keep the whole generated wealth within the family. Otherwise, fruit-sharing brings income to a level that is far from that of a system employing many *sourghas*. This reorientation allows them to make savings again, to invest in pumps and basins in other plots within the family reserve, and to entrust them to a *sourgha* using a fruit-part contract. The third accumulation phase most of the time results in almost total motorization: motor-pump + sprinkler. In such systems, remuneration is the rule. Within motorized farms, the most profitable are the capitalist-type farms of the Niayes' southern peat bogs: these are able to undertake three and even four campaigns a year (PS 3.3.3).

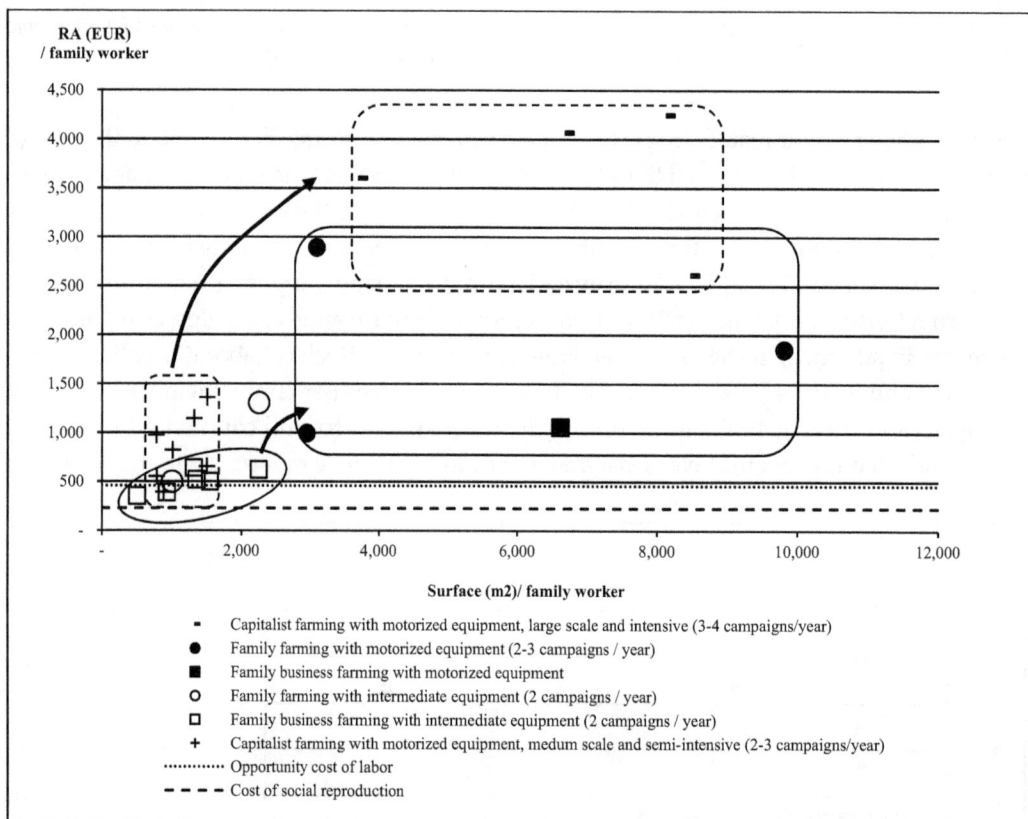

Figure 9. Comparative performance of semi-motorized and motorized production systems.

4. Discussion

In the present agrarian system, we distinguished three main production systems categories (family farming, family business and capitalist agriculture; see Table 7). Within these groups, farms use manual, semi-motorized or motorized cultivating systems. With manual cultivating systems, it is possible for a working person to develop 800 to 1200 m² of Niaye (a piece of fertile land located in depressions between dunes), with at best two yields of vegetables per year. The income varies from 500 to 1500 Euros/working person/year. Systems that combine motorized pumping and manual watering increase that to 2500 m²/working person/year with two plantings per year and an income of 500 to 2600 Euros/working person/year. Complete motorization (pumping and water distribution, using hoses) allows two to four plantings per year and 3000 to 3500 m²/working person. In such a case, income varies between 2000 and 10,000 Euros/working person/year.

Table 7. Synthesis of the main features of the five main production systems identified (N = 79).

Manual Production System (Number of Farms Surveyed)

	Family-Type Farms with Manual Equipment SP1.1 (13)			Family Business/Manual Equipment SP2.1 (37)			Capitalist-Type Farms with Manual Equipment SP3.1 (7)		
	Av.	Max.	Min.	Av.	Max.	Min.	Av.	Max.	Min.
Number of family members	8	17	2	8	25	2	2	3	2
Number of family workers	5	12	1	5	13	1	1	2	1
Number of employees or sharecropper	-	-	-	4	10	1	9	16	5
Arable land (m²/family worker)	733	4000	129	2329	9000	208	5814	8900	3500
Total cultivated area (m²/family worker)	440	750	129	1140	4186	169	4750	8300	3000
Livestock (Tropical Livestock Unit/ha cultivated)	25	125	-	16	146	-	6	13	-
Net annual income (EUR/family worker)									
Average	397			620			2091		
Maximum	751			3025			3653		
Minimum	116			51			1079		

Intermediate and Fully Motorized Irrigation Systems (Number of Farms Surveyed)

	Family-Type Farms/Intermediate Motorizationsp1.2 (2)			Family Business Intermediate Motorization SP2.2 (6)			Family-Type Farms with Machines SP1.3 (3)			Capitalist-Type Farms with Machines SP3.3 (11)		
	Av.	Max.	Min.	Av.	Max.	Min.	Av.	Max.	Min.	Av.	Max.	Min.
Number of family members	15	19	11	10	17	3	31	79	6	12	23	5
Number of family workers	10	12	7	7	11	2	20	51	5	8	13	5
Number of employees or sharecropper	4	8	-	5	11	1	1	3	-	6	16	2
Arable land (m²/family worker)	1627	2254	1000	1313	2250	500	5296	9804	2970	3158	8475	777
Total cultivated area (m²/family worker)	1484	2254	714	1118	2000	500	2152	3113	1431	2358	5556	777
Livestock (Tropical Livestock Unit/ha cultivated)	1	2	1	2	6	-	5	14	-	10	39	1
Net annual income (EUR/family worker)												
Average	902			504			1913			1858		
Maximum	1304			640			2892			4248		
Minimum	499			353			996			393		

Manual family farms (i.e., without employees) or family businesses, which hire few *sourghas*, face difficulties because the income is barely above the survival threshold (an average of 260 to 300 Euros/working person/year, sometimes 100 Euros) on less than 2000 m^2/family working person. When they depend heavily on *sourghas*, farms with manual equipment earn between 1000 and 1800 Euros/working person/year on 4000 m^2 to one ha/family working person. Semi- or completely motorized farms can use between 1000 m^2 (semi-) and 1 ha/family working person (complete), with incomes varying from 1500 Euros/working person/year (family system with motorized pumping and manual watering) to 3500 Euros/working person/year (intensive and motorized capitalist agriculture with four plantings/year).

5. Conclusions

To conclude, we insist that access to capital is one of the main obstacles to market gardening development in the Niayes. Troubled farms are those that have not yet opted for the *mbeye seddo* system due to their inability to cover *sourghas'* salaries and cultivation expenses in advance. Market gardening is very high-value-added production, but requires starting capital, mainly to purchase seeds, organic fertiliser (for those who do not have enough cattle), chemical fertiliser, packing and transportation. Besides, to provide *sourghas'* cost all through the campaign in advance, would weigh on farmers' capital. The main obstacle to family farms' development is access to capital. Family farms are mostly of humble size, and the work is demanding. To increase income, it would be sensible to intensify the cultivation system in a "reasonable way" by optimizing inputs (with respect to fertilisers' quality and doses and those of phytosanitary products), with the introduction of leguminous plants in shifts, using compost from filao trees litter (filao compost is proven to provide nitrogen) and eventually using mulching techniques. This will allow for mobilizing larger surfaces with a good productivity level in the long term. These farms support local employment, which is an essential lever to economic development in the region and to strengthening the social fabric (vs. capitalist ventures that tend to lay off the labour force).

The motorization of drainage would be a first, useful step in these farms' development. Nevertheless, does it lead to sustainable development? Drainage motorization requires use of fossil fuels, of which the price is uncertain. Moreover, if everyone is equipped with motor-pumps, cultivated surfaces will extend and groundwater use will be increasingly required. Risk of salinization will be all the greater, and there will also be the risk of air and water pollution. From the general interest point of view, with these risks taken into account, it would be more efficient to maintain a labor intensive activity. If work conditions in the Niayes region are too harsh, *sourghas* choose to take their chances somewhere else, even if the unemployment rate is very high. It would be more reasonable, in general, to support employment in the Niayes. In doing so, intermediate solutions improving water pumping (mainly in areas where water ground is deep) without going to extremes would be useful. To mitigate the impacts of motorized farms on the water table, two techniques are being tested by NGOs: sprinklers and drip irrigation. The first technique's objective is to control the flow of water by replacing the irrigation hoses with more professional sprinklers: this seems to be one of the best first steps towards sustainability improvement within motorized farms and increase the productivity of manual farms step by step. The second solution's efficiency is proven in terms of low water consumption (diminution by half) and labour productivity (800 to 1200 m^2 for manual farmers versus 10,000 for drip irrigation) [33]; however, the NGO's field teams have faced many failures due to a lack of training given to farmers and high investment costs. These two difficulties need to be addressed in the future [34,35].

The history of the Niayes shows a progressive enhancement of the cultivation of vegetable gardening basins per farmer in the Niayes, which has led, mainly since FCFA devaluation (1994), to a huge transformation in farms' social organization. Today, however, most agricultural enterprises have remained manual, with a cultivation ability of one to two plantings per year. The accumulation process is based on an income differential inherent to the added value sharing method: related to the production obtained with the *sourgha* labour force, the employer got 50% of the added value. Due to the accommodation of migrants escaping from drought, *mbeye seddo* indeed fostered remarkable job creation, but also increased income inequality. The salaries for *sourghas* is quite unattractive and they have to wait until the end of harvesting to receive their income. However, the system urges them to terminate their contract as quickly as possible, at the risk of selling the product for a rather low price (vegetables are not yet mature, therefore weighing less and sold cheaper). Would it not be possible to think of a system where the *sourgha* receives monthly salaries (which would be superior to the labour opportunity cost)? If *sourghas* have better conditions, work would also be improved, and adapted practices (fertilization, pesticide, harvesting calendar) would also be respected. A new system of value-added sharing should be negotiated between the two concerned parties, farmers and *sourghas*, on whom the sustainability of the system depends.

Employers often talk about high living expenses. Yet, it is not the expenses related to *sourghas'* food that gives farmers an incentive to get rid of *sourghas* and replace them with hoses. The price of inputs has risen for some years, mainly that of seeds and imported fertilisers (those containing nitrogen and potassium). Production costs have increased to a greater extent, with decreasing returns (due to unsuitable practices): this places farmers under the obligation to dismiss their labour. Apart from improving workers' conditions, it is also important to improve production quality in order to maintain high returns, and therefore preserve a high employment level in the Niayes area.

All this requires a new approach coupled with "tailor-made" technical monitoring, seeking as much sustainability as possible, and a new pre-financing method for campaigns. Savings and loans may also play a key role.

Acknowledgments: This work was performed within the project "Filao" (implemented by the NGO SOS SAHEL. It was made possible thanks to the availability of the teams (France and Senegal) of the NGO, the members of the Association des Union Maraîchères des Niayes (AUMN, Senegal), as well as the investment of four trainees in the field (Myriam Loloum, Fanny Miss, Akhmat Khastalani and Allassane Pouye). We also thank Ibrahima Sow (expert in agribusiness) for his precious advice.

Author Contributions: Yohann Fare and Pr. Marc Dufumier conceived and designed the research; Myriam Loloum, Fanny Miss, Akhmat Khastalani and Allassane Pouye performed surveys at the rural municipality level; Yohann Fare performed surveys at the regional level, analysed the data and is the lead author.

Conflicts of Interest: The authors declare no conflict of interest. The founding sponsors had no role in the design of the study; in the collection, analyses, or interpretation of data; in the writing of the manuscript, and in the decision to publish the results.

References

1. Loloum, M. *Analyse-Diagnostic de L'Agriculture de la Zone des Niayes au Nord-Ouest du Sénégal: Etude de la Zone des Niayes de la communauté Rurale de Léona, Département et Région de Louga*; Mémoire de Mastère en Agriculture Comparée et Développement Agricole, AgroParisTech: Paris, France, 2010.

2. Mazoyer, M.; Roudart, L. *Histoire des Agricultures du Monde. Du Néolithique à la Crise Contemporaine*; Seuil: Paris, France, 1997; p. 534.

3. Dufumier, M. *Les Projets de Développement Agricole. Manuel d'Expertise*; Karthala and CTA: Paris, France; Wageningen, The Netherlands, 1996; p. 354.

4. Cochet, H. *Comparative Agriculture*; Springer: Dordrecht, The Netherlands; Quae Editions: Versailles, France, 2015; p. 154.

5. Dufumier, M. *Agriculture et Paysanneries Des Tiers Mondes*; Karthala: Paris, France, 2004; p. 600.

6. Brossier, J. Système et système de production, note sur les concepts. In *Cahiers des Sciences Humaines*; ORSTOM: Paris, France, 1987; Volume 23, pp. 377–390.

7. Cochet, H.; Devienne, S. Fonctionnement et performances économiques des systèmes de production agricole: Une démarche à l'échelle régionale. *Cah. Agric.* **2006**, *15*, 578–583.

8. Khastalani, A.K. Études des transformations agraires dans la communauté rurale de Thieppe dans le département de Kébémer. In *Mémoire d'Ingénieur Agronome de l'ENSA*; Ecole Nationale des Sciences Agronomiques: Thiès, Sénégal, 2010.

9. Pouye, A. Etudes des transformations agraires dans la Communauté rurale de Bandègne Wolof, dans le département de Kébémer (région de Louga). In *Mémoire d'Ingénieur Agronome de l'ENSA*; Ecole Nationale des Sciences Agronomiques: Thiès, Sénégal, 2010.

10. Miss, F. La bande de filaos: Analyse de la gestion mise en œuvre par les groupements maraîchers des Niayes et perspectives de développement de la filière. In *Mémoire Pour Obtenir le Diplôme d'ingénieur du GÉNIE Rural Des Eaux et Forêts, Spécialisation Forêt-Nature-Société*; AgroParisTech-ENGREF: Montpellier, France, 2010.

11. Gaye, A.T.; Lo, H.M.; Sakho-Djimbira, S.; Fall, M.S.; Ndiaye, I. *Sénégal: Revue du Contexte Socioéconomique, Politique et Environnemental*; IED: Dakar, Senegal, 2015; p. 87.

12. Ministère de L'Environnement et de la Protection de la Nature. *Rapport National Sur L'état de L'environnement Marin et Côtier*; MEPN: Dakar, Senegal, 2002.

13. Trochain, J. *Contribution à L'étude de La Végétation du Sénégal*; Larose: Paris, France, 1940; p. 434.

14. Fall, M. Environnements sédimentaires quaternaires et actuels des tourbières des Niayes de la Grande Côte du Sénégal. In *Thèse du 3e Cycle Géologie: Sédimentologie*; Université Cheikh Anta Diop: Dakar, Senegal, 1986.

15. Monciardini, C. *La Sédimentation Éocène au Sénégal*; Mémoire B.R.G.M: Paris, France, 1965; p. 105.

16. Michel, P. Les Bassins des Fleuves Sénégal et Gambie: Etude Géomorphologique. Ph.D. Thesis, Université Louis Pasteur, Strasbourg, France, 1973.

17. Noël, Y. *Étude Hydrogéologie Des Calcaires Lutétiens Entre Bambey et Louga (2ème phase)*; BRGM: Dakar, Sénégal, 1978; p. 82.

18. Claude, A. Sensibilité d'Hydrosystèmes Côtiers Particuliers Aux Variations Millénaires du Niveau Marin au Sénégal. Mémoire de Master 2 Sciences de l'Univers, Environnement, Écologie, Parcours Hydrologie-Hydrogéologie. Master's Thesis, Université Pierre et Marie Curie, École des Mines de Paris & École Nationale du Génie Rural des Eaux et des Forêts, Saclay, France, 2007.

19. Saos, J.L.; Fall, M. *Sédimentologie et Variations Climatiques Dans Les Tourbières Holocènes Sénégalaises, SEMINAIRE Paléolacs et Paléoclimats en Amérique Latine et en Afrique (20 000 ans B.P.-Actuel): Résumés des Communications Présentées au Séminaire Géodynamique*; ORSTOM: Bondy/Montpellier, France; Dakar, Sénégal, 29–30 January 1987.

20. Lézine, A.M. *Paléo-Environnements Végétaux d'Afrique Occidentale Nord-Tropicale Depuis 12 000 B.P.: Analyse Pollinique de Séries Sédimentaires Continentales (Sénégal-Mauritanie), Séminaire Paléolacs et Paléoclimats en Amérique Latine et en Afrique (20 000 ans B.P.-Actuel): Résumés des Communications Présentées au Séminaire Géodynamique*; ORSTOM: Bondy/Montpellier, France; Dakar, Sénégal, 1987.

21. Ministère de l'Environnement et de la Protection de la Nature. *Plan D'Action National D'Adaptation Aux Changements Climatiques*; MEPN: Dakar, Sénégal, 2006; p. 84.

22. Pezeril, G.; Châteauneuf, J.J.; Diop, C.E. *La Tourbe des Niayes au Sénégal: Genèse et Gitologie, Symposium International INQUA/ASEQUA du 21 au 28 avril 1986: Changements Globaux en Afrique*; ORSTOM: Paris, France, 1986.

23. Eswaran, M.; Kotwal, A. Access to Capital and Agrarian Production Organization. *Econ. J.* **1986**, *96*, 482–498. [CrossRef]

24. Bélières, J.F.; Bonnal, P.; Bosc, P.M.; Losch, B.; Sourisseau, J.M. *Les Agricultures Familiales du Monde: Définitions, Contributions et Politiques Publiques*; AFD: Paris, France, 2013; p. 281.

25. Raynal, A. *Flore et Végétation des Environs de Kayar (Sénégal): De la Côte au Lac Tanma*; Faculté des Sciences, Université de Dakar: Dakar, Senegal, 1963; pp. 121–231.

26. Maley, J.; Vernet, R. *Peuples et Evolution Climatique en Afrique Nord-Tropicale, de la Fin du Néolithique à l'aube de l'époque Moderne*; Afriques: Paris, France, 2013; p. 50.

27. Boulègue, J. *Les Anciens Royaumes Wolof (Sénégal). Volume 1: Le grand Jolof. XIII–XVIème Siècle*; Editions Façades: Paris, France, 1987; p. 208.

28. Camara, M. Approche Participative Dans la Gestion Intégrée des Ressources en Eau De la Zone des Niayes (de Dakar à Saint-Louis). Master's Thesis, Faculté des Lettres et des Sciences Humaines, Département de Géographie, Université Cheikh Anta Diop, Dakar, Sénégal, 2010.

29. Thiam, E.H.I. Les Terroirs Périphériques de la Reserve Spéciale de Faune de Gueumbeul. Master's Thesis, Université Cheikh Anta DIOP (UCAD), Dakar, Sénégal, 2004.

30. Coly, B. Dynamique Des Ressources Naturelles Dans la côte Nord du Sénégal, de Dakar à Saint Louis. L'exemple de la communauté rurale de Mboro. Master's Thesis, Université Cheikh Anta Diop, Dakar, Sénégal, 2000.

31. Dione, D. Problèmes de Développement Des Activités du Secteur Primaire Dans la Banlieue De Dakar. Ph.D. Thesis, Université de Limoges, Limoges, France, 1986.

32. Dieng, N.M. l'Impact du Maraichage Dans la Dégradation Des Ressources Naturelles Dans Les Niayes de la Bordure du lac Tanma. Master's Thesis, Université Cheikh Anta Diop, Dakar, Sénégal, 2008.

33. Mbengue, A.A. Analyse des Stratégies de Commercialisation de l'Oignon Local Dans Les Niayes. Master's Thesis in Rural economy, ENSA, Thiès, Sénégal, 2007; p. 73.

34. Institut Sénégalais de Recherches Agricoles. *Exploitations Horticoles Avec Irrigation Goutte-à-Goutte Dans le Bassin Arachidier*; Institut Sénégalais de Recherches Agricoles: Dakar, Sénégal, 2013; p. 120.

35. Groupe de Travail Désertification. *Agroécologie, une Transition Vers des Modes de Vie et de Développement Viables Paroles d'Acteurs*; Groupe de Travail Désertification: Montpellier, France, 2013; p. 60.

10

Evaluation of Pectin Extraction Conditions and Polyphenol Profile from *Citrus x lantifolia* Waste: Potential Application as Functional Ingredients

Teresa del Rosario Ayora-Talavera [1], **Cristina A. Ramos-Chan** [1], **Ana G. Covarrubias-Cárdenas** [1], **Angeles Sánchez-Contreras** [1], **Ulises García-Cruz** [2] and **Neith A. Pacheco L.** [1,*]

[1] Centro de Investigación y Asistencia en Tecnología y Diseño del Estado de Jalisco CIATEJ Unidad Sureste. Parque Científico Tecnológico de Yucatán, Km 5.5 Carretera Sierra Papacal-Chuburná puerto, Mérida-CP 97302, México; tayora@ciatej.mx (T.d.R.A.-T.); arapachecol@gmail.com (C.A.R.-C.); covarrubias.ana@hotmail.com (A.G.C.-C.); msanchez@ciatej.mx (A.S.-C.)

[2] Centro de Investigación y Estudios Avanzados del Instituto Politécnico Nacional. Unidad Mérida Km. 6 Antigua carretera a Progreso Apdo. Postal 73, Cordemex, Mérida Yuc. 97310, México; norbertoulisesg@gmail.com

* Correspondence: npacheco@ciatej.mx

Academic Editors: Johnselvakumar Lawrence and Karunanithy Chinnadurai

Abstract: The citrus by-products pectin and polyphenols were obtained from *Citrus x lantifolia* residues. The use of acid type, solute-solvent ratio, temperature, and extraction time on pectin yield recovery was evaluated using a factorial design 3^4; pectin physicochemical characterization, polyphenol profile, and antioxidant activity were also determined. Results indicated a total polyphenol content of 3.92 ± 0.06 mg Galic Acid Equivalents (GAE)/g of citrus waste flour in dry basis (DB), with antioxidant activity of 74%. The presence of neohesperidin (0.96 ± 0.09 mg/g of citrus flour DB), hesperidin (0.27 ± 0.0 mg/g of citrus flour DB), and ellagic acid (0.18 ± 0.03 mg/g of citrus flour DB) as major polyphenols was observed. All of the factors evaluated in pectin recovery presented significant effects ($p < 0.05$), nevertheless the acid type and solute-solvent ratio showed the greatest effect. The highest yield of pectin recovery (36%) was obtained at 90 °C for 90 min, at a ratio of 1:80 (w/v) using citric acid. The evaluation of pectin used as a food ingredient in cookies elaboration, resulted in a reduction of 10% of fat material without significant texture differences ($p < 0.05$). The pectin extraction conditions and characterization from these residues allowed us to determine the future applications of these materials for use in several commercial applications.

Keywords: citrus polyphenols; citrus pectin; *Citrus x lantifolia*; pectin extraction conditions

1. Introduction

The citrus industry in Mexico represents an important economic and social activity. Worldwide, Mexico occupies the fifth place in citrus production; particularly, lemons represent the second most commonly produced citrus fruit in this country, where *Citrus x lantifolia* represents 50% of the citrus crop cultivated in Mexico [1]. Most of the citrus production is intended for fresh consumption; the rest is industrially transformed for the elaboration of juices, pulps, and fruit concentrates. Nevertheless, around 45% of the fruit is wasted; thus as production increases, the generation of solid and liquid waste also increases, which represents significant amounts of by-products not fully industrially exploited [2]. Nowadays, in order to reduce the environmental impact caused by waste materials from food industry, alternatives to obtain added value products that could be exploited in different areas are necessary; as an example, the extraction of biological molecules from citrus residues is an important part of an integrated system that could finish in bioethanol production with the residues free from compounds as polyphenol and pectins, that can be used to favor fermentation process. Specifically, lemon peels are rich in polyphenolic compounds such as phenolic acids and flavonoids, which have been reported to be responsible for a variety of important biological effects [3]. Some of the biological properties reported by polyphenolic compounds are the reduction of cholesterol and blood sugar levels, anti-cancer effects, blood pressure lowering, anti-inflammatory properties, antimicrobial effects, antioxidant capacity, neuroprotection and cardioprotection, all of which are also of great interest for many industrial sectors [3–5]. Pectin, an acidic hydrocolloid widely used as a food ingredient for its gelling properties, is also an important biomolecule of high industrial interest found in citrus waste; it is considered to be a metabolite of biotechnological interest due to its stabilizing properties. Additionally, due to its high water content and easily adjustable physical properties [6], research regarding its potential uses in medical applications has emerged, for instance, nasal and oral drug delivery [7,8], cancer-target drug and gene delivery [9,10], and tissue engineering and wound healing [10]. Most of the commercial pectin is extracted from apple and citrus peels by the use of chloride acid, nevertheless, the trend of consumers looking to find products obtained in a more environmentally friendly way, searching for the lowest chemical residues generation as well as integrated citric waste utilization, led to the study of new extraction methods in order to maintain or improve upon recovery yields [2]. Furthermore, the prolonged commercial success of pectin has shown the importance of using fruit by-products as raw materials to utilize for production [11]. On the other hand, the use of pectin as an ingredient in bakery products has been suggested as an effective fat replacer [12]; pectin from *Yuja* (*Citrus junos*) pomace has been evaluated as an effective fat replacement (up to 10%) in cakes without volume losses. For that reason, the main objectives of this work were: (i) to determine the polyphenol profile from *Citrus x lantifolia* flour residues, (ii) to evaluate the principal factors affecting pectin extraction (solvent type, time, temperature, and solute/solvent radio), (iii) to analyze the physicochemical characteristics of the pectin extracted and compare it to a commercial one, and (iv) to evaluate the pectin obtained as a functional ingredient in a bakery product.

2. Materials and Methods

2.1. Biological Materials and Reagents

The raw material (Persian lime *Citrus x latifolia* peel, bagasse, and seed) was collected at the Akil Juicer from Union de Ejidos Citricultores, Akil, Yucatan. After juice and essential oils were obtained by mechanical methods (scraping), samples were transported to the research center (CIATEJ) and stored at −4 °C until further analysis. Folin-Ciocalteu's phenol, 2,2-diphenyl-1-picrylhydrazyl (DPPH), analytical standards of gallic acid, caffeic acid, ellagic acid, naringin, hesperidin, neohesperidin, morin, quercetin, genistein, kaempferol, methanol and acetonitrile (chromatographic grade), and commercial citrus pectin were purchased in Sigma Aldrich (San Luis, MI, USA). Ultra-pure water was prepared in a Milli-Q water filtration system (Millipore, Bedford, MA, USA).

2.2. Flour Waste Preparation and Characterization

Waste products were oven dried at 65 °C for 48 h before polyphenol extraction. Moisture content was performed according to NMX-F-428-1982 [13], pH and acidity percentage were determined according to the Association of Official Agricultural Chemists (AOAC) method. Color was measured using a MiniScan Ez colorimeter, and L, a, and b parameters were obtained.

2.3. Polyphenol Extraction, Total Polyphenol Quantification (TPC), and Antioxidant Activity

Polyphenol extraction was performed according to the cryogenic methanolic extraction reported in MX/a/2012/014554 patent solicitude and by Sánchez-Contreras et al. [14]. Residual waste after polyphenol extraction was used for pectin recovery. The TPC was determined by Folin-Ciocalteu's phenol method [15], the absorbance was measured in a spectrophotometer (Thermo Scientific, Biomate 3S, Madison, WI, USA) at 760 nm. Estimation of TPC was carried out using gallic acid as a standard and the results were expressed in mg of gallic acid equivalent per gram of citrus waste flour in dry weight (mg GAE/g DW). Antioxidant activity was evaluated using the 2,2-diphenyl-1-picrylhydrazyl (DPPH) scavenging method [16], the absorbance was measured in a spectrophotometer (Thermo Scientific, Biomate 3S, Madison, WI, USA) at 517 nm. The DPPH scavenging activity was evaluated based on the percentage of DPPH radical scavenged according to Equation (1):

$$S_{DPPH} = S_b - (S_c - S_s)/S_b \times 100 \tag{1}$$

where S_{DPPH} is DPPH radical scavenging activity expressed as a percentage, S_b is the A517 nm of blank treatment, Sc is the A517 nm of sample solution, and Ss is the A517 nm of the background of sample.

2.4. Polyphenol Identification and Quantification by HPLC

A mixture of polyphenol standards were considered, containing: gallic acid, ellagic acid, caffeic acid, naringin, hesperidin, neohesperidin, morin, quercetin, genistein, and kaempferol. The retention time of each of the standards was taken as criterion to identify the polyphenol contents in the samples analyzed. To calculate the retention time of each standard, flavonoids were injected individually and the average of 15 individual determinations were taken as retention time value. To quantify the polyphenols identified in the samples, calibration curves of the standards at different concentrations (1, 5, 20, 50, and 100 ppm) were performed; each concentration was injected in triplicate in order to calculate the time of average retention of each of the flavonoids in the mixture. A Finnigan Surveyor Autosampler Plus Equipment was used, with Finnigan Surveyor PDA Plus Detector, the column used was a Phenomenex, 00F-4435-E0, Gemini 5 μ, C 18 110 A, 150 mm × 4.60 mm, 5 microns. The injection volume was 25 μL and the mobile phases used were A:HPLC water with formic acid (0.1%) solvent B: Acetonitrile with formic acid (0.1%), flow rate of 1 mL/min and gradient method starting with a minute at 90% of mobile phase (MP) A, then 40 min at 74% of MP A, 30 min at 35% of MP A, 5 min at 100% of MP B, and lastly, 5 min at 90% of MP A was performed to equilibrate the system. The determination time was 80 min at λ = 290 nm and λ = 350 nm detection.

2.5. Pectin Extraction

A factorial design 3^4 was used to evaluate the independent variables: acid type (hydrochloric, acetic and citric acid), solute-solvent ratio (1:30, 1:50, 1:80), temperature (60, 75, and 90 °C), and time extraction (30, 60, and 90 min) on the pectin yield recovery. Acids were adjusted to pH 2.2. The different acid solutions were placed in flasks of 100 or 250 mL according to the evaluated solute-solvent ratio, then flasks were heating individually to obtain the temperature proposed in the factorial design, and 2 g of citrus waste flour free of polyphenols was immersed into the solutions and kept in agitation during the periods of time evaluated. The resulted extracts were cool at ambient temperature and centrifuged at 5300 rpm for 15 min at 4 °C. The supernatants were further used for pectin recuperation, using ethanol (96% v/v) at 1:2 (v/v) ratio. The response variable of pectin yield was calculated according to the methods previously used by Baltazar et al. [2] (see Equation (2)), where g of recuperated pectin in DB indicates the weight in grams of the product recuperated after ethanol precipitation and oven drying for 24 h at 45 °C. The g of initial flour waste in D, represents the weight in grams of the raw material (*Citrus x latifolia* residues) used in the form of flour.

$$Pectin\ YIELD = \left(\frac{g\ of\ recuperated\ pectin\ in\ DB}{g\ of\ initial\ flour\ waste\ in\ DB} \right) \times 100 \tag{2}$$

2.6. Pectin Physicochemical Characterization

Three different pectin recuperation conditions were evaluated: two stages of acid hydrolysis extraction and alcohol precipitation (A), three stages of acid hydrolysis extraction, alcohol precipitation, and pectin washing (B), two stages of acid hydrolysis extraction with pH adjustment (pH = 6.5) before alcohol precipitation (C), then the resulting pectins were physicochemically characterized. Free Acidity

(FA) and Equivalent Weight (EW) were calculated according to [17], FA was expressed as meq of free carboxyl/g and EW was calculated with sample weight (mg) divided by the meq of the NaOH used for titration. Methoxy Content, Esterification Degree (ED), and Uronic Acid (UA) were calculated according to [18], Methoxy percentage was reported using Equation (3), ED was calculated using Equation (4), and UA was calculated according Equation (5). Identification and pectin conformation were determined according to [19] in order to obtain pectin gels.

$$Methoxy\ (\%) = \frac{\text{meq. of } NaOH \times MW \text{ of methoxy} \times 100}{sample\ weight\ (mg)} \tag{3}$$

$$ED\ (\%) = \frac{meq.\ of\ NaOH\ (methoxy\ content)}{meq.\ NaOH\ (free\ acidity) + meq.\ NaOH\ (methoxy\ content)} \times 100 \tag{4}$$

$$UA\ (\%) = \frac{meq.\ NaOH\ (free\ acidity) + meq.\ NaOH\ (methoxy\ content)}{sample\ weight\ (mg)} \times 176 \times 100 \tag{5}$$

2.7. Cookie Elaboration, Water Activity (a_W), Water Content (%), Physical and Textural Determinations

Cookies were prepared according to [20] with a slight modification, using wheat flour (22%), sugar (32%), vegetal fat (22%), egg powder (10%), coconut (7%), ammonium bicarbonate (0.5%), sodium bicarbonate (0.5%), raisins (6%), and water, then three percentages of the pectin previously obtained were evaluated for fat substitution (2.5%, 7%, and 10%). Water activity and water content (%) were measured using an a_W equipment Novasina AG (Lachen, Switzerland) and a thermobalance OHAUS MB45-2A0 (Switzerland), respectively. Physical characteristics of the cookies in terms of diameter (mm), thickness (mm), and spread ratio (diameter/thickness) were determined as the average of three measurements, using an electronic digital caliper (Truper) (Jilotepec, Mexico) [21]. Texture analyses were performed as maximal straight (textural hardness) using an EZ-SX Shimadzu Corporation, Japón equipment. Textural hardness was quantified using a three flexion points test. The results were expressed as the average of four measurements [21].

2.8. Statistical Analysis

Analysis of Variance (ANOVA) and Tukey comparison test ($p < 0.05$) of the results were determined using the software Statgraphics® Centurion, version XVI (Manugistic, Inc., Rockville, MD, USA). All the results analyzed by comparison test were the average of three independent determinations.

3. Results

3.1. Flour Citrus Waste Characterization

Previous to polyphenol and pectin extraction the Persian lime *Citrus x latifolia* waste was oven dried at 65 °C for 48 h; this procedure was performed in order to reduce water content of the residues to favor material preservation. The results of the physicochemical characterization of the flour obtained after the drying process are shown in Table 1. The moisture content was around 12%; this value favors preservation of the material during storage, avoiding microbial and enzymatic degradation. The pH obtained was 3.38 lower than the value obtained in the raw material (data not shown), the total titratable acidity obtained was 7.43%, reported as citric acid meq due to the greatest presence of this acid in lemon [22]. The color determination could provide information about possible sugar caramelization during the drying process, expressed as a brawn color, or degradation of the material with a low luminosity value [22]. As it can be observed in Table 1, color parameter results were related to a yellow color and luminosity was higher than 60%, indicating a well preserved flour.

Table 1. Physicochemical characterization of the Persian lime *Citrus x latifolia* flour waste.

Parameter		Determination
Moisture content %		12.12 ± 0.007
pH		3.38 ± 0.000
Titratable acidity % (citric acid meq)		7.43 ± 0.288
	Luminosity (L)	62.63 ± 0.08
Color	Parameter a	6.65 ± 0.04
	Parameter b	23.28 ± 0.070

3.2. TPC, Antioxidant Activity and Polyphenol Profile

As is shown in Table 2, the TPC of citrus waste was 3.92 ± 0.06 mg of GAE/g of citrus waste flour (DB), and an antioxidant activity of $73.2\% \pm 4.2\%$ of DPPH$^+$ radical inhibition was observed. The major compound identified was neohesperidin with a concentration of 0.969 ± 0.099 mg/g of waste flour in DB, followed by hesperidin, ellagic acid, caffeic acid, morin, and, in lower concentrations, gallic acid, quercetin, kaempferol, and genistein (Table 2).

Table 2. Total polyphenol content, antioxidant activity, polyphenol identification, and concentration determination from citrus waste flour residues.

Type of Analysis	Determination	mg/g of Waste Flour in Dry Basis
Spectrophotometer analysis	Total Polyphenol content	3.92 ± 0.06
	Antioxidant activity (DPPH + Radical inhibition)	$73.2\% \pm 4.2\%$
HPLC * analysis	Gallic acid content	0.074 ± 0.003
	Caffeic acid content	0.1560 ± 0.007
	Ellagic acid content	0.186 ± 0.0292
	Naringin content	0.003 ± 0.0001
	Hesperidin content	0.278 ± 0.011
	Neohesperidin	0.969 ± 0.099
	Morin	0.134 ± 0.004
	Quercetin	0.058 ± 0.0001
	Genistein	0.00045 ± 0.0001
	Kaempferol	0.015 ± 0.0001

* High pressure liquid chromatography; Values are expressed as mean $\pm$ standard deviation.

3.3. Pectin Extraction

Based on the factorial experimental design 3^4 the results of the different parameters evaluated on the yield of pectin extraction conditions are shown in Figure 1. Higher pectin yield values were observed at 90 °C with solute:solvent ratio of 1:80 with the three acids evaluated. Extractions with citric acid resulted in the best pectin yield compared to the others; the maximal pectin yield attained was 36% using citric acid. The multifactorial analysis of variance (ANOVA) indicated that the interaction of acid type and temperature as well as the four independently evaluated factors presented significant effect ($p < 0.05$) on the pectin extraction yield. According to the Pareto analysis (Figure 2), a major effect was related to acid type being higher with citric acid; solute:solvent ratio also presented a high effect, followed by temperature and time. For all of the factors evaluated, pectin extraction yield increased when the higher levels were evaluated.

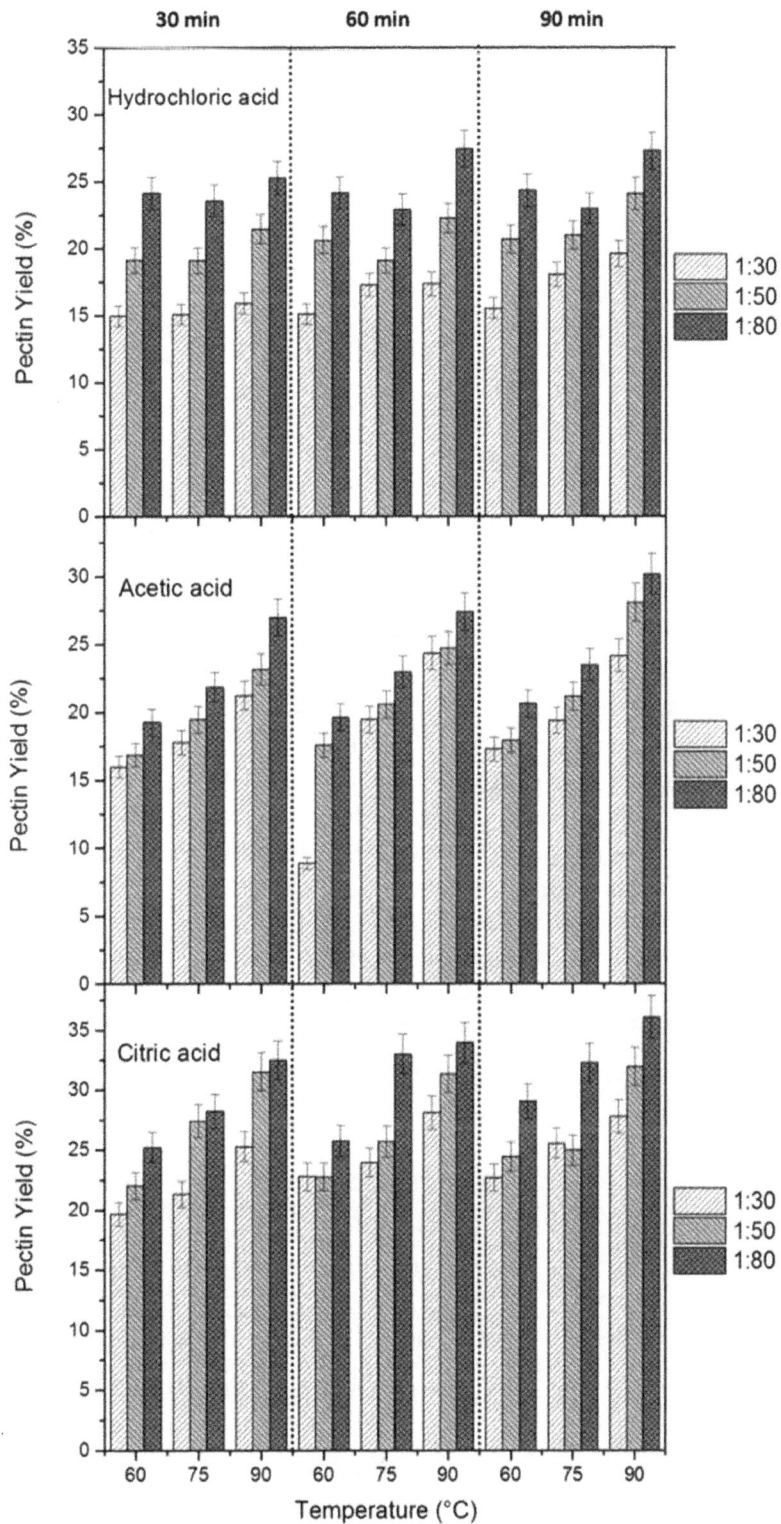

Figure 1. Pectin yield (%) under different extraction conditions: time (30, 60, and 90 min), temperature (60, 75, and 90 °C), solute:solvent ratio (1:30, 1:50, and 1:80), acid type (hydrochloric, acetic, and citric acid).

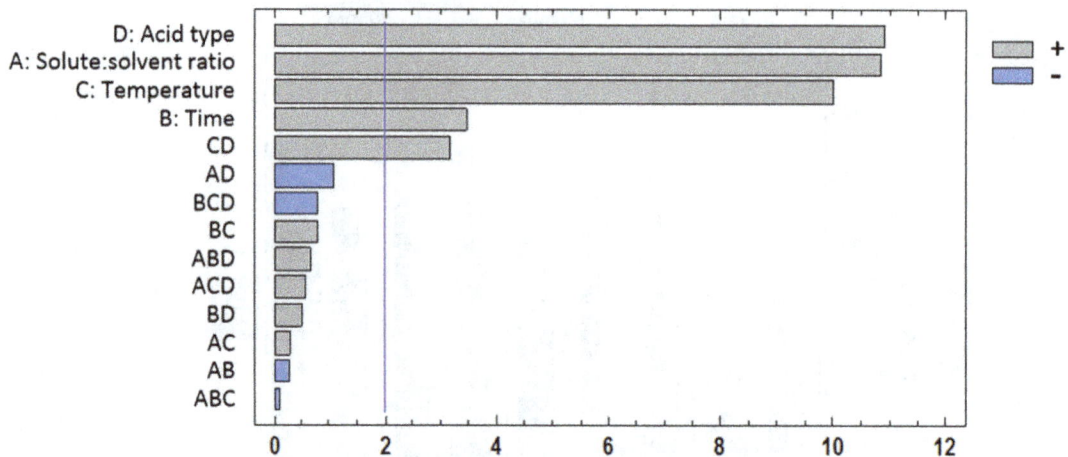

Figure 2. Standardized Pareto analysis of the pectin yield obtained from citrus waste.

3.4. Physicochemical Pectin Characterization

The higher pectin yield was obtained with pectin recuperation process A, using the two stage hydrolysis process, and C, with pH neutralization before alcohol precipitation. Moisture content values obtained with treatments A and B did not show significant differences ($p < 0.05$) between them, but were lower than the values presented in commercial citric pectin and treatment C (Table 3). The titratable acidity expressed as meq of free carboxyl/g of sample indicated that the recuperation procedure A exhibited the higher value; this could be due to the residual citric acid presented in the sample. The equivalent weight was lower using the recuperation procedure A and higher with procedure C when solution was neutralized before precipitation.

Table 3. Physicochemical characterization of pectins obtained from flour citric waste.

Physicochemical Characteristic	Pectin According to Recuperation Process			Commercial Pectin
	A *	B *	C *	
Yield (%)	36.45 ± 0.27 [a]	34.86 ± 0.21 [a]	36.21 ± 1.4 [a]	-
Moisture (%)	7.9 ± 0.01 [a]	8.17 ± 0.16 [a]	10.72 ± 0.27 [b]	10.49 ± 0.04 [b]
FA (meq free carboxyl/g)	3.01 ± 0.34 [d]	2.04 ± 0.05 [c]	1.73 ± 0.29 [b]	0.733 ± 0.0 [a]
Equivalent weight (mg)	400.37 ± 2.98 [a]	622.25 ± 0.0 [b]	706.74 ± 0.0 [c]	1364.63 ± 0.0 [d]
Methoxy (%)	10.12 ± 0.12 [b]	11.29 ± 0.29 [c]	9.00 ± 0.78 [a]	10.56 ± 0.2 [b]
ED * (%)	52.05 ± 0.59 [a]	64.09 ± 0.1 [b]	62.41 ± 1.52 [b]	82.29 ± 0.16 [c]
UA * (%)	DD *	DD *	81.59 ± 5.04 [a]	72.87 ± 0.16 [b]

Values are expressed as mean ± standard deviation. Similar letters in same line indicated no significant differences ($p < 0.05$). * A: two stages of extraction acid hydrolysis and alcohol precipitation, B: three stages of extraction acid hydrolysis alcohol precipitation and pectin washing, C: two stages of extraction with pH adjustment before alcohol precipitation, FA: Free acidity, ED: Esterification degree, UA: Uronic acid, DD: Difficult to determine.

The methoxy content and the ED are chemical parameters related to the gelification rate and pectin solidification. Procedure A showed similar methoxy % as commercial pectin; values higher than 8% are considered to be high methoxy pectin. ED (%) was significantly ($p < 0.05$) lower with the different pectin recuperation processes in comparison to commercial pectin. Nevertheless, all treatments showed high values of ED. The purity of the material can be determined by the UA content; values higher than 65% are accepted as high purity by the FAO, as is the case of the pectin obtained by procedure C.

3.5. Pectin Identification and Conformation

The qualitative test for pectin identification and conformation was performed in the sample obtained with procedure C and compared to the commercial one; all the tests indicated positive results to pectin (Table 4).

Table 4. Qualitative tests for the identification and conformation of pectin.

Test	Laboratory Pectin		Commercial Pectin	
	Description	Result	Description	Result
Pectin solution + ethanol	Yellow gelatinous pp	+	Sandy color gelatinous pp *	+
Pectin solution + NaOH 2N	Yellow gel	+	Sandy color gel	+
Precipitated gel + HCl 3N	Colorless gelatinous pp	+	Colorless gelatinous pp *	+

* pp: precipitate.

3.6. Evaluation of Pectin as Functional Ingredient

The addition of pectin in cookie elaboration was performed at 2.5%, 7%, and 10% of fat substitution. The results of water content, a_W, physical characterization, and texture of the resulting cookies are shown in Table 5. The water content in samples was between 5.3% and 8.3%; this parameter is related to the capability of the cookie to absorb water. The control and treatment with 2.5% of pectin substitution presented the lower water % content (0.45) and increased ($\approx$0.48) for the treatments with 7% and 10% of pectin substitution. Diameter and thickness measurements of the cookies indicated that higher substitution significantly decreased these values, nevertheless, the largest spread ratio was obtained with the highest pectin substitution. Textural hardness indicated that there were not significant differences ($p < 0.05$) among the substitution concentrations and control treatments. The result values were in the interval of 18.3–20.6 N, which indicates the presence of a soft cookie that did not feature drastic fracturing of the components.

Table 5. Water content, A_W, physical characterization, and texture analysis of the resulting cookies with added citric pectin.

Treatments	Water Content (%)	A_W	Physical Characterization			Hardness (N)
			Diameter (mm)	Thickness (mm)	Spread Ratio	
Control	5.39 ± 0.3 [a]	0.44 ± 0.02 [a]	50.12 ± 0.15 [b]	10.45 ± 0.15 [c]	4.80	19.9 ± 3.3 [a]
2.5% of PS *	5.42 ± 0.5 [a]	0.45 ± 0.01 [a]	50.05 ± 0.12 [b]	10.35 ± 0.17 [c]	4.83	20.6 ± 2.1 [a]
7% of PS	8.23 ± 0.4 [b]	0.49 ± 0.02 [b]	47.50 ± 0.1 [a]	9.92 ± 0.11 [b]	4.78	18.3 ± 3.2 [a]
10% of PS	7.65 ± 0.1 [b]	0.48 ± 0.01 [b]	48.3 ± 0.11 [a]	9.68 ± 0.02 [a]	4.99	18.9 ± 3.2 [a]

Values are expressed as mean ± standard deviation. * PS: pectin substitution. Similar letters in the same line indicate no significant differences ($p < 0.05$).

4. Discussion

4.1. Flour Citrus Waste Characterization

Moisture % of the citrus waste flour is in concordance to the Mexican normativity for flour materials NOM 247-SSAI-2008 [23], which indicates a maximal value of 15%; this is an important parameter to ensure the preservation of the material, reducing the risk of microbial contamination and enzymatic degradation that could reduce polyphenols and pectin yields. pH is related to the acidity presented in citrus residues that could be reduced by the dehydration process due to the salts dissociation as reported by Badillo [22], who obtained pH values of 5.5 for fresh fruits and 3.2 after dehydration. Total titrable acidity is related to the citric acid presence that might remain in the peel waste after juice extraction [22]. The slight dark coloration of citrus waste flours could be related to the natural pigments and sugars present in the products that could be partial degraded during waste drying, nevertheless the luminosity value higher than 60% indicates that the elaborated flour presented the possibility of preserving the quality of the by-products (polyphenols and pectin) that can be extracted from it.

4.2. TPC, Antioxidant Activity, and Polyphenol Profile

The importance of the polyphenol extraction method used in this study is related to the use of solid carbon dioxide, which favors a higher disruption of the vegetal matrix, which combines with the solvent, and promotes faster liberation of the phytochemicals compounds. A faster extraction process reduces the possibility of degradation to the polyphenols biological activity occurring due to heating and time consuming methods [24]. Furthermore, the total polyphenol content was higher than the results reported by Li et al. [25] and similar to the values obtained by Wang et al. [26] in citrus residues; in both cases, they used methanol as an extraction solvent. However, the TPC value was lower than the values mentioned by Papoutsis et al. [27]; this can be explained due to the different conditions used by these authors for drying peel preparation prior to TPC extraction with hot water, furthermore, the species of the citrus residue evaluated was also different (*C. limon*). Additionally, in our case, a lower TPC content could also be attributed to a previous essential oil extraction that could take part of the polyphenols presents in the peel and increase polyphenolic degradation before flour preparation. Although it has been reported [24] that a soft heat pre-treatment of the vegetal matrix (like the drying process) could enhance polyphenol extractions due to the previous disruption of the vegetal structures that favor polyphenol liberation, there are different conditions such as: pre-treatments methods, and different solvents and extraction conditions that can reduce or improve TPC yield [24]. Related to the antioxidant activity, the values reported herein are promising due to the preservation of more than 70% of the antioxidant activity. It has been reported that principal polyphenol compounds present in lemon residues, such as the flavonoids hesperidin and eriocitrin, may also have a major part in the antioxidant effect [28]. In this study, hesperidin is the second major compound detected, although the presence of eriocitrin was not determined due to the lack of the standard for the quantification analysis, the presence of unidentified peaks in the chromatogram was observed. If the methanolic polyphenol extraction is performed for further applications in the food industry, in all the cases a total elimination of the methanol content by evaporation is required for the restriction of methanol presence in food products, as reported by FDA. Phenolic compounds extracted from citrus waste could be excellent functional ingredients in the food industry and especially in bakery products due to the presence of antioxidant activity, nevertheless the sensitivity of the compounds suggests the use of encapsulation matrix to preserve their biological activities.

4.3. Pectin Extraction

For pectin extraction, all the factors evaluated presented a significantly effect. The solubilization of protopectin depends on a control acidic medium during the extraction process as well as the temperature and time [29]. A better pectin yield was obtained with citric acid than acetic or hydrochloric acid. Stronger acids could breakdown the polysaccharide bonds and reduce pectin yield; furthermore, the hydroxyl groups of the citric acid benefits the formation of hydrogen bonds between pectin and the citric acid favoring extraction [29]. As is indicated in results section, the Pareto analysis showed that conditions with higher values resulted in higher extraction yields; hence 90 min of time extraction and a ratio of 1:80 resulted in higher pectin yields regardless of the type of acid used. Pectin yields obtained in this work are comparable with those obtained from apple and citrus (10%–40%) at laboratory scale [29]. The use of citric acid that favors pectin yield also promotes the application of green technologies to pectin production, reducing the generation of hydrochloric acid residues.

4.4. Physicochemical Pectin Characterization

Pectin characteristics depend principally on the vegetal source as well as extraction conditions. Pectin moisture content along with all the physicochemical parameters of pectin is related to the quality of the product as it indicates the water absorption capability. Thus, pectin obtained with treatment C presented similar pectin moisture as the commercial pectin, indicating similar water absorption capability during storage at ambient temperature. Titratable acidity values were higher than those obtained in the literature [17]. This might be due to the residue evaluated and the ripening stage, as well as the residues of the citric acid used during extraction. The equivalent weight is associated with the maturity stage of the fruit and is inversely proportional to the free acidity; this parameter is related to the gelling power and viscosity of the resulted pectin. Ferreira et al. [30] determined the equivalent weight of different citric fruits, obtaining values between 528 and 1130 mg; thus, the values obtained herein between 400 and 750 mg are similar to the values reported elsewhere. Madhav and Pushpalatha [31] characterized pectins from different fruits, and found high methoxy pectins for citric sources of around 9%, as was found in this work. This value is similar to the values obtained for pomelo pectins [32]. Characterization of pectin from citrus residues (*Citrus x latifolia*) requires further analysis to study rheological parameters that can influence pectin application. However, considering its characteristic of forming gels, these pectins could also be used for medical applications [5] or to fabricate new entrapment bio-composites for probiotic delivery in the food industry [33].

4.5. Evaluation of Pectin as Functional Ingredient

As reported in the results section, the addition of pectin increased water content, although up until a 3% addition of pectin the conditions evaluated did not cause a significant difference in texture analysis. Thus, these results suggest the use of pectin from *Citrus x latifolia* flour residues could be used as a fat replacement in bakery products due to the capability of this pectin to trap water and to give weight to the product without increasing calories. Further analyses to increase substitution values and sensorial analyses are needed in order to obtain a healthy and high quality product, nevertheless an approximation of a functional application is given in this work.

5. Conclusions

The interaction between the temperature and acid type, as well as the individual factors: extraction time, solute-solvent ratio, acid type, and temperature presented a significant effect ($p < 0.5$) on the yield of pectin extracted from Persian lime *Citrus x latifolia* waste flour. According to the multifactorial variance analysis, the best extraction conditions were: citric acid at 90 °C for 90 min at a ratio of 1:80 (w/v) with a yield of 36% (g of pectin recovered per g of flour used). The evaluation of pectin recuperation process on the pectin physicochemical characteristics indicated that the best treatment was obtained with neutralization before precipitation. Using these conditions, a pectin with moisture content of 10.72%, free acidity of 1.73 meq free/g carboxyl equivalent weight of 706.74 mg, methoxy content of 9.0%, esterification degree of 62.41%, uronic acid of 81.59%, and yield of 36.21% can be obtained. The pectin was also categorized as having high methoxy, slow gelation, and a high degree of purity. Polyphenols profile determination indicated the major presence of the flavonoids neohesperidin and hesperidin in the residue with a TPC of 3.9 mg of GAE/g of citrus waste flour DB and a value higher than 73% for antioxidant activity. The application of the citrus pectin as fat replacer in the cookies elaboration indicated a potential functional use of the pectin extracted using the best conditions observed. The characterization of pectin allowed for the determination of the characteristics of pectins that might be useful in other commercial applications.

Acknowledgments: The authors wish to thank Proyecto Fomix Yuc-2011-C09-0169165 for the financial support to the experimental work and for Cristina A. Ramos-Chan's scholarship.

Author Contributions: Neith A. Pacheco L. and Teresa del Rosario Ayora-Talavera conceived and designed the experiments, the corresponding author Neith A. Pacheco L. also analyzed and wrote the paper, Cristina A. Ramos-Chan and Ana G. Covarrubias-Cárdenas performed the experiments, Angeles Sánchez-Contreras and Ulises García-Cruz contributed to the discussion and analysis of data.

Conflicts of Interest: The authors declare no conflict of interest.

References

1. SIAP-Agri-Food and Fisheries Information Service. (2014). SIAP gob.mx. Available online: http://www.siap.gob.mx/agricultura-produccion-anual/ (accessed on 16 July 2016).

2. Baltazar, R.; Carbajal, D.; Baca, N.; Salvador, D. Optimization of the conditions of pectin extraction from lemon rind french (*Citrus medica*) using response surface methodology. *Agroind. Sci.* **2013**, *2*, 77–89. [CrossRef]

3. Casquete, R.; Castro, S.M.; Martín, A.; Ruiz-Moyano, S.; Saraiva, J.A.; Córdoba, M.G.; Teixeira, P. Evaluation of the effect of high pressure on total phenolic content, antioxidant and antimicrobial activity of citrus peels. *Innov. Food Sci. Emerg. Technol.* **2015**, *31*, 37–44. [CrossRef]

4. Benavente-García, O.; Castillo, J. Update on uses and properties of citrus flavonoids: New findings in anticancer, cardiovascular, and anti-inflammatory activity. *J. Agric. Food Chem.* **2008**. [CrossRef] [PubMed]

5. Gil-Izquierdo, A.; Riquelme, M.T.; Porras, I.; Ferreres, F. Effect of the Rootstock and Interstock Grafted in Lemon Tree (*Citrus limon* (L.) Burm.) on the Flavonoid Content of Lemon Juice. *J. Agric. Food Chem.* **2004**, *52*, 324–331. [CrossRef] [PubMed]

6. Sungthongjeen, S.; Sriamornsak, P.; Pitaksuteepong, T.; Somrisi, A.; Puttipipatkhachorn, S. Effect of degree of esterification of pectin and calcium amount on drug release from pectin-based matrix tablets. *AAPS PharmSciTech* **2009**, *5*, 50–57.

7. Luppi, B.; Bigucci, F.; Abruzzo, A.; Corace, G.; Cerchiara, T.; Zecchi, V. Freeze-dried chitosan/pectin nasal inserts for antipsychotic drug delivery. *Eur. J. Pharm. Biopharm.* **2010**, *75*, 381–387. [CrossRef] [PubMed]

8. Sriamornsak, P. Application of pectin in oral drug delivery. *Expert Opin. Drug Deliv.* **2011**, *8*, 1009–1023. [CrossRef] [PubMed]

9. Glinsky, V.V.; Raz, A. Modified citrus pectin anti-metastatic properties: One bullet, multiple targets. *Carbohydr. Res.* **2009**, *344*, 1788–1791. [CrossRef] [PubMed]

10. Munarin, F.; Tanzi, M.C.; Petrini, P. Advances in biomedical applications of pectins gels. *Int. J. Biol. Macromol.* **2012**, *51*, 681–689. [CrossRef] [PubMed]

11. Chan, S.Y.; Choo, W.S.; Young, D.J.; Loh, X.J. Pectin as a rheology modifier: Origin, structure, commercial production and rheology. *Carbohydr. Polym.* **2017**, *161*, 118–139. [CrossRef] [PubMed]

12. Lim, J.; Ko, S.; Lee, S. Use of *Yuja* (*Citrus junos*) pectin as a fat replacer in baked foods. *Food Sci. Biotechnol.* **2014**, *23*, 1837–1841. [CrossRef]

13. Norma Oficial Mexicana NMX-F-428. Determinación de Humedad (Método Rápido de la Termobalanza). 1982. Available online: http://www.colpos.mx/bancodenormas/nmexicanas/NMX-F-428-1982.PDF (accessed on 10 September 2015).

14. Sánchez-Contreras, A.; Rufino-Gonzalez, Y.; Ponce-Macotela, M.; Sánchez-Garcia, S.; Jiménez-Estrada, M.; Rodriguez-Buenfil, I.; Chel-Guerrero, L. Psidium guajava and Tagetes erecta flavonoids: Isolation, identification and biological activity. In *Nutraceuticals Funct Foods Conv Non-Conventional Sources*; Studium Press: Houston, TX, USA, 2011; pp. 64–71.

15. Vasco, C.; Ruales, J.; Kamal-Eldin, A. Total phenolic compounds and antioxidant capacities of major fruits from Ecuador. *Food Chem.* **2008**, *111*, 816–823. [CrossRef]

16. Chen, M.L.; Yang, D.J.; Liu, S.C. Effects of drying temperature on the flavonoid, phenolic acid and antioxidative capacities of the methanol extract of citrus fruit (*Citrus sinensis* (L.) *Osbeck*) peels. *Int. J. Food Sci. Technol.* **2011**, *46*, 1179–1185. [CrossRef]

17. Gamboa, B.M.; Universidad de Oriente, Núcleo de Anzoátegui, Puerto la Cruz, Venezuela. Aprovechamiento de los Residuos Obtenidos del Proceso de Despulpado del Mango (*Mangifera indica* L.) Como Materias Primas Para la Obtención de Pectinas. Personal Communication, 2009.

18. Mohd, N.; Ramli, N.; Meon, H. Extraction and characterization of pectin from Dragon Fruit (*Hylocereus polyhizus*) using various extraction conditions. *Sains Malays.* **2012**, *41*, 41–45.

19. Pomeranz, Y.; Meloan, C.E. *Food Analysis: Theory and Practice*, 3rd ed.; Springer Science & Business Media, An Aspen Publication, Aspen Publishers, Inc.: Gaitherburg, MD, USA, 2000.

20. Handa, C.; Goomer, S.; Siddhu, A. Effects of Whole-Multigrain and Fructoligosaccharide incorporation on the Quality and Sensory Attributes of cookies. *Food Sci. Technol. Res.* **2011**, *17*, 45–54. [CrossRef]

21. Sudha, M.L.; Srivastava, A.K.; Vetrimani, R.; Leelavathi, K. Fat replacement in soft dough biscuits: Its implications on dough rheology and biscuit quality. *J. Food Eng.* **2007**, *80*, 922–930. [CrossRef]

22. Badillo, M.D.; Escuela Superior Politécnica de Chimborazo, Riobamba, Ecuador. Estudio Comparativo del Potencial Nutritivo de Limón Persa (*Citrus latifolia tanaka*) Deshidratado en Secador de Bandejas y en Microondas. Personal Communication, 2011.

23. Norma Oficial Mexicana NOM-247-SSA1. Productos y Servicios. Cereales y sus Productos. Cereales, Harinas de Cereales, Sémolas o Semolinas. 2008. Available online: http://depa.fquim.unam.mx/amyd/archivero/NOMcereales_12434.pdf (accessed on 8 September 2015).

24. Baas-Dzul, L.V.; Centro de Investigación y Asistencia en Tecnología y Diseño del Estado de Jalisco A.C. Mérida México. Obtención de Extractos Polifenólicos con Actividad Biológica a Partir de Harinas Elaboradas con Subproductos de Limón Italiano. Personal Communication, 2015.

25. Li, B.; Smith, B.; Hossain, M. Extraction of phenolics from citrus peels I. solvent extraction method. *Sep. Purif. Technol.* **2006**, *48*, 182–188. [CrossRef]

26. Wang, Y.; Chueh, Y.; Hsing, W. The flavonoid, carotenoid and pectin content in peels of citrus cultivated in Taiwan. *Food Chem.* **2008**, *106*, 277–284. [CrossRef]

27. Papoutsis, K.; Pristijono, P.; Golding, J.B.; Stathopoulos, C.E.; Bowyer, M.C.; Scarlett, C.J.; Vuong, Q.V. Effect of vacuum-drying, hot air-drying and freeze-drying on polyphenols and antioxidant capacity of lemon (*Citrus Limon*) pomace aqueous extracts. *Int. J. Food Sci. Technol.* **2017**. [CrossRef]

28. Del Río, J.A.; Fuster, M.D.; Gómez, P.; Porras, I.; García-Lidón, A.; Ortuño, A. *Citrus limon*: A source of flavonoids of pharmaceutical interest. *Food Chem.* **2004**, *84*, 457–461. [CrossRef]

29. Canteri, M.H.; Ramos, H.C.; Waszczynskyj, N.; Wosiacki, G. Extraction of pectin from Apple pomace. *Braz. Arch. Biol. Technol.* **2005**, *48*, 259–266. [CrossRef]

30. Ferreira, S. Aislamiento y caracterización de las pectinas de algunas variedades de frutos cítricos colombianos. *Rev. Colomb. Cienc. Quím. Farm.* **1976**, *3*, 5–25.

31. Madhav, A.; Pushpalatha, P. Characterization of pectin extracted from different fruit wastes. *J. Trop. Agric.* **2002**, *40*, 53–55.

32. Methacanon, P.; Krongsin, J.; Gamonpilas, C. Pomelo (*Citrus maxima*) pectin: Effects of extraction and its properties. *Food Hydrocoll.* **2013**, *30*, 1–9. [CrossRef]

33. Khorasani, A.C.; Shojaosadati, S.A. Pectin-non-starch nanofibers biocomposites as novelgastrointestinal-resistant prebiotics. *Int. J. Biol. Macromol.* **2017**, *94*, 131–144. [CrossRef] [PubMed]

11

Development and Evaluation of Poly Herbal Molluscicidal Extracts for Control of Apple Snail (*Pomacea maculata*)

Guruswamy Prabhakaran *, Subhash Janardhan Bhore and Manikam Ravichandran

Department of Biotechnology, Faculty of Applied Sciences, AIMST University, Bedong-Semeling Road, Semeling, 08100 Bedong, Kedah, Malaysia; subhashbhore@gmail.com (S.J.B.); mravichandran08@gmail.com (M.R.)
* Correspondence: prabhakaran@aimst.edu.my

Academic Editor: Les Copeland

Abstract: Golden Apple Snail (GAS) is the most destructive invasive rice pest in Southeast Asia. The cost of synthetic molluscicides, their toxicity to non-target organisms, and their persistence in the environment have propelled the research of plant-derived molluscicides. Most research efforts have focused on individual plant extracts for their molluscicidal potency against GAS and have not been proven to be entirely effective in rice field conditions. Selective combination of synergistically acting molluscicidal compounds from various plant extracts might be an effective alternative. In this direction, ethanolic extracts from six different plants (Neem, Tobacco, Nerium, Pongamia, Zinger, and Piper) were evaluated against *Pomacea maculata* Perry. Of the various combinations studied, a binary extract (1:1) of nerium and tobacco (LC_{90} 177.71 mg/L, 48 h), and two tri-herbal extract formulations (1:1:1) of (nerium + tobacco + piper) and (nerium + tobacco + neem) were found to be most effective, with LC_{90} values of 180.35 mg/L and 191.52 mg/L, respectively, in laboratory conditions. The synergistic effect of combined herbal extracts resulted in significant reduction in LC_{90} values of the individual extracts. The findings of this study demonstrate that the selective combinations of potent molluscicidal herbal extracts are effective for management of *P. maculata* under laboratory conditions.

Keywords: golden apple snail; plant molluscicides; poly herbal extracts; *Pomacea maculata*

---

1. Introduction

Rice is a staple food and an essential crop grown worldwide, with Asia being the largest producer and consumer. Golden Apple Snail (GAS), a major rice pest, has a voracious appetite for the young rice seedlings of both transplanted and direct-seeded rice [1]. Infestations by GAS are responsible for a huge economic loss of $1.47 billion per annum in rice production [2]. *Pomacea canaliculata* Lamarck is the most pervasive; it is destructively invasive in most rice growing regions of the world and amongst the world's top 100 worst invasive alien species [3]. The distribution of apple snails is more pronounced in Peninsular Malaysia due to the method of paddy cultivation, which is predominately (90%) by wet paddy method. Among the two species of apple snails, *Pomacea maculata* is more abundant than *P. canaliculata* in the states of Kedah and Perlis, the two major paddy growing regions which produce more than 50% of rice in Malaysia [4,5]. Hence, the present study was focused on *P. maculata*. The Integrated Pest Management (IPM) strategies for the control of GAS are based on the use of cultural approaches, biological methods, snail predators (ducks and fish), by water management and chemical control measures [6,7].

The application of synthetic molluscicide formulations based on niclosamide and metaldehyde is widely practiced for effective control of snails. Niclosamide has become the mainstay of GAS control and eradication programs [8]. Synthetic molluscicides are known for their knock-down effect; however, their negative impact on the environment and the high costs of application have stimulated interest in the search for plant-derived molluscicides which are target specific, and environmentally and toxicologically safe [9–11]. Botanical pesticides have long been explored as effective alternatives to synthetic chemical counterparts due to their low toxicity to non-target organisms, their biodegradability, and due to them being less expensive in their crude form, in addition to the prevention of development of resistance against phytochemical mixtures [12–15]. Moreover, botanical pesticides are easily obtainable and more associated with indigenous self-sufficient pest snail control strategies than their imported synthetic pesticides.

Molluscicidal properties have been reported in more than 1400 plant species [16]. In the recent past, most research efforts were focused on individual plant extracts for their molluscicidal potency against GAS. The different plants (*Annona squamosa* seed, *Nerium indicum* leaves, *Stemona tuberose* root, *Cyperus rotundus* corm, and *Derris elliptica* root) were evaluated [17]. The results indicated that *D. elliptica* and *C. rotundus* (LC_{50} 23.68 mg/L and 133.20 mg/L, respectively) showed the highest toxicity on GAS, while Vulgarone B, a sesquiterpene isolated from *Artemisia douglasiana*, showed 100% mortality of GAS at a concentration on par with metaldehyde [18]. The molluscicidal potency of methanol extract from neem seed (*A. indica* A. Juss), oleoresin of *Zingiber officinale*, and cyclotide extracts of *Oldenlandia affinis* and *Viola odorata* on GAS was ascertained [19–22]. The extract of neem leaves and garlic (*Allium sativum*) recorded more than 90% mortality at 1000 mg/L on GAS [23]. The toxicity of different solvent extracts of *Agave sisalana* on *P. canaliculata* was established by Li et al. [11].

The extracts of *Cymbopogon citratus* induced alterations in growth and development of *P. canaliculata* and resulted in high mortality of snails [24,25]. The ethyl acetate soluble fraction of *Aglaia duperreana* was effective against GAS [26]. In rice field conditions, the molluscicidal efficacy of chemically modified quinoa (*Chenopodium quinoa*) saponins [27], methanol extract of *Camellia oleifera* seed meal [28], and *Sapindus saponaria* extract [29] were tested against GAS. The application of tobacco waste in field trials proved its efficacy with 100% mortality of GAS [30]. However, the vast majority of individual plant extracts were not proven to be entirely effective in rice field conditions. Therefore, to leverage the cumulative benefits of synergistically acting phytochemical mixtures, a selective combination of potent molluscicidal compounds from various plant extracts might be an effective alternative against apple snails.

2. Materials and Methods

2.1. Collection of Snails (Pomacea maculata)

Snails with shell length of 10 mm to 30 mm were collected from a snail-infested rice field located in Semeling, Kedah, Malaysia. The snails were identified based on the conchological characters [31,32]. The identification of species was confirmed by officials at the District Agriculture Office, Sungai Petani, Kedah, Malaysia. The shell length of each snail was measured with a Vernier calliper. The snails were acclimated in 20 litre laboratory aquarium tank (60 × 30 × 30 cm) containing dechlorinated tap water at room temperature (25 ± 2 °C). The snails were fed with fresh lettuce (*Lactuca sativa)* which was maintained ad libitum by feeding them two to three times a week according to their consumption, in order to avoid bacterial growth and water fouling [33,34].

2.2. Collection of Plant Materials

A total of 5 kg fresh leaves of *Nerium indicum* and *Nicotiana tabacum* were collected from Sungai Petani and Jitra, Kedah, Malaysia. The plants were taxonomically identified and authenticated by the botanist at the Biotechnology Department of AIMST University, Malaysia. Three kilograms of fresh rhizome of *Zingiber officinale* and 0.5 kg of black piper seeds (*Piper nigrum*) were purchased from the local TESCO supermarket. One kilogram of pongamia oil (*Pongamia pinnata*) and neem oil (*Azadirachta indica)* were obtained from Spic Ltd., Chennai, Tamil Nadu, India. The leaves of *N. indicum* and *N. tabacum* were thoroughly cleaned with running tap water and shade dried for three weeks at room temperature (25 ± 2 °C). The zinger rhizome was chopped into small pieces and shade dried for four weeks. The dried materials were ground into coarse powder using a food blender and stored in air-tight containers. The details of plant materials evaluated for their molluscicidal properties are given below in Table 1.

Table 1. Plant materials evaluated for molluscicidal properties.

No.	Scientific Name	Family	Parts Used
1	*Azadirachta indica* A. Juss.	Meliaceae	Neem kernel oil (cold processed)
2	*Nicotiana tabacum* L.	Solanaceae	Leaves
3	*Nerium indicum* Mill.	Apocynaceae	Leaves
4	*Pongamia pinnata* L.	Fabaceae	Seed oil
5	*Zingiber officinale* L.	Zingiberaceae	Rhizome
6	*Piper nigrum* L.	Piperaceae	Seeds

2.3. Preparation of Plant Extracts

The 100 g of dried and coarse plant material was macerated in 1000 mL of 95% ethanol at room temperature (25 ± 2 °C) for three days. The 95% ethanol was used to maintain the homogeneity in the individual plant extracts and to enhance the compatibility among the plant extracts for development of poly herbal formulations. The crude extract was separated by filtering under vacuum using No. 1 Whatman filter paper (Camlab Limited, Cambridge, UK). The retentate was then added to 1000 mL of fresh 95% of ethanol, macerated for three days further at room temperature, and filtered. Both the filtrates were pooled together and concentrated to dryness using a rotary evaporator at 60 °C until the solvent was completely evaporated. The pongamia oil and neem oil were extracted twice with equal volume of 95% of ethanol, and ethanol layers were pooled and concentrated under vacuum in a rotary evaporator. The ethanolic extract of individual plants obtained was labelled and stored at 5 °C until evaluation of molluscicidal activities and analysis of phytochemicals.

2.4. Evaluation of the Molluscicidal Potency of Plant Extracts

Evaluation of molluscicidal activity of plant extracts against juvenile snails of *P. maculata* was performed as per the World Health Organization (WHO) guidelines [35]. The different concentrations of individual, binary, and poly plant extracts were prepared to test against the snails. A stock solution of individual crude plant extract was prepared by dissolving 1 g of each extract in 1000 mL of distilled water to make it a 0.1% concentration which constitutes 1000 mg/L. From this stock solution, a 0.01% (100 mg/L) test solution was made by tenfold dilution with dechlorinated tap water. Similarly, the required concentrations (200 mg/L to 500 mg/L) were prepared from the 0.1% stock solution.

2.5. Preparation of Binary, Tri, and Poly Herbal Combinations of Plant Extracts

The plants examined in this study were selected on the basis of ethnobotanical information and recognized molluscicidal activity on *P. canaliculata* as reported in previous studies [17,21,22,30]. The following six different binary combinations of plant extracts were evaluated (Table 2). The different concentrations of binary plant extracts (100 mg/L–500 mg/L) at the ratio of 1:1 *v/v* were prepared from 0.1% stock solutions of individual extracts with appropriate dilution with dechlorinated tap water.

Table 2. Binary combination of plant extracts.

No	Plant A	Plant B
1	Nerium (*Nerium indicum*)	Tobacco (*Nicotiana tabacum*)
2	Nerium (*Nerium indicum*)	Piper (*Piper nigrum*)
3	Nerium (*Nerium indicum*)	Neem (*Azadirachta indica*)
4	Tobacco (*Nicotiana tabacum*)	Piper (*Piper nigrum*)
5	Tobacco (*Nicotiana tabacum*)	Neem (*Azadirachta indica*)
6	Piper (*Piper nigrum*)	Neem (*Azadirachta indica*)

The selective combinations of tri-herbal extracts at 1:1:1 ratio (v/v) were prepared by combining individual plant extract test solutions for the evaluation against *P. maculata* (Table 3). Five different concentrations of tri-herbal extracts (50 mg/L to 250 mg/L) were prepared from 0.1% stock solution of individual extract with appropriate dilution with dechlorinated tap water. Similarly, different concentrations (50 mg/L to 250 mg/L) of poly herbal extract test solutions with nerium, tobacco, piper and neem were prepared at the ratio of 1:1:1:1 (v/v) by combining the 0.1% stock solution of appropriate plant extracts.

Table 3. Tri-herbal combinations of plant extracts.

No.	Plant A	Plant B	Plant C
1	Nerium	Tobacco	Piper
2	Nerium	Piper	Neem
3	Tobacco	Piper	Neem
4	Nerium	Neem	Tobacco

2.6. Preparation of Niclosamide Test Solutions

The synthetic chemical molluscicide niclosamide 70% wettable powder (WP) was used as the reference standard for comparison. The Bayluscide® WP 70 (Bayer, AG, Leverkusen, Germany) was purchased from the local market at Sungai Petani (Kedah, Malaysia) and was used as positive control in the experiments. Bayluscide consists of 81.4% w/w active ingredient (niclosamide ethanolamine), as indicated on the label. To prepare different concentrations of niclosamide (0.25 mg/L, 0.50 mg/L, 0.75 mg/L, and 1 mg/L), a stock solution of 10 mg/L was prepared and then diluted appropriately. Since 1 g of Bayluscide contains 0.814 g of niclosamide as active ingredient, 12.3 mg was dissolved in one litre of distilled water to obtain 10 mg/L stock solution. Tenfold dilution of the stock solution with dechlorinated tap water was prepared to obtain 1 mg/L stock solution. From this stock solution, two-fold dilutions were made to prepare 0.5 mg/L test solution.

2.7. Evaluation of Molluscicidal Potency of Plant Extracts

Groups of 10 uninfected juvenile snails (20 to 30 mm of shell length) were placed in plastic trays with 1000 mL of different concentrations of individual, binary, and poly herbal extracts test solutions separately. Five different concentrations of each test solution of the plant extracts were tested, each with three replicates of 10 snails. Three replicates were prepared with 0.5 mg/L niclosamide in dechlorinated water and used as positive control. Control experiments were performed with dechlorinated tap water alone (negative control). A total of 210 snails were used for the evaluation of an extract. All of the molluscicidal evaluations were carried out at room temperature (25 ± 2 °C) under normal diurnal lighting. The plastic trays were individually covered with a fine plastic mesh to prevent the snails from crawling out. Snails exposed to different concentrations of the plant extract were left for observation for 24 h and 48 h. After 48 h, the plant extract suspension was decanted; the snails were rinsed twice with dechlorinated tap water and transferred to a new container filled with dechlorinated tap water and observed for another 24 h, which served as the recovery period, following which the mortality rates were determined. The snails were not fed during the exposure and recovery periods. Upon observation, the dead snails were removed from the containers. Snails were considered dead if they did not move and either had retracted well into their shells or were hanging out of the shells. The death of each snail was further ascertained by the complete opening of operculum and if the head did not respond when pricked with a sharp needle. The mortality of snails was recorded for both exposure periods and the recovery period. The LC_{50} and LC_{90} values with their associated 95% confidence intervals (95% CI) were calculated by probit analysis using SPSS (Version 22, IBM Software Group, Chicago, IL, USA). The mortality of snails exposed to different concentrations of six selected plant extracts was analysed using a one-way analysis of variance (ANOVA) and mean comparisons were performed by Duncan's multiple comparison tests using SPSS (Version 22) at the 5% significant level.

3. Results

3.1. Molluscicidal Activity of Six Individual Plant Extracts

The molluscicidal activities of the six plants at different concentrations (100 mg/L to 500 mg/L) against *P. maculata* were evaluated. Neither behavioural symptoms observed nor death occurred in control groups with water, indicating that no factor other than plant moieties were responsible for the altered behaviour and mortality of snails. An analysis of the molluscicidal activities of the six plant extracts on *P. maculata* showed that there was a linear relationship between the concentrations of the individual extracts and the mortality of snails (Figure 1).

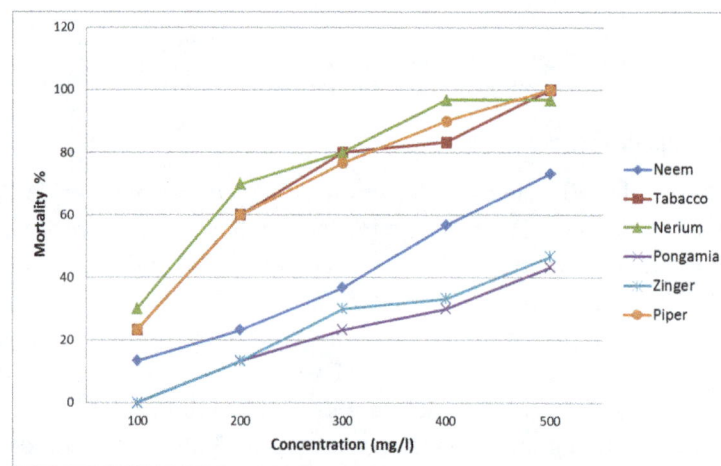

Figure 1. Molluscicidal activity of six plant extracts at different concentrations on *P. maculata*.

This observation was in agreement with the findings of Chauhan and Singh [36] and EI-Din et al. [37] wherein increases in the concentration of plant extracts resulted in high mortality rates of various pest snails. A similar positive correlation between the concentration of plant extracts (*Sandoricum vidalii, Harpulia arborea*, and *Parkia* sp.) and the mortality of *P. canaliculata* was reported by Taguiling et al. [38]. Likewise, a positive correlation between the neem seed crude extract concentrations and the mortality of *P. canaliculata* was observed [22]. In the present study, the LC_{50} and LC_{90} of *N. indicum, P. nigrum, N. tabacum*, and *A. indica* extracts were 179.36 and 341.57 mg/L, 202.01 and 359.89 mg/L, 205.70 and 375.84 mg/L, and 365.10 and 624.67 mg/L, respectively. Compared to the above four plants, *Z. officinale* and *P. pinnata* recorded higher LC_{50} and LC_{90} values (485.48 and 767.63 mg/L, 512.62 and 804.49 mg/L, respectively) and hence were not effective against snails. The results indicate that the snails were more sensitive to *N. indicum, P. nigrum*, and *N. tabacum* extracts, as evidenced by their lower LC_{50} and LC_{90} values indicating the highest potency. Eventually, the crude ethanol extract of *N. indicum* was found to be the most effective, with a LC_{90} value (341.57 mg/L,

48 h), whereas Dai et al. [39] recorded the lowest LC_{50} value (3.71 mg/L, 96 h) with purified cardiac glycosides of *N. indicum* for the control of *P. canaliculata*.

The result of the probit analysis on the mortality data of individual plant extracts on snails at LC_{50} and LC_{90} and fiducial limits are summarized in Table 4. Chi-square tests for probit analysis of Pearson Goodness-Fit test were carried out. A significant molluscicidal potency ($p < 0.001$) of nerium coincides with its LC_{90} value. A similar trend was observed with piper and tobacco. The heterogeneity value of less than 1 indicates that all the plant extracts exhibited a significant effect on the mortality of snails. Similarly, the heterogeneity value of less than 1 was correlated to the chi-squared values by Massaguni and Latip [22]. On comparative analysis, nerium extract was found to be the most effective, followed by piper and tobacco extracts. Niclosamide (0.5 mg/L) was used as a positive control, where 100% mortality was recorded in 24 h.

Table 4. Summary of probit analysis on the mortality data of six different plant extracts on *P. maculata*.

Plant Name	Hetero-Geneity	Chi Square	*p*	LC_{50}	Fiducial Limits (mg/L)		LC_{90}	Fiducial Limits (mg/L)	
					LL	UL		LL	UL
Azadirachta indica	0.999	3.863	0.920	365	321	420	624	541	768
Nicotiana tabacum	0.765	11.68	0.232	205	172	237	375	334	436
Nerium indicum	0.052	26.114	0.001	179	130	222	341	289	430
Pongamia pinnata	0.996	4.904	0.842	512	445	641	804	666	1111
Zingiber officinale	0.973	7.024	0.634	485	425	592	767	643	1028
Piper nigrum	0.704	12.575	0.183	202	170	232	359	320	417

LC: Lethal concentration. LL, UL: Lower and Upper Confidence Limit. Degree of freedom (df) = 9 for each concentration of individual plant extracts tested.

3.2. Molluscicidal Activity of Binary Combinations of Plant Extracts

A plant extract is considered to be effective at a concentration below 100 mg/L [40], whereas, the six different plant extracts investigated in this study were in the order of 341.57 to 804.50 mg/L for 90% mortality of the target snails. Therefore, in order to ascertain potent combinations of plant extracts which could bring down the concentration, different binary combinations of these plant extracts were formulated at 1:1 ratio by *v/v* and tested against *P. maculata*. The molluscicidal activity of six different binary combinations of plant extracts was compared at their LC_{90} values, and the results are depicted in Figure 2.

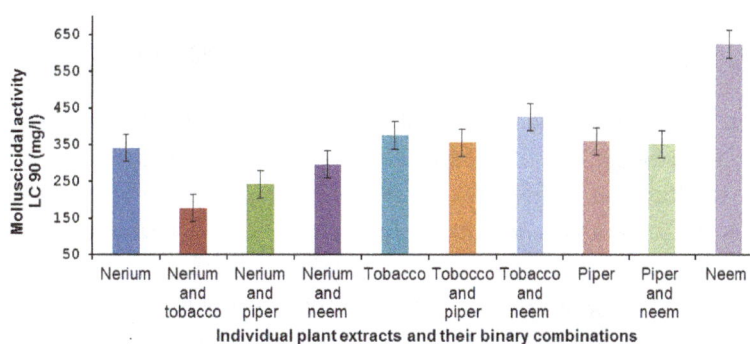

Figure 2. Comparison of molluscicidal activity (LC_{90}) of individual and binary combinations of plant extracts against *P. maculata*.

When two plant extracts were combined at a ratio of 1:1 (*w/w*), the individual plant extract's concentration would reduce to 50%. Therefore, it is expected that in the diluted form their combination would be needed in a high concentration for achieving 90% mortality of the snails. In this aspect, when LC_{90} values of individual extracts were compared with LC_{90} values of their binary combinations, the synergistic effect among four extracts such as (nerium + tobacco), (nerium + piper), (nerium + neem), and (tobacco + neem) was observed. This resulted in significant reductions in the LC_{90} values on the mortality of *P. maculata* in comparison to the individual extracts. Similarly, Chauhan and Singh [36] reported that the binary combination (1:1 ratio) of taraxerol from *Codiaeum variegatum* and acetone extract of *Euphorbia tirucalli* on *L. acuminata* increased thetoxicity by 9.51 times relative to the individual treatments. Furthermore, the binary combination of latex powder of *Euphorbia pulcherima* and *Jatropha gossypifolia* significantly reduced the fecundity, hatchability, and survival of young snails of *L. acuminata* [41]. The molluscicidal activity of six different binary combination of plant extracts on *P. maculata* is shown in Table 5.

Table 5. Molluscicidal activity of six different binary combinations of plant extracts on *P. maculata*.

Concentration (mg/L)	Percentage Mortality of *Pomacea maculata*					
	Binary Combination of Plant Extracts					
	Nerium and Tobacco	Nerium and Piper	Nerium and Neem	Tobacco and Piper	Tobacco and Neem	Piper and Neem
100	63.33 ± 5.7	63.33 ± 5.7	63.33 ± 5.7	36.60 ± 5.7	36.60 ± 5.7	36.60 ± 5.7
200	90.00 ± 10	76.60 ± 15.2	70.00 ± 10	66.60 ± 5.7	56.60 ± 5.7	56.60 ± 5.7
300	100 ± 0.00	93.30 ± 5.7	80.00 ± 10	80.00 ± 10	73.30 ± 5.7	76.60 ± 5.7
400	100 ± 0.00	100 ± 0.00	100 ± 0.00	93.30 ± 5.7	83.30 ± 5.7	93.30 ± 5.7
500	100 ± 0.00	100 ± 0.00	100 ± 0.00	96.60 ± 5.7	93.30 ± 5.7	100 ± 0.00

Each value is Mean ± standard deviation (SD) of three replicates (*n* = 10).

The binary combination of nerium and tobacco resulted in 90% mortality of snails, even at 200 mg/L. In comparison, nerium extract administered on an individual basis at 200 mg/L concentration recorded 70% mortality. At the same concentration, the tobacco extract recorded 60% mortality (Figure 1). The higher kill rate of the binary combination at the same concentration level suggests a synergistic effect. The binary combination of nerium and piper resulted in 93.3% mortality at 300 mg/L, whereas, the individual plant extracts of nerium and piper showed 80% and 76.7% mortality, respectively, at 300 mg/L. Nerium combined with neem at higher concentration (400 mg/L) exhibited an increase in mortality up to 100%. It was observed that in all cases where nerium was one of the components of binary combination, a higher mortality was recorded compared to any of the other combinations.

3.3. Molluscicidal Activity of Tri and Poly Herbal Extracts

Combining different plant species with effective molluscicidal properties is advantageous for the development of molluscicides. In the present study, among the six plant extracts evaluated individually as well as in binary combinations, *N. indicum*, *N. tabacum*, *P. nigrum*, and *A. indica* were found to be effective on the snails in laboratory conditions. When these four plant extracts were tested individually the LC_{90} values ranged from 341.57 mg/L to 624.67 mg/L, whereas, in their binary combinations significant reductions in the LC_{90} values (171.71 to 296.17 mg/L) were recorded. It is evident that the binary combinations of plant extracts resulted in approximately 45% reductions in LC_{90} values based on the mortality of *P. maculata* as well as demonstrating the synergistic effect among these plant extracts. Therefore, combining different plant species with effective molluscicidal properties is advantageous for further reducing the lethal concentrations against snails. In light of this, the selective consortiums of tri-herbal and poly herbal extracts were prepared by combining individual plant extracts at 1:1:1 to 1:1:1:1 ratio (*v/v*) and evaluated at various concentrations on *P. maculata*. The molluscicidal activity of tri and poly herbal combinations of plant extracts on *P. maculata* is shown in Table 6. The results indicate that the combination of three plant extracts (nerium + neem + tobacco) showed the highest mortality rate (93.3%) with the lowest LC_{50} and LC_{90} (62.60 mg/L and 191.52 mg/L, respectively) even at a lower concentration of 50 mg /L. In comparison, at the same concentration, the combined extracts of (nerium+ tobacco + piper) had 86.7% mortality and with LC_{50} (73.91 mg/L) and LC_{90} (180.35 mg/L) (Table 7).

Table 6. Molluscicidal activity of tri and poly herbal combination of plant extracts on *P. maculata*.

Concentration (mg/L)	Percentage Mortality of *Pomacea maculata*				
	Tri and Poly Herbal Combination of Plant Extracts				
	Nerium, Tobacco, and Piper	Nerium, Piper, and Neem	Tobacco, Piper, and Neem	Nerium, Neem, and Tobacco	Nerium, Tobacco, Piper and Neem
50	86.60 ± 5.7	46.66 ± 5.7	36.70 ± 5.7	93.30 ± 5.7	66.60 ± 15.2
100	96.60 ± 5.7	73.30 ± 5.7	70.00 ± 10	96.60 ± 5.7	83.30 ± 5.7
150	100 ± 0.00	86.60 ± 5.7	80.00 ± 10	100 ± 0.00	90.00 ± 10
200	96.60 ± 5.7	96.60 ± 5.7	86.70 ± 5.7	100 ± 0.00	100 ± 0.00
250	100 ± 0.00	96.60 ± 5.7	93.30 ± 5.7	96.60 ± 5.7	96.60 ± 5.7

Each value is Mean ± SD of three replicates ($n = 10$).

Table 7. Comparison of LC_{90} values of individual, binary, and tri-herbal extracts for their synergistic molluscicidal effects on *P. maculata*.

LC (mg/L) 48 h	Nerium	Tobacco	Piper	Neem	Nerium and Tobacco	Nerium, Tobacco, and Piper	Nerium, Tobacco, and Neem
LC_{50}	179.36	205.71	202.02	365.10	100.18	73.91	62.60
LC_{90}	341.57	375.84	359.90	624.67	177.71	180.35	191.52

LC: Lethal concentration.

All the tri-herbal combinations exhibited a significant ($p < 0.05$) increase in mortality from 86.7% to 100% at 200 mg/L concentration. The (nerium + neem + tobacco) and (nerium + tobacco + piper) emerged as the most effective tri-herbal combinations in terms of mortality of snails with LC_{90} 191.52 mg/L and 180.35 mg/L, respectively. The results of this study supported the findings of Taguiling [38] who reported that the activities of plant extracts vary significantly according to the species combinations and dosages against GAS. Further, the variations in the molluscicidal efficacy of plant extracts might be attributed to three major factors: species-tolerance, concentrations used, and the phytochemical constituents [42].

The piper was more effective with LC_{50} (202 mg/L) compared to neem (LC_{50} 365 mg/L) (Table 4). The binary combination of nerium and tobacco was found to be more effective with a LC_{50} value of 100.18 mg/L (Figure 2). Hence, it is obviously expected that a tri-herbal combination containing nerium, tobacco, and piper should be most effective. Therefore, the molluscicidal activities of individual, binary, tri-herbal, and poly herbal extract combinations were compared (Table 7). Notably, it was found that the tri-herbal combination made with nerium, tobacco, and neem proved to be the most effective, with a LC_{50} value of 62.60 mg/L. It is inferred that synergism among nerium, tobacco, and neem is higher than the nerium, tobacco, and piper combination.

4. Discussion

When the stock solutions of three plant extracts are combined at a ratio of 1:1:1 (v/v), there will be a dilution effect in the concentration of individual plant extracts. Therefore, such combinations should result in higher concentration requirement to achieve 90% mortality of snails. Even under the dilution effect if the plant extracts are synergistic, then the concentration of the plant extract required to achieve 90% mortality will be reduced. In the present study, the two tri-herbal extract combinations, namely (nerium + tobacco + neem) and (nerium + tobacco + piper), recorded low LC_{90} values, demonstrating a synergistic effect among the extracts of these plants. It is interesting to note that nerium and tobacco plant extracts are found to be common to all the effective tri-herbal extract combinations. The synergistic effect of these two extracts was also exhibited in their binary combination with LC_{90} of 177.71 mg/L and a similar trend was observed in their LC_{50} values (100.18 mg/L). Our results are in agreement with the research findings reported by Taguiling [38] who reported that the combination of extracts of three species (*S. vidalii* fruit and barks of *H. arborea* and *Parkia* sp.) at a ratio of 1:1:1 (w/w) was the most effective, with a mean mortality time of 6.56 min against adult GAS under laboratory and field trials. Furthermore, the binary combination of (*S. vidalii* + *Parkia* sp.) and (*H. arborea* + *Parkia* sp.)

recorded a mean mortality time of 8.11 min and 10.11 min, respectively [38]. A similar synergistic effect was reported with the binary combination of ferulic acid and azadirachtin against *Fasciola* larva in the snail *Lymnaea acuminata*, which was 64.28 times more effective than a single treatment with ferulic acid [43]. Rao and Singh [44] reported that the synergistic action in binary and tri-herbal combinations of *A. indica* and *Cedrus deodara* oil (1:7 ratio) and *A. indica*, piperonyl butoxide, and *C. deodara* at a ratio of 1:5:7 were found to be more toxic to *L. acuminata* than single treatment. Similarly, the binary (1:1) and tri-herbal (1:1:1) combinations of *Euphorbia pulcherima* latex powder, botulin, and taraxerol decreased significantly the LC_{50} dosages against snail *L. acuminata* [41].

Within five poly extracts studied only two combinations consisting of (nerium + neem + tobacco) and (nerium + tobacco + piper) were effective due to synergistic effects. This synergism may be attributed to the phytochemical constituents of individual plant extracts. Among the binary combinations, the nerium and tobacco extracts (LC_{90} 177.71 mg/L) was found to be the most effective against *P. maculate*, and this may be due to both the nicotine and cardiac glycosides content in these extracts. Glycogen is the primary and intermediate source of energy. The cardiac glycosides of *N. indicum* decreased acetylcholinesterase (AChE) activities and impaired the hepatopancreas tissues of *P. canaliculate*, resulting in the fatal inhibition of the activity of digestive enzymes as well as the feeding rate [39,45]. Nicotine acts by mimicking acetylcholine and it exerts a toxic effect in both vertebrates and invertebrates as a neurotoxin by binding to nicotinic acetylcholine receptors and inhibiting their penetration into the synapse [46]. The combination of lethal action in the active compounds of both nerium and tobacco possibly increased the snail mortality rate.

In summary, our study showed that of all the combinations of plant extracts evaluated, one binary combination (nerium and tobacco) and two tri-herbal combinations (nerium, tobacco, and neem) and (nerium, tobacco, and piper) were found to be the most effective against *P. maculata*. The study demonstrated the synergistic molluscicidal effect among the selectively combined crude extracts of *Nerium indicum*, *Nicotiana tabacum*, *Piper nigrum*, and *Azadirachta indica* for the effective control of *P. maculata* under laboratory conditions.

Acknowledgments: The authors are grateful to the AIMST University for financial support (Grant reference No: AIMST University Internal Grant Scheme—AURGC/20/FAS/2013).

Author Contributions: Guruswamy Prabhakaran conceived this research, designed the experiments, and performed the research work. The research activities were jointly supervised by Manikam Ravichandran and Subhash Janardhan Bhore, and they contributed to the writing and editing of this manuscript.

Conflicts of Interest: The authors declare no conflict of interest.

References

1.   International Rice Research Institute (IRRI), Rice Knowledge Bank. Available online: http://www.knowledgebank.irri.org/step-by-step-production/growth/pests-and-diseases/golden-apple-snails (accessed on 14 November 2016).

2.   Nghiem, L.T.P.; Soliman, T.; Yeo, D.C.J.; Tan, H.T.W.; Evans, T.A.; Mumford, J.D.; Keller, R.P.; Baker, R.H.A.; Corlett, R.T.; Carrasco, L.R. Economic and Environmental Impacts of Harmful Non-Indigenous Species in Southeast Asia. *PLoS ONE* **2013**, *8*, e71255. [CrossRef] [PubMed]

3.   GISD (Global Invasive Species Database); Invasive Species Specialist Group ISSG. The Global Invasive Species Database. Version 2015.1. Available online: http://www.issg.org/database (accessed on 1 June 2016).

4.   Salleh, N.H.M.; Arbain, D.; Daud, M.Z.M.; Pilus, N.; Nawi, R. Distribution and Management of *Pomacea canaliculata* in the Northern Region of Malaysia: Mini Review. *APCBE Procedia* **2012**, *2*, 129–134. [CrossRef]

5.   Arfan, A.G.; Muhamad, R.; Omar, D.; Nor Azwady, A.A.; Manjeri, G. Distribution of two *Pomacea* spp. in rice fields of Peninsular Malaysia. *Annu. Res. Rev. Biol.* **2014**, *4*, 4123–4136. [CrossRef]

6.   Zhang, J.; Zhao, B.; Chen, X.; Luo, S. Insect Damage Reduction while Maintaining Rice Yield in Duck-Rice Farming Compared with Mono Rice Farming. *J. Sustain. Agric.* **2009**, *33*, 801–809. [CrossRef]

7. Liang, K.; Zhang, J.; Song, C.; Luo, M.; Zhao, B.; Quan, G.; An, M. Integrated Management to Control Golden Apple Snails (*Pomacea canaliculata*) in Direct Seeding Rice Fields: An Approach Combining Water Management and Rice-Duck Farming. *Agroecol. Sustain. Food Syst.* **2014**, *38*, 264–282. [CrossRef]

8. Plant Health Australia. *Contingency Plan: Golden Apple Snail*; Plant Health Australia: Deakin, Australia, 2009; p. 15.

9. Duke, S.O.; Cantrell, C.L.; Meepagala, K.M.; Wedge, D.E.; Tabanca, N.; Schrader, K.K. Natural Toxins for Use in Pest Management. *Toxins* **2010**, *2*, 1943–1962. [CrossRef] [PubMed]

10. Singh, S.K.; Singh, A. Metabolic Changes in Freshwater Harmful Snail *Lymnaea acuminata* Due to Aqueous Extract of Bark and Leaf of *Euphorbia pulcherima* Plant. *Am.-Eurasian J. Toxicol. Sci.* **2010**, *2*, 13–19.

11. Li, X.; Deng, F.; Shan, X.; Pan, J.; Yu, P.; Mao, Z. Effects of the molluscicidal agent GA-C13:0, a natural occurring ginkgolic acid, on snail mitochondria. *Pestic. Biochem. Physiol.* **2012**, *103*, 115–120. [CrossRef]

12. Prakash, A.; Rao, J.; Nandagopal, V. Managing resistance to insecticides is a key for the future of crop protection. *J. Biopestic.* **2008**, *1*, 154–169.

13. El-Wakeil, N.E. Botanical pesticides and their mode of action. *Gesunde Pflanz.* **2013**, *65*, 125–149. [CrossRef]

14. Isman, M.B.; Grieneisen, M.L. Botanical insecticide research: Many publications, limited useful data. *Trends Plant Sci.* **2014**, *19*, 140–145. [CrossRef] [PubMed]

15. Pavela, R. Essential oils for the development of eco-friendly mosquito larvicides: A review. *Ind. Crops Prod.* **2015**, *76*, 174–187. [CrossRef]

16. Teixeira, T.; Rosa, J.S.; Rainha, N.; Baptista, J.; Rodrigues, A. Assessment of molluscicidal activity of essential oils from five Azorean plants against Radix peregra (Müller, 1774). *Chemosphere* **2012**, *87*, 1–6. [CrossRef] [PubMed]

17. Ruamthum, W.; Visetson, S.; Milne, J.R.; Bullangpoti, V. Toxicity of botanical insecticides on golden apple snail (*Pomacea canaliculata*). *Commun. Agric. Appl. Biol. Sci.* **2010**, *75*, 191–197. [PubMed]

18. Joshi, R.C.; Meepagala, K.M.; Sturtz, G.; Cagauan, A.G.; Mendoza, C.O.; Dayan, F.E.; Duke, S.O. Molluscicidal activity of vulgarone B from *Artemisia douglasiana* (Besser) against the invasive, alien, mollusc pest, *Pomacea canaliculata* (Lamarck). *Int. J. Pest Manag.* **2005**, *51*, 175–180. [CrossRef]

19. Shukla, S.; Singh, V.K.; Singh, D.K. The effect of single, binary, and tertiary combination of few plant derived molluscicides alone or in combination with synergist on different enzymes in the nervous tissues of the freshwater snail *Lymnaea* (Radix) *acuminata* (Lamark). *Pestic. Biochem. Physiol.* **2006**, *85*, 167–173. [CrossRef]

20. Plan, M.R.R.; Saska, I.; Cagauan, A.G.; Craik, D.J. Backbone Cyclised Peptides from Plants Show Molluscicidal Activity against the Rice Pest *Pomacea canaliculata* (Golden Apple Snail). *J. Agric. Food Chem.* **2008**, *56*, 5237–5241. [CrossRef] [PubMed]

21. Keni, M.F.; Latip, S.N.H.M. *Azadirachta indica* seed as potential biopesticides for controlling golden apple snail, *Pomacea canaliculata* in rice cultivation. In Proceedings of the 2013 IEEE Business Engineering and Industrial Applications Colloquium (BEIAC), Langkawi, Malaysia, 7–9 April 2013; pp. 251–256.

22. Massaguni, R.; Latip, S.N.H.M. Assesssment the Molluscicidal Properties of Azadirachtin against Golden Apple Snail, *Pomacea Canaliculata*. *Malays. J. Anal. Sci.* **2015**, *19*, 781–789.

23. Benchawattananon, R.; Boonkong, U. The toxicity of leave crude extract from neem tree (Azadirachta indica Juss.) and Garlic (*Allium sativom* L.) on mortality rate of golden apple snail (*Pomacea* sp.). In Proceedings of the 32nd Congress on Science and Technology of Thailand, Bangkok, Thailand, 10–12 October 2006.

24. Musman, M.; Kamaruzzaman, S.; Karina, S.; Rizqi, R.; Arisca, F. A preliminary study on the anti hatching of freshwater golden apple snail *Pomacea canaliculata* (*Gastropoda: Ampullariidae*) eggs from *Barringtonia racemosa* (*Magnoliopsida: Lecythidaceae*) seeds extract. *Aquac. Aquar. Conserv. Legis. Int. J. Bioflux Soc.* **2013**, *6*, 394–398.

25. Demetillo, M.T.; Baguio, M.L.; Limitares, D.E.; Madjos, G.G.; Abrenica-Adamat, L.R. Effect of *Cymbopogon citratus* (lemon grass) crude leaf extracts on the developmental stages of *Pomacea canaliculata* (golden apple snail). *Adv. Environ. Sci.* **2015**, *7*, 460–467.

26. Zhang, H.; Xu, H.H.; Song, Z.J.; Chen, L.Y.; Wen, H.J. Molluscicidal activity of Aglaia duperreana and the constituents of its twigs and leaves. *Fitoterapia* **2012**, *83*, 1081–1086. [CrossRef] [PubMed]

27. Martín, R.S.; Ndjoko, K.; Hostettmann, K. Novel molluscicide against *Pomacea canaliculata* based on quinoa (*Chenopodium quinoa*) saponins. *Crop Prot.* **2008**, *27*, 310–319. [CrossRef]

28. Kijprayoona, S.; Toliengb, V.; Petsoma, A.; Chaicharoenpongb, C. Molluscicidal activity of *Camellia oleifera* seed meal. *Sci. Asia* **2014**, *40*, 393–399. [CrossRef]

29. Quijano, M.; Riera-Ruiz, C.; Barragan, A.; Miranda, M.; Orellana, T.; Manzano, P. Molluscicidal activity of the aqueous extracts from *Solanum mammosum* L., *Sapindus saponaria* L. and *Jatropha curcas* L. against *Pomacea canaliculata*. *Emir. J. Food Agric.* **2014**, *26*, 871–877. [CrossRef]

30. Tangkoonboribun, R.S.S. Molluscicide from Tobacco Waste. *J. Agric. Sci.* **2009**, *1*, 76. [CrossRef]

31. Cowie, R.H.; Hayes, K.A. Invasive ampullariid snails: Taxonomic confusion and some preliminary resolution based on DNA sequences. In Proceedings of the APEC Symposium on the Management of the Golden Apple Snail, Pingtung, Taiwan, 6–11 September 2005.

32. Hayes, K.A.; Cowie, R.H.; Thiengo, S.C.; Strong, E.E. Comparing apples with apples: Clarifying the identities of two highly invasive Neotropical Ampullariidae (Caenogastropoda). *Zool. J. Linn. Soc.* **2012**, *166*, 723–753. [CrossRef]

33. Burela, S.; Martín, P.R. Evolutionary and functional significance of lengthy copulations in a promiscuous apple snail, *Pomacea canaliculata* (Caenogastropoda: Ampullariidae). *J. Molluscan Stud.* **2011**, *77*, 54–64. [CrossRef]

34. Seuffert, M.E.; Martín, P.R. Juvenile growth and survival of the apple snail *Pomacea canaliculata* (Caenogastropoda: Ampullariidae) reared at different constant temperatures. *SpringerPlus* **2013**, *2*, 312. [CrossRef] [PubMed]

35. World Health Organization. *Snail Control in Prevention of Bihariasis Monograph*; World Health Organization: Geneva, Switzerland, 1965; Volume 50, pp. 124–138.

36. Chauhan, S.; Singh, A. Impact of Taraxerol in combination with extract of Euphorbia tirucalli plant on biological parameters of *Lymnaea acuminata*. *Rev. Inst. Med. Trop. São Paulo* **2011**, *53*, 265–270. [CrossRef] [PubMed]

37. El-Din, A.S.; El-Sayed, K.; Mahmoud, M. Effect of ethanolic extract of Dalbergia sissoo plant parts on Biomphalaria alexandrina snail, the intermediate host of *Schistosoma mansoni*. Medical Malacology Laboratory. *J. Evolut. Biol. Res.* **2011**, *3*, 95–100.

38. Taguiling, N.K. Effect of Combined Plant Extracts on Golden Apple Snail (*Pomacea canaliculata* (Lam.)) and Giant Earthworm (*Pheretima* sp). *Int. J. Agric. Crop Sci.* **2015**, *8*, 55–60.

39. Dai, L.; Wang, W.; Dong, X.; Hu, R.; Nan, X. Molluscicidal activity of cardiac glycosides from Nerium indicum against *Pomacea canaliculata* and its implications for the mechanisms of toxicity. *Environ. Toxicol. Pharmacol.* **2011**, *32*, 226–232. [CrossRef] [PubMed]

40. World Health Organisation. Guidelines for evaluation of plant molluscicides. In *Phytolacca Dodecandra (Endod) Dublin*; Lemma, A., Heyneman, D., Silangwa, S., Eds.; Tycooly International Publishing Limited: Dublin, Ireland, 1983; pp. 121–124.

41. Yadav, R.; Singh, A. Combinations of binary and tertiary toxic effects of extracts of *Euphorbia pulcherima* latex powder with other plant derived molluscicides against freshwater vector snails. *Internet J. Toxicol.* **2008**, *7*, 2.

42. Olofintoye, L.K. Comparative evaluation of Molluscicidal effects of *Securidaca longepedunculata* (Fres.) and *Tephrosia bracteolata* (Guilland Perr) on *Bulinus globosus*. *J. Parasitol. Vector Biol.* **2010**, *2*, 44–47.

43. Sunita, K.; Kumar, P.; Singh, V.K.; Singh, D.K. In vitro phytotherapy of vector snails by binary combinations of larvicidal active components in effective control of fascioliasis. *Rev. Inst. Med. Trop. Sao Paulo* **2013**, *55*, 303–308. [CrossRef] [PubMed]

44. Rao, I.G.; Singh, D.K. Combinations of *Azadirachta indica* and *Cedrus deodara* oil with *piperonyl butoxide*, MGK-264 and *Embelia ribes* against *Lymnaea acuminata*. *Chemosphere* **2001**, *44*, 1691–1695. [CrossRef]

45. Dai, L.; Qian, X.; Nan, X.; Zhang, Y. Effect of cardiac glycosides from *Nerium indicum* on feeding rate, digestive enzymes activity and ultrastructural alterations of hepatopancreas in *Pomacea canaliculata*. *Environ. Toxicol. Pharmacol.* **2014**, *37*, 220–227. [CrossRef] [PubMed]

46. Okwute, S.K. Plants as Potential Sources of Pesticidal Agents: A Review. In *Pesticides—Advances in Chemical and Botanical Pesticides*; Soundararajan, R.P., Ed.; InTech: Rijeka, Croatia, 2012.

12

State Support in Brazil for a Local Turn to Food

Ana Paula Matei [1], Paul Swagemakers [2,*], Maria Dolores Dominguez Garcia [3], Leonardo Xavier da Silva [4], Flaminia Ventura [5] and Pierluigi Milone [5]

[1]   Secretaria de Desenvolvimento Tecnológico (SEDETEC/UFRGS), Porto Alegre 90040-020, Brazil; ana.matei@ufrgs.br
[2]   Department of Applied Economics, University of Vigo, Ourense 32004, Spain
[3]   Department of Applied Economics, Complutense University of Madrid, Pozuelo de Alarcón 28223, Spain; mariaddo@ucm.es
[4]   Departamento de Economia e Relações Internacionais, Universidade Federal Do Rio Grande Do Su, Porto Alegre 90040-000, Brazil; leonardo.xavier@ufrs.br
[5]   Dipartimento di Ingegneria Civile ed Ambientale (DICA, Department of Civil and Environmental Engineering), University of Perugia, Perugia 06125, Italy; flaminia.ventura@unipg.it (F.V.); pierluigi.milone@unipg.it (P.M.)
*   Correspondence: paul.swagemakers@uvigo.es

Academic Editors: Giaime Berti, Moya Kneafsey, Larry Lev, Irene Monaserolo and Sergio Schneider

Abstract: The local turn to food is often claimed to be a way to increase the value-added component retained by primary producers and to provide healthy, fresh and affordable food to consumers. Rio do Grande do Sul in Brazil has several governmental support programs that aim to empower family farmers and open up new market opportunities for them. This article examines these programs, investigates how small-scale farmers engage with them and the resultant changes in farming and marketing practices that ensue. The article uses cluster and content analysis to identify and interpret the extent, and the different ways, in which these farmers engage with and make use of the local knowledge and innovation system. The results provide useful insights into how policy instruments improve the performance of family agribusinesses, helping them to make better use of the resources available to them, encouraging farm diversification, and strengthening local interrelations between producers and consumers.

Keywords: farm diversity; public programs and policies; regional development; rural entrepreneurship

1. Introduction

A holistic understanding of the term 'local', and its inherent relevance for connectivity, raises the notion that food systems are not external to their users but that they are created in, and through, practice [1,2]. A local turn to food, therefore, implies increasing the use of locally available resources and relating this to farmers' on-farm diversification strategies, the development of new markets that connect producers and consumers with public and private sectors and knowledge institutions [3,4]. In such markets, 'producers and consumers are linked through specific networks and commonly shared frames of reference' [5] (p. 171). They also involve new social practices in which producers, production places, and consumers link (new) food qualities. These markets have been conceptualized as 'nested markets' characterized by a mix of connections between existing markets historically associated with the creation of new wealth and acceleration of development; there is the creation of new markets that are nested within wider (global) markets and the creation of new governance structures for both existing and new markets [5]. From this theoretical perspective, a local turn to production and consumption patterns should enhance the autonomy of farmers and local control over farming [6–8].

Transforming local raw materials into products, and connecting them to the supply of services is increasingly seen as a viable way for farmers to improve their incomes, quality of life and strengthen their autonomy. The adoption of such re-territorialized practices (based on endogenous or locally available resources) implies a local turn to food and contributes to sustainability as both producers and consumers are giving more consideration to 'the use of environment and resources to meet the needs of the present without compromising the ability of future generations to meet their own needs (...) maintaining the capacity of ecological systems to support social and economic systems' [9] (p. 2). Such approaches are often based on farmers' personal interests and agency [10,11] and are mostly adopted by farmers who want to create added value, optimize the use of local resources [12,13] and develop active new producer-consumer relationships [14–16].

As such, a local turn to food holds the promise of promoting sustainable territorial development. However, it also requires changes in the social organization of the knowledge and innovation system. In Brazil, there are a number of different public policies that promote family farming and create a supportive context for farmers' innovation, encouraging local sustainable development and entrepreneurship.

Following the proposition that a local turn to food can generate revenues for family farmers, but that farmers' business strategies are based on their own personal interests and agency, the central question that this article explores is the extent to which family farmers make use of a supportive institutional environment in order to become involved in (different) processes of innovation. As such, we evaluate the support measures that promote different processes of innovation (product innovation, process innovation, and marketing innovation) for a local turn to food. In so doing, this article draws on case study research from Rio do Grande do Sul, Brazil, that identifies and evaluates the implementation of policy programs aimed at improving farmers' incomes, providing consumers with fresh and healthy food and promoting new relationships between family farms and markets. The case study involved developing portraits of farmers (the 'beneficiaries' of the support measures) who have different interrelations with the knowledge and innovation system. The analysis combines local and regional levels, since this is where the 'interactions between the socio-economic fabric, institutions and the creation and diffusion of knowledge take place' [17] (p. 5). It is hoped that the results will inspire researchers, innovation brokers and policymakers to design and implement policy instruments that effectively contribute to the development of entrepreneurial skills, farm diversification and strengthen new (short) food chains. Together, these activities can help to sustain farming that is rooted in endogenous capital and generate sustainable economic growth. The next section presents the data collection and analysis, and introduces the case study area. This is then followed by the analysis of how farmers are differently involved in, and respond to, policy programs. Finally, we summarize the lessons learnt about how policy instruments can contribute to sustain farming and the provision of healthy, fresh and affordable food to consumers.

2. Materials and Methods

This article draws upon data derived from a larger research project that examined the potential of family farming to improve farmers' incomes, their quality of life and autonomy. It identifies the actual innovations adopted by 19 family farmers in Rio Grande do Sul and how they interact with a supportive institutional environment [18]. It describes the implementation of different policy programs and regulations intended to promote dynamism and innovation in family farming.

Institutional support for family farmers in Brazil is provided through a number of policy programs. The longest established of these is PRONAF (the National Program to Strengthen Family Farming) which, since the 1990s, has supported family farming. Other programs include, the Food Acquisition Program (PAA), the National School Nutrition Program (PNAE) and the Program for Family-based Agroindustry (PEAF-RS) (see Table 1 for details). These programs aim to sustain family farms and family agro-food businesses either nationally or within the region.

Table 1. State programs in Rio Grande do Sul.

The 'Programa Nacional de Fortalecimento da Agricultura Familiar' (PRONAF, National Program to Strengthen Family Farming) is an investment program, established in 1990, to support farmers' individual and collective projects to construct, expand or modernize their agricultural or non-agricultural production infrastructure and services. The program provides support for modernization and/or acquiring new machinery and equipment related to soil improvements, milk coolers, the genetic improvement of production factors (plants and animals), irrigation systems, orchards, greenhouses and storage as well as investments in tourism activities, such as craft workshops and cottages. The investment should strengthen the capital assets of family farmers and improve their standard of living.

The 'Programa de Aquisição de Alimentos' (PAA, Food Acquisition Program) is a federal program, established in 2003, that aims to reduce hunger and poverty in Brazil by strengthening family farming. The program promotes the creation of added value to fresh produce through on-farm processing of raw materials and organizes markets for these food products in the private sector and through public food procurement programs. Part of its remit involves building strategic food stocks and distributing these to socially vulnerable groups.

Since 1955, the 'Programa Nacional de Alimentação Escolar' (PNAE, National School Nutrition Program) has been acquiring food for school meals, usually from large suppliers. Since 1994, when the federal program was decentralized, there has been a move towards preferentially purchasing locally produced food. With a potential target group of 47 million students the further development of this local preference program has the potential to create new markets for small- scale farmers and to improve the nutritional quality of school meals.

The 'Programa de Agroindústria Familiar do Estado do Rio Grande do Sul' (PEAF-RS, Rio Grande do Sul's Program for Family-based Agroindustry, also known as Programa Sabor Gaúcho) has been operational since 1999. Firmly grounded in the principles of agroecology, it supports small-scale farmers (individuals and farmers' cooperatives) in the state by promoting diversification of farm activities (including on-farm processing of raw materials) and selling these farm products. Measures in the program include access to low interest credit rates and offering support to farmers engaged in the PAA and PNAE programs. It offers extension services on health and sanitary regulations on production and on-farm processing activities and promotes the delivery of end products to public entities and/or fairs and private market chains.

Despite the expansion of milk and beef production, and particularly soy production, a number of different diversification strategies have been developed in Rio Grande de Sul, including organic production and direct selling [19]. Yet, the *'colonos'* (colonists), *'posseiros'* (squatters), *'parceiros'* (partners), *'assentados'* (settlers), peasants and other small-scale farmers [20] are also in competition for public resources and social legitimacy with the larger players in food production. Family farmers (90% of the 435,000 farm units in the region of Rio Grande do Sul) [21], work an average of 6.1 ha per person, and are widely seen as having the potential to use productive resources more efficiently than large mono-cultural farms.

Policy programs are increasingly supporting these types of farms by stimulating the adoption of small-scale food processing technologies and improving access to local markets as these measures increase the value added at the farm level. Through providing knowledge and expertise about on-farm processing and developing linkages with local and regional markets, these policy programs aim to make farmers and family members more entrepreneurial [22]. At the family and community level, the changes consist of adapting their on-farm practices (e.g., adopting organic production methods) and the processing and selling of fresh produce through new supply chains (mostly characterized by more direct market relations) [23,24].

The research methodology is rooted in the tradition of performance story reporting [25] and grounded case study research [26], and combines quantitative and qualitative methods. It applies multivariate techniques of data analysis: cluster analysis (using SPSS statistical software package) and content analysis [27,28] from primary and secondary sources (interviews and document analysis).

The study focuses on 19 family agribusiness farms (or family-based agroindustries that participated in the policy programs set out in Table 1) within the Regional Councils of Development (Corede) in Rio Grande do Sul. These councils were created by Law 10.283, 17 October 1994, and form a "discussion and decision-making forum regarding policies and actions that aim at regional development" [18,24]. Locations where the study was implemented are Serra, Vale do Caí, and Vale do Rio Pardo (see Figure 1). Serra consists of 32 municipalities. In 2012, it had 878,500 inhabitants spread over an area of 6949 km^2 (125.1 habitants/km^2). The field research included 10 family-based agro-industries in Antônio Prado, Carlos Barbosa, Flores da Cunha, Garibaldi, Nova Roma do Sul and Santa Tereza. Vale do Caí has 19 municipalities. In 2012, it had 172,400 inhabitants spread over an area of 1854 km^2 (92 habitants/km^2). The field research included 5 family-based agro-industries in Bom Princípio, Harmonia, Montenegro, Pareci Novo e São. Vale do Rio Pardo consists of 23 municipalities. In 2012, it had a population of 65,946 people spread over an area of 773.24 km^2 (85.29 habitants/km^2). The field research included 4 family-based agro-industries, all located in Venâncio Aires. The farms in the sample share a similar socio-economic, environmental, political and cultural context, and were selected as farms contributing to the main objectives of the Corede: to promote regional, sustainable and harmonious development, improving the quality of life of the population, the equitable distribution of produced wealth, attracting new settlers to the region and preserving and restoring the environment [18,24]. These objectives are also consistent with the goals established by the State's policy programs.

Family agribusiness are defined according to Article 2 of State Law No. 13.921/2012 [29], as a family enterprise, managed individually or collectively, located in a rural or urban area, that aims to use and/or transform (either artisanally or not) on-farm raw materials, which can (the agribusinesses in our sample do) but do not necessarily, include geographical, historical-cultural, local or regional aspects. These agro enterprises can process locally produced raw materials derived from vegetables, livestock, fisheries, aquaculture, or forest fruits. These operations can range from simple processes to more complex ones, including physical, chemical and/or biological operations. The raw materials are processed into an end product in small-scale, artisanal production units operated directly by the farmer and/or family member(s) using their own means of production and labor input or through a partnership contract.

The main criterion for our selection of municipalities was a high number of family family-based agro-industries involved in the PEAF-RS program. The second criterion was proximity of farms to Porto Alegre, giving them relatively easy access to a large market, with low transportation costs. The family-based agro-industries mostly produced vegetables and drinks, and there were no farmers processing animal produce due to the difficulties of meeting sanitary regulations. A potential sample of 36 farms that met these criteria was identified, though time and money restrictions meant slimming down the sample to 21 farms of which two declined to be interviewed.

The research consisted of four phases. In the first phase, secondary data (including scientific books and articles, reports, regulations and historical data trends on the number of farms, agro-food businesses, and farming area) were gathered and analyzed. The second phase consisted of meetings with stakeholders, participation in seminars, fairs and events related to farmers, and interviews with key-informants. The third stage involved collecting primary, quantitative and qualitative data. This included the analysis of the cadaster, the collection of farm data of farms involved in the PEAF-RS, and the definition of criteria for selecting the cases to be included in the sample. During this stage, 19 semi-structured interviews were held with farmers, as well as 15 with representatives of organizations involved in local and regional innovation systems, such as the Rural Development Secretary (SDR), Emater (the rural extension service), and the Union of Rural Workers. These representatives were asked about their opinions on the development of the regional councils (Corede), their budgetary allocations and the State actions to promote family farming. These interviews with non-farming actors provided qualitative information on the programs and helped us identify the variables used for selecting a sample. The semi-structured interviews with farmers were done on the farms, recorded and later transcribed. The fourth phase consisted of data analysis and interpretation, which included qualitative (content analysis) and quantitative techniques (cluster analysis within-linkage groups, Jaccard measure). The objective of content analysis is to infer knowledge from a large amount of text (the transcripts of the interviews, documents), turning qualitative contents into indicators for further quantitative analysis [27]. Content analysis was carried out in three stages [27,28]: (1) pre-analysis; (2) exploration of the materials; and (3) data treatment, inferences and interpretation. The pre-analysis included the organization, compilation and preliminary evaluation of the primary data, organizing them in individual matrixes according to three dimensions: (1) the organizational and production characteristics of the family agribusiness (9 variables); (2) innovation processes (27 variables); and (3) interactions with the institutional environment (26 variables that were organized into two groups relating to how and why interactions took place). The exploratory analysis of the field materials resulted in the categorization of variables, on which further qualitative data evaluation was based. The data treatment involved applying a cluster analysis, which permitted us to group farms with similar innovation processes, and describe and interpret the results.

Nine variables were used for characterizing the family agribusinesses. These included the number of people involved in the farm, and the number of families involved in the production and processing of raw matter. Other variables included the type of production (vegetable, drink or both), the model of production (organic, integrated or conventional), the business model (mono-activity, pluri-activity) and the organizational structure (cooperative, family, association). Different governance structures (hierarchical or hybrid, based on the contractual relations among those involved) were also considered. These variables allowed us to make an initial qualitative analysis of the type and features of the 19 family agribusinesses.

Innovation processes were delineated according to 27 variables that covered three different types of innovation. First, for product innovation, variables on the use of new or diversified raw matter, the certification of raw materials, new end products or lines of products, new packaging or presentation of final products, and services (tourism, gastronomy) were used. Second, process innovation was framed by variables that included changes in the process of production of raw materials, new processing techniques, the construction of new facilities, the enlargement or adaption of the farm area, new equipment, new technologies, certified processing, waste treatment and the use of renewable energies. Market innovation included variables showing different ways of selling (direct from the farm, through fairs, in institutional markets, local markets and shops, street fairs, and e-selling); different sorts of communication (mouth-to-mouth, fairs and events, use of site, folders, journals, magazines, specialized catalogues, newspapers, radio programs, participation in contests, involvement in tourist routes); and the use of different types of certification (quality labels, organic, linked to family farming cooperatives, linked to specific programs, and institutional labels).

The third dimension, on institutional interactions, was defined by 26 variables. A group of 8 variables indicating how interactions took place (i.e., participation in training courses, technical and productive assistance, the use of consulting or other specialized services, participation in networks, associations and unions, the exchange of experiences and practice, the promotion in fairs, events, and participation in seminars). Another group of 18 variables delineated the aim of the interactions, that were either directly related to markets and marketing, or to technical issues (certification, compliance with different regulations, raw matter processing, etc.)

The primary data to run the cluster analysis originated from the answers given by farmers during the semi-structured interviews. These were transcribed and the data organized, processed and analyzed, first in Excel files, and then in SPSS. The answers were transformed in a binary code signaling the presence (1) or absence (0) of the variable. A hierarchical cluster was then run (using within-group linkage method, Jaccard measure) to identify groups of respondents with a similar behavior in terms of innovation processes. This method enabled us to group the presence of variables ($1 \times 1; 0 \times 1; 1 \times 0$), ignoring those where these variables were absent (0×0) [30]. These were then interpreted, allowing us to identify different portraits of innovation pathways, which were then analyzed to identify any patterns in farmers' institutional interactions.

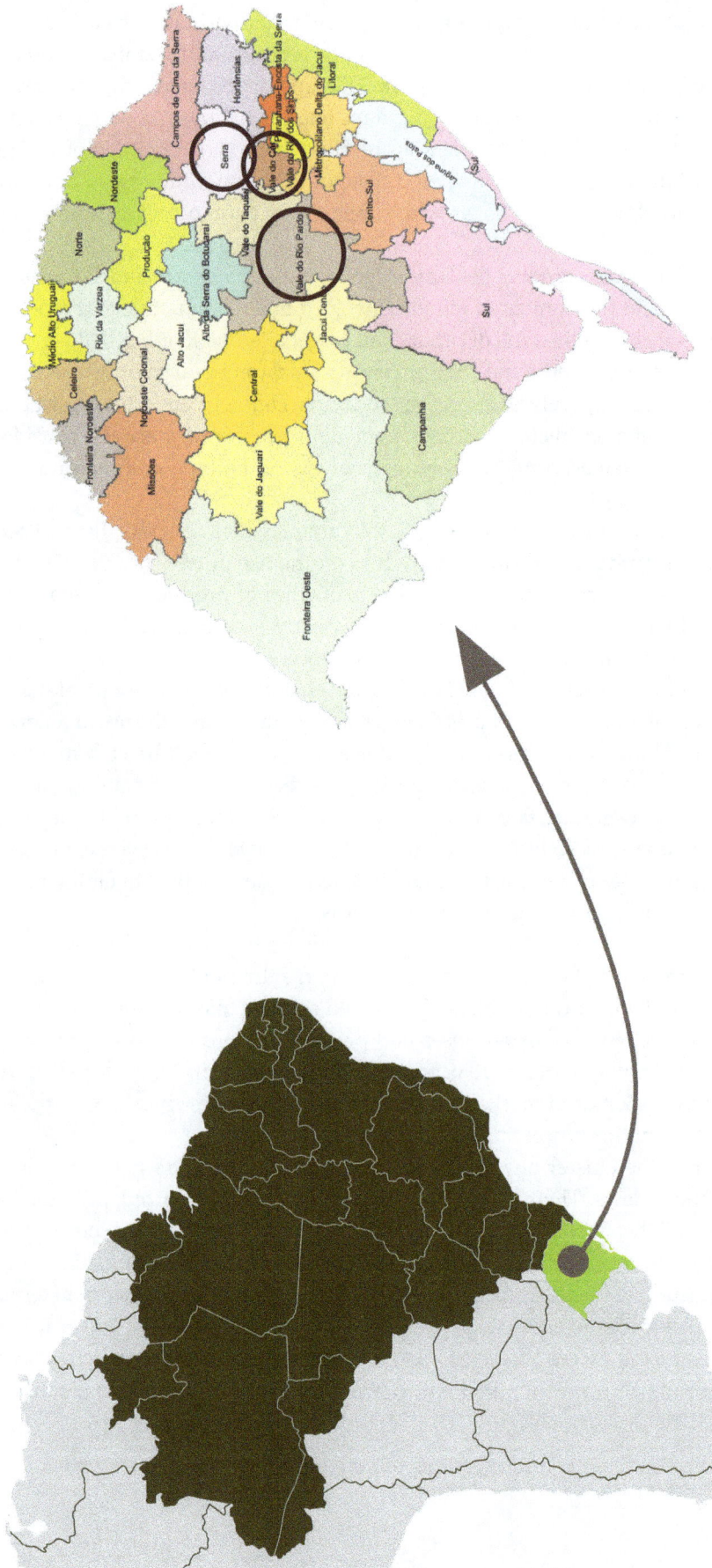

Figure 1. Rio Grande do Sul in Brazil and the location of the study areas.

A cluster can be considered as an exploratory, non-inferential, technique [31] and the results obtained should be strictly restricted to the specific sample and cannot be extrapolated or used to make more general inferences. The small sample size does not pose a problem as long as it is representative. In this case, the representativeness is assured by the researcher's contextual and conceptual knowledge, and by the definition and variables selected for the content analysis. At the end of the day, the clusters represent a consolidation of the qualitative information provided by the variables of the first dimension, the contextual knowledge of the researchers, and the results obtained by using descriptive statistics (frequency) to analyze the features and behavior of the farmers within dimensions 2 and 3.

The three 'portraits' of family-based agro-industries can be seen as prototypes within a typology [32] of a specific story of an imaginary farmer. As such, they are representations of the different practices and strategies pursued by the family farmers in the sample. To construct these portraits, we built upon a set of well-elaborated research methods that have been commonly used in farming styles analysis [33–35]. Our qualitative analysis of the interview materials focused on the policy programs to enable us to develop an understanding of the different ways in which family farmers, for practical reasons, respond differently to available public support measures.

3. Results and Discussion

The identification and definition of portraits (Section 3.1, below) is based on the different innovation processes and forms of interaction that farmers engaged in and captures the level of farmers' involvement in the supportive institutional environment framed by the various State programs.

The departure point for outlining those portraits is the grouping obtained by the cluster analysis that considered the variables related to different processes of innovation (product, processing, and commercial). By adding a description of state facilitation embodied in the different measures within the different policy programs, we can understand better how farmers within each portrait have been able to unleash their capacity for innovation. For each portrait, we describe how the programs encouraged entrepreneurship, and how farmers have benefitted from, and prioritized, elements from within these programs. Quotes from the respondents (the abbreviation AF is an abbreviation of family agribusiness) provide illustrations of, and evidence for, this. Finally, we show the qualitative impact of the policy programs within each portrait in providing financial, institutional, technical, operational, management and marketing support (Section 3.2).

3.1. Portraits: Innovation Proccesses and Institutional Interaction

The first result of the cluster analysis, run with the variables that show different innovation processes, enabled us to distinguish three portraits within the 19 family agribusinesses. Table 2 shows the frequency of the different types of innovation within every portrait. Next, Table 3 illustrates each portrait's level of participation in the different State programs. This shows the level of institutional interaction, i.e., the frequency with which farms within each portrait made use of the different measures within the four State programs. This helps us better define and understand each portrait.

The variables for institutional interaction are based on 16 measures contained in these various State programs (in the original work on which this article is based on, this set of variable was wider and included other more specific goals). These variables are: (1) providing training support to acquire specific knowledge about practice and/or management; (2) providing specialized consulting services, free of charge; (3) providing financial and informational support for promoting and increasing the market profile of small scale agribusiness products (i.e., assisting them to participate in fairs, events, workshops, etc.); (4) providing (subsidized) consulting and specific technical services that reduce financing costs; (5) policy articulation; (6) providing support for participating in cooperatives and non-profit associations; (7) financing access to loans and credit (8) providing management training; (9) providing financial support for improving old or constructing new physical infrastructure; (10) providing administrative support (information) to access environmental licenses; (11) facilitating market access (product promotion, access to new markets and commercialization circuits); (12) providing technical and technological support for product development and improvement (possibly including laboratory tests); (13) providing technical information on specific sectors of production; (14) providing technical support on raw matter production and/or processing (Raw Matter P & P); (15) providing technical support to comply with sanitary controls; and (16) providing support for strengthening the image of products by stressing their quality and/or product differentiation.

Table 2. Portraits of family agribusinesses based on different types of innovation. Frequencies (%) of presence (Sample 19 Brazilian farms, 27 variables of innovation).

Types of Innovation	Portrait 1 (14)	Portrait 2 (5)	Portrait 3 (1)
Product innovation			
New or diversified raw matter	46%	80%	0%
Certified raw matter production	62%	20%	0%
New final products or line of products	100%	60%	100%
New packaging or presentation of final products	77%	40%	100%
Services (tourism, gastronomy)	69%	0%	0%
Process innovation			
Changes in the process of producing raw materials	85%	60%	0%
New process of production (processing)	92%	100%	100%
New construction of facilities	54%	0%	100%
Enlargement or adaption of the farming area	77%	100%	0%
New equipment	85%	80%	100%
New technologies	54%	0%	100%
Certified processing	46%	0%	0%
Waste treatment, use of renewable energies	69%	20%	0%
Market innovation			
Direct selling from the farm	79%	75%	0%
Selling at street fairs	71%	75%	0%
Selling to institutional markets	71%	0%	0%
Selling in local markets and shops	93%	100%	100%
Internet sales	14%	100%	0%
Mouth-to-mouth communication	93%	50%	0%
Communication in fairs and events	71%	75%	0%
Use of a web site, folders, journals, catalogues, newspapers, etc.	64%	75%	0%
Communication and participation in contests, prizes	29%	75%	0%
Communication and involvement in tourist routes	50%	25%	0%
Organic certification	64%	0%	0%
Certification link to family farming cooperatives	79%	25%	0%
Certification linked to programs (e.g., Sabor Gaúcho)	100%	50%	100%
Institutional certification (INPI)	36%	25%	0%

Table 3. Participation of family agribusiness (by portrait = P) in the different State programs as beneficiaries of different measures.

Programs/Portraits	PEAF-RS			PRONAF			PAA		PNAE	
Interactions	P 1	P 2	P 3	P 1	P 2	P 3	P 1	P 2	P 1	P 2
Training (courses, seminars)	4	1	0	0	0	0	0	0	0	0
Free consulting	3	0	0	0	0	0	0	0	1	0
Participation in fairs/events	12	2	0	0	0	0	0	0	0	0
Subsidized consulting	1	0	0	12	5	1	0	0	0	0
Policy articulation	1	0	0	0	0	0	0	0	0	0
Promotion of cooperatives	2	0	0	0	0	0	2	2	5	3
Financing Loans and Credits	7	0	0	12	5	1	0	0	0	0
Manag training	3	1	0	0	0	0	0	0	0	0
Physical infrastruc	0	0	0	11	4	1	0	0	0	0
Envir licences	5	1	0	0	0	0	0	0	0	0
Market and commerc	13	2	1	0	0	0	3	2	9	3
Product development	0	0	0	0	0	0	0	0	1	0
Tech support on sector	1	0	0	0	0	0	0	0	0	0
Tech Raw Matter P & P	2	0	0	12	5	0	0	0	0	0
Tec. sanitary control	1	1	0	0	0	0	0	0	0	0
Image strength	13	5	1	0	0	0	3	2	9	3

From the results shown in Table 2, we can say that, in general terms, Portrait 1 is very dynamic regarding all types of innovation (product, processing, and market), but is mostly engaged in product innovation. In comparison with the other two portraits, farms within this group engaged more in developing new products or lines of final products (100%). This was linked to a high level of process innovation in terms of raw matter production, both in changing from conventional to organic (85%) and market innovation (69% organic certification). This group is typified by an interest in meeting new societal demands, such as organic produce, but also by providing services beyond traditional agricultural production (e.g., tourism, gastronomy) (69%). Their participation in tourist routes (54%) is also indicative of a diversification in their activities. This group is also more inclined to opt for more sustainable approach to farming than the other groups with 69% choosing to use renewable energy and take over waste treatment (on farm compost production) (69%) with the aim of reducing costs. This group is also quite dynamic in terms of market innovation, preferring word of mouth communication and short distribution circuits especially small local shops and supermarkets (92%) and street fairs (71%). They are also more engaged in seeking recognition for their products and participate more than the other two groups in contests and prizes (46%). They also participate more in institutional markets than the other two groups (62% participation rate). In addition to organic certification, they also engage in other quality certification schemes (69%) and other types of brand protection (62%). In summary, this group represents an organic and diversified farming typology.

Farms in Portrait 2 are more focused on processing and marketing innovations than on product innovation. In product innovation, they pay more attention to new raw materials (80%), and less to product diversification (60%). Their only interest in 'moving beyond' farming is to engage in processing (100%). All members of this group invested in farming infrastructure, enlarging and/or adapting their farm area and most (80%) combined this with investments in new equipment (80%). While this group is involved in developing new processing facilities, they do not pay much attention to seeking organic certification although many (80%) participate in other quality label schemes such as Sabor Gaucho. They all sell their produce via small local shops and supermarkets and 75% sell directly from their farms, again showing a clear preference for short commercialization circuits. We consider them as a type of farmer that is looking to enhance product quality, but not wishing to convert to organic production. These farmers are aware of the opportunities of communicating the virtues of their produce to a broad range of consumers: all of them use a website, brochures and journals to promote themselves and 60% are engaged in e-marketing to consumers. This portrait then consists of conventional farmers who are aware of and profit from the opportunities of multifunctional farming, and can be described as a portrait of, conventional farming of quality products that are distributed through local short, circuits.

Portrait 3 only contains one farm, which makes difficult to justify any sort of representation in statistical terms. However, we have kept it as a separate typology because of its specificity: it is the only one out of the 19 family agribusinesses to adopt a totally new line of product and processing (including freezing the end-product) and therefore a totally different line of commercialization. The characteristics of this new activity required the farmer to establish a new business and to build new processing facilities. Apart from this, this farm is not interested in making other market innovations, except for the necessary use of new packaging and the presentation of the final product. This shows a rather different strategy than the other two profiles, which involves paying less attention to market segmentation (in terms of quality and demand) although this farm also only sells through local shops and markets. Thus, this farm represents a portrait in which the goal is to look for product differentiation based on a specific technological innovation and product processing and packaging rather than quality (however defined).

In the next section, we supplement the information provided by Tables 2 and 3 with some statements from farmers made during the semi-structured interviews, in which they directly expressed their motivations and their opinions about the impact of the programs on their businesses. This allows us to build a more nuanced picture of each portrait and shows how the different programs have helped the farmers' expand and diversify their activities.

3.1.1. Portrait 1: Farming Organically and Activity Diversification

This portrait represents farm-based agro-industries that combine organic production with valorizing product qualities through participation in nested markets created by the policy programs. The 13 respondents include 11 families that produce organically, 12 that are involved with associations and 1 that belongs to a cooperative. They mostly produce drinks (juice and wine) and vegetables. The families are located in Serra (9 families), and Vale do Caí (4 families).

All farms in this group participate in the PEAF-RS program, 12 out of the 13 business initiatives in this group also receive support from PRONAF and 9 from the PNAE, but only 3 from the PAA.

More than two-thirds have some form of product certification often through the Rio Grande do Sul Organic Program, linked to PEAF-RS, which supports the commercialization of organic produce. They have made this transition in response to new demands for healthier products, which has led them to actively engage in product innovation. While they have benefited from public programs, they have also had to confront some problems, such as adapting to bureaucracy and complying with organic standards. Sometimes they have also had to make an effort to explain the value added of their products—in terms of nutritional and or environmental aspects, which also implies often a higher price.

> *There was an increase in demand for organic products and sales of the products from the PNAE. Our farm income improved and the PNAE contributed around 30% of the cost. However, the contracts we have with some public sectors have some requirements that we have difficulty meeting. They are not well adapted to the realities facing family farmers. The public tenders are sometimes quite a long way from the particularities at our farm and it is difficult to adapt to the application forms they send us. We always try to contact them in order to present and explain the nutritional values of our products so they are aware of their value and quality.* (Cooperative member AF13)

Farmers in this group are also quite engaged in processing innovations, especially making changes to raw material production and introducing processing. Most have applied adaptation strategies that improve their use of locally available resources and have adopted new, innovative, processing methods that involved installing new equipment. Most received assistance from PRONAF to make these changes, which alongside PEAF-RS, contributed to the investments by making low interest loans available.

> *I could only start my business after getting a loan of R$ 10,000 (around 3000 Euro) through one of the funds of PEAF-RS.* (AF06)

> *I received R$ 40,000 (around 10,000 Euro) to buy new equipment through PRONAF to improve my agro-industrial processes. It was very important to transform my business.* (AF18)

> *Our agribusiness started after getting a loan through PRONAF and with (technical) support from Emater. In the beginning, our aim was to plant a new vineyard but when we realized the difficulties of selling the grapes we decided to go into processing instead and so the agro-business evolved.* (AF17)

Certification is an important support measure, whether it is for organic standards, the Family Support Label coordinated by PRONAF or the Sabor Gaucho label created by PEAF-RS. This group's products are mostly sold in local markets and small shops, directly from the farm, and at street fairs and are promoted through personal contacts and at fairs. State programs, such as PEAF-RS, have promoted market innovation through direct selling, establishing small markets and fairs and promoting existing ones. PNAE has provided new market opportunities for selling to schools and public cantinas. While this opportunity is mostly welcome, some farmers also pointed out the problems associated with their dependency on public contracts and public calls.

I started my activity with the support of the Centro Ecológico de Ipê (Ipe Organic Center). I always liked to produce herbal remedies, and used to do so and sell them directly in fairs (green and undried), but I was losing money. With the support of the Ipe and Emater I introduced more varieties and PEAF-RS's support allowed me to improve my commercialization which made my business profitable, allowing me to continue farming and to stay in the countryside. (AF10)

We sell our products to a large number of municipally-run schools (there are 340 in the RS). PNAE has had a strong impact in RS but PNAE also depends on the public calls and this creates uncertainty. (AF13)

Produce from farms in this group is mainly sold in local markets and shops and through the internet but also directly from the farm or at fairs. PEAF-RS has played an active role in supporting and facilitating access to specialized fairs and events and promoted participation in competitions for quality wines, juices and cachaças (a Brazilian spirit), while PAA and the PNAE have provided support for market development and commercialization.

(...) We used to produce wine and sell it to another enterprise that bottled it, so we lost part of the benefit. We decided to sell the wine directly. PRONAF and PNAE helped us to access new markets and PEAF-RS helped us to attend relevant events, such as fairs and to fulfill the requirements for certification. (AF08)

Wine production is a tradition in our family, but we sold through an intermediary. About 10 years ago, we decided to take control over the whole process of commercialization (...) and in 2005 we started to bottle our own wine brand, creating two new high-quality brands. PRONAF and PEAF-RS were of vital importance in helping us to access new markets and new market circuits and to differentiate ourselves. (AF07)

3.1.2. Portrait 2: Conventional Farming of Quality Products and Distribution in Local Short Circuit Channels

This portrait represents farm-based agro-industries that are involved in conventional, but quality, production. Four out of the five respondents that fit this portrait produce vegetables (ranging from seeds, to horticulture and avocadoes, some of which are processed into oil), and one produces drinks (wine and grape juice). The initiatives are located in Serra (1 initiative), and Vale do Rio Pardo (4 initiatives).

All of them have received support from the PEAF-RS and PRONAF programs, and 3 from the PAA and PNAE programs. The main features of their innovations, which received policy support from state programs, are new processing and marketing processes. The support mostly took the form of specific consulting services (support for identifying solutions and recommendations related to the specific demands of the farmers, including financial support for providing these services), technical support to change and/or introduce production processes (PRONAF) and financial support for investments in infrastructure.

The production of avocado oil required a large investment to change our production process and making changes to our infrastructure. The support of State programs like PRONAF for consulting and investments was required to overcome the instability of production processes, to adapt our production cycles and to be able to process without making a financial loss. (AF05)

Farmers within this portrait stressed the support of programs such as PEAF-RS and PRONAF in supplying training to support them with their new processes of production and the adaptation of their business as well as the provision of financing.

(...) I had to follow some training to improve management, especially to reduce some costs because I need to enlarge and adapt my farm and invest in new equipment. (...) PEAF-RS and PRONAF were a big help. (AF02)

In terms of marketing innovation, these farmers mostly sell their products directly from their farms, or via small local shops and supermarkets, showing a clear preference for short commercialization circuits (that reduce the involvement of intermediaries and means that the produce is sold close to the locus of production). Farmers in this group have pursued these commercialization strategies largely without calling on state agencies' assistance. However, some have made use of the publicly supported schemes for image labeling, such as Sabor Gaucho and INPI (national industrial property).

Although this group is less oriented to product innovation than the first group, product innovation still plays a significant role, with many (figures) producing new raw materials and diversifying their primary production, indicating a growing awareness about quality and consumer demands.

> *I produce 90% of my raw material. It is a priority to me to control the quality of my product and eventually to move from conventional to organic production.* (AF04)

> *Although the production of wine is a tradition in my family, we did not care about the variety of grapes (…) we used to buy from other farmers to complete our production. But we realized that customers did not value the quality of our wine. So we decided to follow some training in enology and asked for technical support to improve our wine, grape juice, looking for better (grape) varieties. We also got training to improving the management and administration of our business, in order to also to improve our product commercialization. The PRONAF program played a supportive role in all.* (AF11)

3.1.3. Portrait 3: Technological Innovation in Product Conservation and Packaging

The main characteristic that distinguishes this portrait from the others is the focus on product conservation and the packaging and presentation of products. This helps increase the value added and allows the products to find their way to more distant consumers in more 'anonymous' consumption circuits. This portrait contains just one farm (AF19), producing yucca. The farm radically reoriented its production processes by installing new processing facilities and equipment. PRONAF supported the investment in a new refrigeration plant, while PEAF-RS supported the certification of the product under Sabor Gaúcho, which helped with market access.

> *Thanks to the PEAF-RS and the Sabor Gaucho certification I could enter new markets. Although I mostly sell to big retailers, I also sell my product directly to a local supermarket network.* (AF19)

We have maintained this sole example as a portrait in its own right as it shows a different development trajectory based around innovation in processing and marketing that could be followed by others.

3.2. Policy Programs, the Provision of Opportunity for Innovation

The qualitative analysis, based on the institutional performance through the implementation of the programs, and interviews with key actors close to those programs, allows us to evaluate the ways in which these programs encourage family farm innovation.

Institutional recognition of the significance of family farming started in Brazil in the 1990s. Social movements played a large part in this, but State regulation (starting with PRONAF) was also an important driving force for a turn towards supporting small-scale, and more sustainable, family farms [18].

Figure 2 illustrates that public support programs working at different levels (State and Federal) that are designed to improve infrastructure, markets and market access, sanitary control and management. These various measures enable family farmers to deliver fresh surplus produce (mainly fruits and vegetables) to markets with different target groups, and create opportunities for them to add value to their primary production, increase their incomes and to foster socio-economic development at the local, regional and state levels.

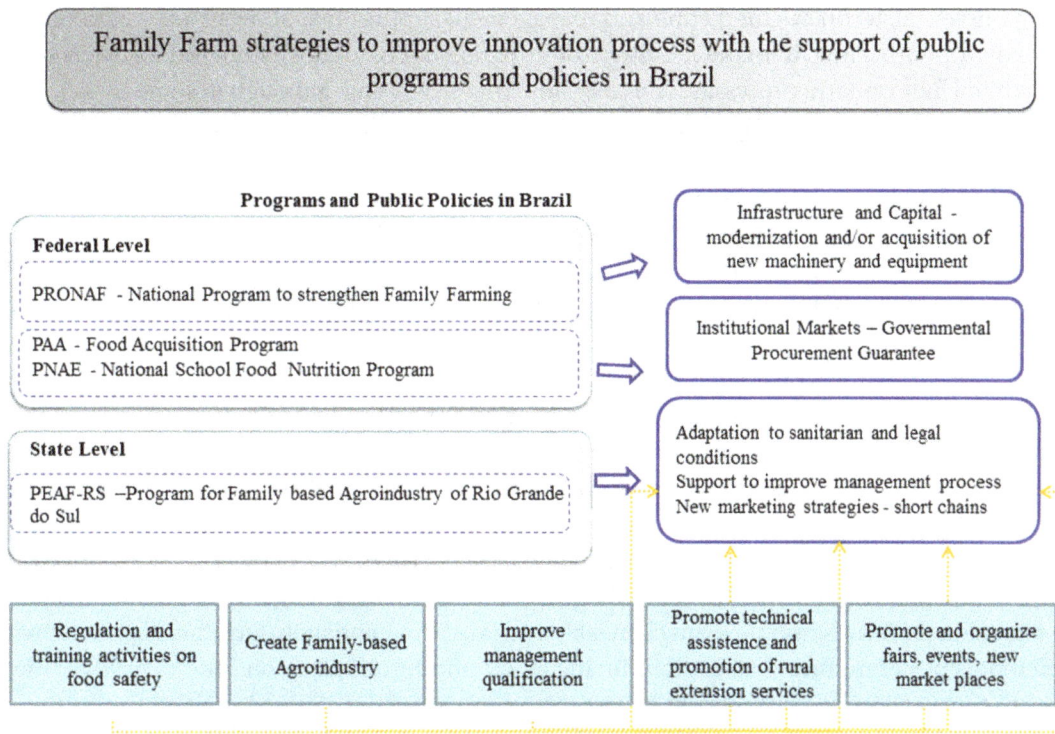

Figure 2. State programs: Support for small-scale farmers and public food procurement.

Some of these programs provide farmers with financial support to develop their infrastructure and or make capital investments in modern equipment. Others (especially the PAA and the PNAE) facilitate access to markets either by providing quality guarantees or opening up new public sector markets (school meals, etc.) At the state level, PEAF-RS helps farmers to comply with legal and sanitary requirements through educational programs, management support and training, and also opens access to new markets (fairs and food-related events) and new (short) circuits. While there are other programs and contributions from private and public institutions, these four programs played a central role in promoting and enabling innovations at the farm level among our sample.

From the farmers' responses on their participation in public programs (Table 3), we observe that the PEAF-RS and PRONAF programs provide more support to promoting innovation than the PNAE and the PAA. Counting the times that the agribusinesses in our sample made use of each different measure (their institutional interactions), we see that market innovations were the most commonly utilized: specifically in the domains of image strengthening, market and commercialization, and enabling participation in special events and food fairs. Support for processing innovation was also significant, mainly through technical raw matter production and processing. The provision of finance, on favorable terms, to enable innovation (of all types) has also been highly significant. Farms within Portrait 1 most frequently interacted with the PEAR-RS program, which they used to help them promote the image of their produce, to access markets and other aspects of commercialization, such as enabling them to participate in specific events and fairs. PRONAF was also widely used, though more by farmers in Portrait 2, who found its subsidies for consulting services, provision of cheaper loans and credits and technical support over raw material production, processing and physical infrastructure, particularly useful.

4. Conclusions

This case study shows how supportive governmental frameworks in Rio Grande do Sul have encouraged and motivated farmers to engage in innovation processes so as to improve the family agribusiness. State programs, such as PEAF-RS, PRONAF, PNAE, and PAA, have played a key role in providing financial, technical, marketing and management support that allows family agribusinesses to find new opportunities to make a living from the farm. These mostly consist of extending the business to include on-farm processing activities and local marketing, although in some cases they also involve moving 'beyond farming'. Both strategies enhance farm livelihoods and strengthen the rural socio-economy. Farmers and society at large both benefit: consumers from access to healthy, fresh and affordable food, politicians through prestige and, technicians who are usefully employed in supplying services to farmers and processing cooperatives. The local food chains that emerge help producers valorize their products. At the same time, the programs help producers improve and develop their knowledge and skills so they are able to produce and process final goods from raw materials which are locally available.

The State programs considered in this article have promoted family agribusiness dynamics by facilitating access to technical knowledge in production and processing, in the possibilities and opportunities of product diversification and differentiation, and have helped farmers to access new markets. They also have reduced investment risks: investing in capital assets (such as constructing a processing plant) is costly and many farmers might otherwise consider it to be too risky. The financial and marketing support these programs offer make these decisions less risky and costly and ease the pressure of paying back the investments, as these are amortized over a longer time period than normal, thus stabilizing the farm's cash flow. Investments in product differentiation, facilitated by the provision of different types of quality certification, financial aid and information on how to access different sources of communication, have helped farmers to communicate key messages about product quality, which is often associated with place of origin. The support provided by the programs allows farmers to improve the balance between the resources available to them and market opportunities.

Furthermore, in interacting with various support agencies, members, and associations (promoted by support measures to cooperatives and creation of associations) farmers increasingly shared their knowledge with, and supported, each other. This has helped them to make better decisions and conduct their farm businesses from a more informed position. Participating in these support networks and making more contact with neighbors, friends, experts, technicians and politicians leads farmers to become more socially engaged whilst making their family-based agro-industry more visible.

Acknowledgments: This article results from collaboration with the University of Perugia for the 'Plan Galego de Investigación, innovación e crecemento 2011–2015' of the Xunta de Galicia, financial support provided by the project 'POS-B/2016/028' of the Xunta de Galicia, and case study research by Ana Paula Matei, bolsista da CAPES—Proc. No. BEX 8103/13-5, Servidora Pública Federal, Secretaria de Desenvolvimento Tecnológico e Doutora em Desenvolvimento Rural pela Universidade Federal do Rio Grande do Sul—SEDETEC-PGDR-UFRGS. We are grateful to all who commented on earlier versions of this article. Our special thanks go to the anonymous reviewers who encouraged us to improve this article and for their useful suggestions. Many thanks go to Nicholas Parrott (TextualHealing.eu) for English language editing and editorial advice. The responsibility for the views and the argumentation provided in this article remain the authors.

Author Contributions: Ana Paula Matei performed the case study research and cluster analysis; Leonardo Xavier da Silva, Flaminia Ventura and Pierluigi Milone supervised the case study research and content and cluster analysis; Paul Swagemakers made suggestions on how to frame the case study materials and helped draw out the conclusions; M. Dolores Dominguez Garcia and Paul Swagemakers developed the materials and methods and results and discussion sections in collaboration with Ana Paula Matei; M. Dolores Dominguez Garcia and Paul Swagemakers wrote the final draft of the article.

Conflicts of Interest: The authors declare no conflict of interest.

References

1. Berkes, F.; Colding, J.; Folke, C. Introduction. In *Navigating Social-Ecological Systems: Building Resilience for Complexity and Change*; Berkes, F., Colding, J., Folke, C., Eds.; Cambridge University Press: Cambridge, UK, 2003; pp. 1–20.

2. Moore, M.L.; Westley, F. Surmountable chasms: Networks and social innovation for resilient systems. *Ecol. Soc.* **2011**, *16*, 5. [CrossRef]

3. Wilson, G.A. From 'weak' to 'strong' multifunctionality: Conceptualising farm-level multifunctional transitional pathways. *J. Rural Stud.* **2008**, *24*, 367–383. [CrossRef]
4. Marsden, T.; Smith, E. Ecological entrepreneurship: Sustainable development in local communities through quality food production and local branding. *Geoforum* **2005**, *36*, 440–451. [CrossRef]
5. Darnhofer, I. Strategies of family farms to strengthen resilience. *Environ. Policy Gov.* **2010**, *20*, 212–222. [CrossRef]
6. Sevilla Guzmán, E.; Martínez Alier, J. New rural social movements and agroecology. In *Handbook of Rural Studies*; Cloke, P., Marsden, T., Mooney, P.H., Eds.; Sage: London, UK, 2006; pp. 472–483.
7. Holloway, L.; Kneafsy, M.; Venn, L.; Cox, R.; Dowler, E.; Tuomainen, H. Possible food economies: A methodological framework for exploring food production-consumption relationships. *Sociol. Rural.* **2007**, *47*, 1–19. [CrossRef]
8. Brunori, G.; Malandrin, V.; Rossi, A. Trade-off or convergence? The role of food security in the evolution of food discourse in Italy. *J. Rural Stud.* **2013**, *29*, 19–29. [CrossRef]
9. Mahon, M.; Fahy, F.; Cinneide, M.O. The significance of quality of life and sustainability at the urban-rural fringe in the making of place-based community. *GeoJournal* **2012**, *77*, 265–278. [CrossRef]
10. Jongerden, J.; Swagemakers, P.; Barthel, S. Connective storylines: A relational approach to initiatives in food provisioning and green infrastructures. *Span. J. Rural Dev.* **2014**, *5*, 7–18. [CrossRef]
11. Kneafsy, M. Tourism, place identities and social relations in the European rural periphery. *Urban Reg. Stud.* **2000**, *7*, 35–50. [CrossRef]
12. Horlings, L.G. Place branding by building coalitions; lessons from rural-urban regions in the Netherlands. *Place Brand. Public Dipl.* **2012**, *8*, 295–309. [CrossRef]
13. Van der Ploeg, J.D. The food crisis, industrialized farming and the imperial regime. *J. Agrar. Chang.* **2010**, *10*, 98–106. [CrossRef]
14. Marsden, T. *The Condition of Rural Sustainability*; Van Gorcum: Assen, The Netherlands, 2003.
15. McDonagh, J.; Nienaber, B.; Woods, M. *Globalization and Europe's Rural Regions*; Ashgate: Aldershot, UK, 2015.
16. Van der Ploeg, J.D. *The New Peasantries: Struggles for Autonomy and Sustainability in an Era of Empire and Globalization*; Earthscan: London, UK, 2008.
17. Rodríguez-Pose, A. The rise of the "city-region" concept and its development policy implications. *Eur. Plan. Stud.* **2008**, *16*, 1025–1046. [CrossRef]
18. Matei, A.P. Innovation Processes and Interactions of Family-Based Agroindustries in Brazilian and Italian Regions. Ph.D. Thesis, Universidade Federal do Rio Grande do Sul (UFRGS), Porto Alegre, Brazil, 2015.
19. Vander Vennet, B.; Schneider, S.; Dessein, J. Different farming styles behind the homogenous soy production in southern Brazil. *J. Peasant Stud.* **2016**, *43*, 396–418. [CrossRef]
20. Schneider, S.; Niederle, P.A. Resistance strategies and diversification of rural livelihoods: The construction of autonomy among Brazilian family farmers. *J. Peasant Stud.* **2010**, *37*, 379–405. [CrossRef]
21. Schneider, S. (Universidade Federal do Rio Grande do Sul (UFRGS), Porto Alegre, Rio Grande do Sul, Brazil). Personal communication, 2015.
22. Schneider, S.; Gazolla, M. Seeds and Sprouts of Rural Development: Innovations and Nested Markets in Small Scale on-Farm Processing by Family Farmers in South Brazil. In *Rural Development and the Construction of New Markets*; Hebinck, P., van der Ploeg, J.D., Schneider, S., Eds.; Routledge: London, UK, 2015; pp. 127–156.
23. Oliveira, D.; Gazolla, M.; Schneider, S. Producing novelties in family farming: Adding value and sustainable agriculture for rural development. *Cad. Ciênc. Tecnol.* **2011**, *28*, 17–49.
24. Gazolla, M. Knowledge, Novelty Production and Institutional Actions: Short Chains of Family Agroindustries. Ph.D. Thesis, Universidade Federal do Rio Grande do Sul (UFRGS), Porto Alegre, Brazil, 2012.
25. Vanclay, F. The potential application of qualitative evaluation methods in European regional development: Reflections on the use of performance story reporting in Australian natural resource management. *Reg. Stud.* **2015**, *49*, 1326–1339. [CrossRef]
26. Yin, R.K. *Case Study Research: Design and Methods*; Bookman: Porto Alegre, Brazil, 2001.
27. Bardin, L. *Análise de Conteúdo*; Edições 70: Lisboa, Portugal, 2006.
28. Martins, G.A. *Estudo de Caso: Uma Estratégia de Pesquisa*; Atlas: São Paulo, Brazil, 2008.
29. Estado Do Rio Grande Do Sul Assembleia Legislative. Available online: http://www.sdr.rs.gov.br/upload/20170111116231313.921_institui_a_pol_utica_estadual_de_agroind__stria_familiar.pdf (accessed on 13 January 2017).

30. Everitt, B.S.; Landau, S.; Leese, M.; Stahl, D. *Cluster Analysis*, 5th ed.; King's College London: London, UK, 2011.

31. Hair, J.F.; Black, W.C.; Babin, B.J.; Anderson, R.E.; Tatham, R.L. *Análise Multivariada de Dados*, 6th ed.; Tradução Adonai Schlup Sant'Anna; Porto Alegre Bookman: Porto Alegre, Brazil, 2009.

32. Wiskerke, J.S.C. Arable Farming in Zeeland between Change and Continuity: A Sociological Study of Diversity in Farm Practices, Technological Development, and Renewing the Countryside. Ph.D. Thesis, Wageningen Agricultural University, Wageningen, The Netherlands, 1997.

33. Van der Ploeg, J.D. Styles of Farming: An Introductory Note on Concepts and Methodology. In *Born from Within: Practice and Perspectives of Endogenous Rural Development*; van der Ploeg, J.D., Long, A., Eds.; Van Gorcum: Assen, The Netherlands, 1994; pp. 7–30.

34. Domínguez García, M.D. The Way You Do It Matters: A Case Study: Farming Economically in Galician Dairy Agroecosystems in the Context of a Co-Operative. Ph.D. Thesis, Wageningen University, Wageningen, The Netherlands, 2007.

35. Van der Ploeg, J.D.; Ventura, F. Heterogeneity reconsidered. *Curr. Opin. Environ. Sustain.* **2014**, *8*, 23–28. [CrossRef]

13

Practices for Reducing Greenhouse Gas Emissions from Rice Production in Northeast Thailand

Noppol Arunrat [1,2] and Nathsuda Pumijumnong [2,*]

[1] Laboratory of Soil Science, Graduate School of Agriculture, Hokkaido University, Kita 9 Nishi 9, Kita-ku, Sapporo 060-8589, Japan; n_noppol@hotmail.com
[2] Faculty of Environment and Resource Studies, Mahidol University, Salaya, Phutthamonthon, Nakhon Pathom 73170, Thailand
* Correspondence: nathsuda.pum@mahidol.ac.th or nathsuda@gmail.com

Academic Editor: Ryusuke Hatano

Abstract: Land management practices for rice productivity and carbon storage have been a key focus of research leading to opportunities for substantial greenhouse gas (GHG) mitigation. The effects of land management practices on global warming potential (GWP) and greenhouse gas intensity (GHGI) from rice production within the farm gate were investigated. For the 13 study sites, soil samples were collected by the Land Development Department in 2004. In 2014, at these same sites, soil samples were collected again to estimate the soil organic carbon sequestration rate (SOCSR) from 2004 to 2014. Surveys were conducted at each sampling site to record the rice yield and management practices. The carbon dioxide (CO_2), methane (CH_4), and nitrous oxide (N_2O) emissions, Net GWP, and GHGI associated with the management practices were calculated. Mean rice yield and SOCSR were 3307 kg·ha^{-1}·year^{-1} and 1173 kg·C·ha^{-1}·year^{-1}, respectively. The net GWP varied across sites, from 819 to 5170 kg·CO_2eq·ha^{-1}·year^{-1}, with an average value of 3090 kg·CO_2eq·ha^{-1}·year^{-1}. GHGI ranged from 0.31 to 1.68 kg·CO_2eq·kg^{-1} yield, with an average value of 0.97 kg·CO_2eq·kg^{-1} yield. Our findings revealed that the amount of potassium (potash, K_2O) fertilizer application rate is the most significant factor explaining rice yield and SOCSR. The burning of rice residues in the field was the main factor determining GHGI in this area. An effective way to reduce GHG emissions and contribute to sustainable rice production for food security with low GHGI and high productivity is avoiding the burning of rice residues.

Keywords: land management practices; rice field; net global warming potential; greenhouse gas intensity; Northeast Thailand

1. Introduction

Rice fields have been a concern of scientists worldwide because they emit the three most potent and long-lived greenhouse gases (GHGs), carbon dioxide (CO_2), methane (CH_4), and nitrous oxide (N_2O) [1,2], because of their positive increases in radiative forcing and their contribution to global warming [3]. Flooded rice fields emit CH_4 due to a methanogenesis process that occurs in anaerobic conditions, during which organic matter (OM) undergoes decomposition [4]. Factors affecting CH_4 emissions, such as weather conditions [5], the water regime [6], soil properties [7], land practices, i.e., irrigation [8], organic amendments [9], fertilization [10], and rice varieties [11], have been considered. Most N_2O emissions occur from nitrogen (N) fertilizer application [12], for which the N application rate is the main driver of N_2O production for either wet or dry soil [13]. However, rain-fed areas are more comparable and have stronger N_2O emissions from rice fields than other areas [14] because of changes in soil oxygen status, soil redox potential, soil moisture, and soil temperature [15]. With regard to CO_2 emissions, the main sources are the activities of farmers on their land, particularly when crop residues are burned and machines use energy either for cropping operations (i.e., tillage, harvesting, and so on) or stationary operations (i.e., water pumping, land preparation, and application of insecticides and herbicides) [16]. Furthermore, the burning of crop residues not only emits CO_2 but is also a major source of gaseous pollutants such as carbon monoxide (CO), CH_4, N_2O, and hydrocarbons in the troposphere [17]. However, soils have the potential to mitigate increasing CO_2 concentrations through carbon sequestration, with the maximum potential global sequestration varying from 0.45 to 0.9 Pg·C·year^{-1} [18]. Therefore, understanding the effects of management practices on GHG emissions and soil organic carbon (SOC) is necessary to improve management practices to reduce GHG emissions from rice fields.

Thailand's rice production area is 13.28 million ha, which is approximately 55.6% of the country's total agricultural area [19]. Geographically, Thailand is divided into four main regions, the North, Northeast, Central, and South. Rice is grown throughout the country, but the Northeast, North, and Central are the most important rice-growing regions with 49.1%, 25.4%, and 22.3%, respectively, of the country's total rice growing area. The Northeast has a majority of the rain-fed lowland areas with around 4.8 million ha [20] and shallow drought-prone areas [21]. Traditional rice varieties, particularly Jasmine rice or Khao Dawk Mali 105 (KDML 105), are grown in the Northeast. Although the rice quality is high, the rice yield is low, and the farmers in this area are the poorest compared to the other regions [19]. To obtain sustainable management of the sandy soil in this area, it is necessary to understand the land management practices, which include the farmers' actual activities and practices, and determine the appropriate practices in terms of the soil characteristics and farmers' capability. By understanding these elements, the pros and cons of each land management practice in relation to GHG emissions, SOC, and rice yield can be thoroughly estimated. To estimate the overall effects of rice fields, the concept of net global warming potential (GWP) was proposed based on the radiative properties of CO_2, CH_4, and N_2O emissions and SOC variations, expressed as kg·CO_2eq·ha^{-1}·year^{-1} [22]. Moreover, the agricultural practices can be related to GWP by estimating net GWP per ton of crop yield, which is referred to as greenhouse gas intensity (GHGI) [23]. Net GWP reflects the balance between SOC storage and GHG emissions. A negative net GWP value means that the system is taking GHGs out of the atmosphere, whereas a positive net GWP value means that GHGs are being added to the atmosphere and net GWP increases [24]. In addition, a positive GHGI value indicates a net source of GHGs per kilogram of yield per year, whereas a negative value indicates a net sink of GHGs in soil [25]. No studies to date have estimated net GWP and GHGI in this area under different land management practices, including irrigated versus rain-fed fields. The aim of this study was to estimate the effect of land management practices on net GWP and GHGI.

2. Materials and Methods

2.1. Description of the Study Area

The study area is situated in Thung Kula subdistrict, Suwannaphum district, Roi-Et province, Thailand (15°28′ N, 103°48′ E) and covers 59.45 ha, 22. 29 ha of which was irrigated and 37.16 ha was rain-fed. Roi Et soils are derived from washed deposits of sandstone and occur on the lower parts of peneplains. The elevation ranges from 100 to 200 m above sea level. This area has a tropical monsoon climate (Köppen 'Aw'). The average annual precipitation in 2014 ranged from 800 to 2900 mm, and the mean annual air temperature ranged from 26 to 28 °C. The major soil type in Roi-Et Province is Ultisol with more than 60% sand content; low SOC, ranging from 0.40% to 1.29%; and medium acid surface soil of pH 5.0–6.0 [26]. In general, the soils are deep, and are characterized by different colors; however, the dominant colors are a grayish-brown or light brown sandy loam A horizon overlying a light brown grading to pinkish-gray sandy clay loam or loam kandic B horizon, which, in turn, overlies a light gray or whitish clay loam or clay C horizon. The soils are mottled throughout the profile, with strong brown or yellowish brown or dark brown and some yellowish red or red mottles being common in the subsoil. The reaction is medium acidic over strong to very strongly acidic [27].

Rice in the study area refers to the major rice crop, which is grown during the rainy season between July and December, and the second rice crop, which is grown during the dry season between January and April of the following year [28]. Most rice fields use rain-fed cultivation, in which rice is grown only once a year because precipitation is a major limiting factor. Farmers in some irrigated areas are able to grow rice twice a year, for both major and second rice crops. Jasmine rice (*Oryza sativa*) is most commonly grown in this area. The dominant rice varieties recorded in this study were Khao Dawk Mali 105 (KDML 105), RD 6, and Suphanburi 60. KDML105 and RD 6 are strongly photoperiod sensitive and flower in late October, regardless of sowing time, whereas Suphanburi 60 is a non-photosensitive rice variety.

The main conventional management practices used in the study area are as follows. First, during the growing period, both major and second rice crops are cultivated using the broadcast method and harvested by machine. Second, for tillage management, conventional tillage to a depth of 20 to 30 cm is performed by a machine. Third, after harvesting, farmers apply one of two forms of rice straw and stubble management incorporation into the soil or burning in the field. The farmers usually burn rice residue after major rice harvesting (dry season) because of the ease and convenience of tillage to prepare for the next crop. Fourth, for water management, continuous flooding and shallow flooding are used for the irrigated and rain-fed areas, respectively. In irrigated areas, fields are inundated with 10 to 15 cm of standing water throughout the growing period and drained or naturally dried 7 to 10 days before harvesting. In rain-fed areas, the soil was temporarily flooded depending on rainfall, or water pumping when rain water was unavailable. Fifth, for manure and chemical fertilizer application, cattle manure was often added to the soil as a basal fertilizer once a year, usually after the previous crop was harvested or at the beginning of planting the next crop. The following chemical fertilizer types were found in the study area: 46-0-0, 16-16-8, 16-20-0, 0-0-60, 15-15-15, and 16-8-8.

According to the survey of soil nutrient status in Thailand from 2004–2008, soil samples of 13 sites in the study area were collected during the dry season after the rice harvest by the laboratory of the Office of Science for Land Development, Land Development Department, Ministry of Agriculture and Cooperatives, Thailand. Data from the 13 sites, nine of which were irrigated areas and four rain-fed areas, in Thung Kula subdistrict, were obtained from the soil pH, bulk density, OM, organic carbon (OC), total nitrogen, available P, available K, electrical conductivity, and lime requirement in 2004. We, therefore, estimated SOC in 2004 based on the soil bulk density, OC, and depth (30 cm) in 2004. In 2014, we again collected soil samples from the same 13 study sites to estimate the soil organic carbon sequestration rate (SOCSR) from 2004 to 2014.

2.2. Data Collection

The data were obtained over a five-year period (2010–2014). Questionnaires were conducted at each sampling site to record the crop and management practices by farm owners. At 13 sites, 13 farms provided crop and land management data. Rice yields and management practice data (i.e., dates of planting and harvesting; rates of application of fertilizers, manure, pesticides, and irrigation; and field operations performed) were collected from the questionnaire survey in 2014 and from the record book for the standards for good agricultural practices (GAP) of farm owners over the five-year study period (2010–2014). The record books were disseminated to the farmers by the Department of Agricultural Extension, Ministry of Agriculture and Cooperatives, Thailand to record their agricultural activities, which helped this study to obtain precise data on operational practices.

2.3. Soil Sampling

Soil samples were collected during the dry season after the rice harvest (November 2014). At each site, the soil horizons from 0 to 40 cm depth were identified by considering specific physical features, namely color and texture. Three soil samples (replications) of each soil horizon were then collected. Any visible roots, stones, or organic residues were removed manually after the samples were air-dried at room temperature (31–33 °C). The samples were passed through a 2-mm sieve. The SOC content (fine fraction < 2 mm) was determined using the wet oxidation method with $K_2Cr_2O_7$ and concentrated H_2SO_4 as described by Walkley and Black [29]. Soil bulk density was taken using the soil core. After a 24 h drying period in an oven at 105 °C, the soil bulk density was determined as the dry weight per unit volume of the soil core.

2.4. Estimation of GHG Emissions

The GHG emissions were calculated within the farm gate. Therefore, the GHG emissions of raw materials production and the transportation of agricultural inputs to the farm were not included. The emission factors for the calculation of GHG emissions, which were provided by Arunrat et al., are presented in Table 1 [30].

Table 1. Emissions factors used for the calculation of GHG emissions within the farm gate (utilization phase) [30].

Activity	Emissions Factor	Unit	Source
Agriculture Input			
Diesel used (stationary combustion) for farm operation	2.7446	$kg \cdot CO_2eq \cdot L^{-1}$	[31]
Gasoline used (stationary combustion) for farm operation	2.1896	$kg \cdot CO_2eq \cdot L^{-1}$	
Diesel used (mobile combustion) for farm operation	Tractor = 3.908 / Harvester = 2.645	$kg \cdot CO_2eq \cdot L^{-1}$	[32] (calculated with diesel density of 0.832 $kg \cdot L^{-1}$)
Gasoline used (mobile combustion) for farm operation	2.319	$kg \cdot CO_2eq \cdot L^{-1}$	[33]
Insecticide	5.1	$kg \cdot CO_2eq \cdot kg^{-1}$	[34]
Herbicide	6.3	$kg \cdot CO_2eq \cdot kg^{-1}$	[34]
CH_4 Emission from Rice Cultivation			
EF_c	3.12	$kg \cdot CH_4 \cdot ha^{-1} \cdot day^{-1}$	[35]
SF_w	0.52 in all systems		
SF_p	$Rw = 0.68, Lw, Ld = 1$		
ROA_i	2.5	$ton \cdot ha^{-1}$	[31]
$CFOA_i$	$Rw = 0.29, Lw, Ld = 1$		
SF_0	$Rw = 1.4, Lw, Ld = 2.1$		
Direct and Indirect N_2O Emission from Managed Soils (Chemical and Organic Fertilizer)			
EF_1	0.01	$kg \cdot N_2O\text{-}N \cdot kg^{-1}$ N input	[31]
EF_{1FR}	0.003	$kg \cdot N_2O\text{-}N \cdot kg^{-1}$ N input	
EF_2	0.01	$kg \cdot N_2O\text{-}N \cdot (kg \cdot NH_3\text{-}N + kg \cdot NO_x\text{-}N$ volatilized$)^{-1}$	
EF_3	0.0075	$kg \cdot N_2O\text{-}N \cdot kg$ leaching per runoff	
$Frac_{GASF}$	0.1	$kg \cdot NH_3\text{-}N \cdot + NO_x\text{-}N \cdot kg^{-1}$ N applied	
$Frac_{LEACH\text{-}(H)}$	0.3	$kg \cdot N \cdot kg^{-1}$ N additions	

Table 1. *Cont.*

Activity	Emissions Factor	Unit	Source
Burning Crop Residue			
CH_4	2.7	$g \cdot kg^{-1}$ dry matter burned	
N_2O	0.07	$g \cdot kg^{-1}$ dry matter burned	
Dry matter fraction	1		[31]
Fraction burned	0.29		
Fraction oxidized	0.9		
Rice residue to crop ratio	Irrigated areas: major rice = 1.06; second rice = 0.65		[36]
	Rain-fed areas: major rice and second rice = 0.55		

2.4.1. CO_2 Emissions from Fossil Fuel Utilization

The CO_2 emissions for the diesel and gasoline usage of stationary combustion were also taken from the IPCC [31]. The CO_2 emissions from the mobile combustion of the diesel fuel of farm tractors and harvesters were estimated from the emission factors of Maciel et al. [32], and CO_2 emissions from gasoline fuel were estimated from the EPA [33]. Figures for insecticides and herbicides were provided by the emissions factors from Lal [34]. The equations are detailed as following details:

(1) Diesel fuel

$$CO_2 \text{ emissions from diesel fuel utilization} = \text{Total amount of diesel fuel} \times \text{emissions factor of diesel fuel combustion.} \tag{1}$$

(2) Gasoline fuel

$$CO_2 \text{ emissions from gasoline fuel utilization} = \text{Total amount of gasoline fuel} \times \text{emissions factor of gasoline fuel combustion.} \tag{2}$$

2.4.2. CO_2 Emissions from Insecticide and Herbicide Utilization

The calculation for CO_2 emissions from insecticide and herbicide utilization was calculated as follows:

$$CO_2 \text{ emissions from insecticide and herbicide utilization} = \text{Total amount of insecticide and herbicide application} \times \text{emissions factor of insecticide and herbicide utilization.} \tag{3}$$

2.4.3. CH_4 Emissions from Rice Production

Field CH_4 emissions from rice cultivation were used as the model for the calculations according to the 2006 IPCC Guidelines for National Greenhouse Gas Inventories [31]. The baseline emission factor was taken from Yan et al. [35], who adjusted the region-specific emission factors for rice fields in east, southeast, and south Asian countries, and all scaling factors used were from the IPCC [31].

The basic equation to estimate CH_4 emissions from rice cultivation is based on the IPCC Guidelines [31] (Tier 2) (Equation (4)). CH_4 emissions were estimated by multiplying the daily emissions factor by the cultivation period of rice:

$$CH_4 = EF \times t, \tag{4}$$

where CH_4 is the methane emissions from rice cultivation (kg·CH$_4$·ha^{-1}), EF is the adjusted daily emissions factor (kg·CH$_4$·ha^{-1}·day^{-1}), and t is the cultivation period of rice (days).

Emissions from each different region were calculated by multiplying a baseline default emissions factor by the various scaling factors, as shown in Equation (5):

$$EF = (EF_c \times SF_w \times SF_p \times SF_o \times SF_{s,r}), \tag{5}$$

where EF is the adjusted daily emissions factor for a particular harvested area, EF_c is the baseline emissions factor for continuously flooded fields without organic amendments, SF_w is the scaling factor to account for the differences in water regime during the cultivation period, SF_p is the scaling factor to account for the differences in water regime in the season before the cultivation period, SF_0 is the scaling factor that accounts for differences in both type and amount of organic amendment applied source, and $SF_{s,r}$ is the scaling factor for soil type, rice cultivar, etc., if available.

Meanwhile, Equation (6) and the default conversion factor for farmyard manure presented an approach to vary the scaling factor according to the amount of farmyard manure applied:

$$SF_0 = \left(1 + \sum_i ROA_i \times CFOA_i\right)^{0.59}, \tag{6}$$

where ROA_i is the application rate of organic amendment i in dry weight for straw and fresh weight for others in tons·ha^{-1}, and $CFOA_i$ is the conversion factor for organic amendment i in terms of its relative effect with respect to straw applied shortly before cultivation.

2.4.4. N$_2$O Emissions from Managed Soils

The direct and indirect N$_2$O emissions were estimated using the methodology proposed by the IPCC [31].

The methodology for estimating direct N$_2$O emissions from chemical fertilizer application is given by IPCC Guidelines [31] (Tier 1) as follows:

$$\text{Direct N}_2\text{O emissions} = [F_{SN} \times EF_1 + (F_{SN})_{FR} \times EF_{1FR}] \times 44/28, \tag{7}$$

where F_{SN} and $(F_{SN})_{FR}$ are the annual amount of synthetic fertilizer N applied to dry land and rice fields respectively, and EF_1 and EF_{1FR} are the emissions factors of N$_2$O caused by fertilizer N input in the two types of fields respectively.

The calculation formula for indirect N$_2$O emissions caused by chemical fertilizer application is listed below:

$$\text{Indirect N}_2\text{O emissions} = [F_{SN} \times Frac_{GASF} \times EF_2 + F_{SN} \times Frac_{LEACH\text{-}(H)} \times EF_3] \times 44/28, \tag{8}$$

where $Frac_{GASF}$ is the fraction of synthetic fertilizer N that volatilizes as NH$_3$ and NO$_X$, EF_2 is the emissions factor for N$_2$O emissions from atmospheric deposition of N on soil and water surfaces, $Frac_{LEACH\text{-}(H)}$ is the fraction of all N which is lost when added to/mineralized in managed soil in regions where leaching/runoff occurs, and EF_3 is the emissions factor for N$_2$O emissions from N leaching and runoff.

2.4.5. GHG Emissions from Field Burning

The calculation for GHG emissions from field burning was calculated as follows [37]:

$$\text{The quantity of rice straw} = \text{Rice production} \times \text{Residue to crop ratio} \tag{9}$$

$$\text{Amount of burned residues} = \text{Quantity of rice straw} \times \text{fraction of area burned} \times \text{dry matter fraction} \times \text{fraction burned} \times \text{fraction oxidized} \tag{10}$$

$$\text{CH}_4 \text{ emissions from field burning} = \text{Amount of burned residues} \times \text{CH}_4 \text{ emissions factor} \tag{11}$$

$$N_2O \text{ emissions from field burning} = \text{Amount of burned residues} \times N_2O \text{ emissions factor.} \quad (12)$$

where fraction of area burned is proportion of rice straw subject to open field burning, which is based on the field survey (fraction of area burned = 1 for rice straw in the whole area was burned). The amount of burned residue was estimated "0" if no burning rice straw.

2.5. SOC Calculation

SOC stock was calculated by:

$$SOC = (BD \times OC \times D) \times 100, \quad (13)$$

where SOC is soil organic carbon stock (kg·C·ha^{-1}), BD is soil bulk density (kg·m^{-3}), OC is organic carbon content (%), and D is soil sampling depth (m).

The soil organic carbon sequestration rate (SOCSR) was calculated as follows:

$$SOCSR \text{ (kg·C·ha}^{-1}\cdot\text{year}^{-1}) = (SOC_t - SOC_0)/10, \quad (14)$$

where SOC_t and SOC_0 are the SOC contents measured in 2014 and 2004, respectively (kg·C·ha^{-1}), and 10 is the number of years from 2004 to 2014.

2.6. Net Global Warming Potential

The GWP based on the CH_4, CO_2, and N_2O emissions was used to account for the climatic impact on rice yield under different land management practices. To assess the combined GWP, CH_4 and N_2O were calculated as CO_2 equivalents over a 100-year time scale using a radiative forcing potential relative to CO_2 of 28 for CH_4 and 265 for N_2O [38]. The net GWP of a rice field equals the total CO_2 emissions equivalents minus the SOCSR in the rice field [39,40]:

$$\text{Net } GWP = (CO_2 \text{ emissions} \times 1) + (N_2O \text{ emissions} \times 265) + (CH_4 \text{ emissions} \times 28) - (SOCSR \times 44/12). \quad (15)$$

2.7. Greenhouse Gas Intensity

The GHGI is calculated as a ratio of net GWP and rice yield, as described in Shang et al. [25]:

$$GHGI = \text{net } GWP/\text{rice yield.} \quad (16)$$

2.8. Statistical Analysis

Statistical analyses of the data were carried out using SPSS (Version 20.0, Chicago, IL, USA). The mean and standard deviation values were used to represent rice yield, SOC, emissions of CO_2, N_2O and CH_4, GWP, SOCSR, net GWP, and GHGI in each site. Differences in rice yield, SOC, emissions of CO_2, N_2O and CH_4, GWP, SOCSR, net GWP, and GHGI between irrigated and rain-fed areas were analyzed with t-test and least significant difference (LSD) test ($p < 0.05$). Simple linear regression analysis was used to find the relationship between two variables by fitting a linear equation. Stepwise multiple regression analysis was conducted to evaluate the relationships of rice yield and SOCSR with the pertinent management practice (manure, fertilizer application rates, and amount of burned rice residues). Pearson's correlation analysis was conducted to evaluate the relationships among the GHG emissions and pertinent factors.

3. Results

3.1. Pertinent Management Practices, Rice Yield, and SOC

There were no significant differences in the manure and fertilizer application rates, the amount of burned rice residues, rice yield, and SOC between irrigated and rain-fed areas (Table 2). The manure application rate varied across sites, from 0 to 4830 kg·ha^{-1}·year^{-1}, with averages of 2864 and

1888 kg·ha^{-1}·year^{-1} for irrigated and rain-fed areas, respectively. The N fertilizer application rate ranged from 38 to 98 kg·ha^{-1}·year^{-1} across sites, with averages of 72 and 56 kg·ha^{-1}·year^{-1} for irrigated and rain-fed areas, respectively. The P_2O_5 and K_2O application rate ranged from 14 to 46 and 6 to 46 kg·ha^{-1}·year^{-1} across sites for irrigated and rain-fed areas, respectively. The average P_2O_5 and K_2O application rates were 27 and 22 kg·ha^{-1}·year^{-1} for irrigated areas and 26 and 15 kg·ha^{-1}·year^{-1} for rain-fed areas. Meanwhile, the amount of burned rice residue was also variable, ranging from 0 to 593 kg·ha^{-1}·year^{-1}, with averages of 263 and 148 kg·ha^{-1}·year^{-1} for irrigated and rain-fed areas, respectively.

The average rice yield, SOC_t, SOC_0, and SOCSR were 3,307 kg·ha^{-1}·year^{-1}, 53,884 kg·C·ha^{-1}, 42,151 kg·C·ha^{-1}, and 1,173 kg·C·ha^{-1}·year^{-1} (Table 2) respectively. There was a significant correlation between rice yield and SOC_t ($R^2 = 0.51$, $p < 0.01$), SOC_0 ($R^2 = 0.46$, $p < 0.01$) (Figure 1), and SOCSR ($R^2 = 0.52$, $p < 0.01$) (Figure 2). Although, the manure, N, P_2O_5, and K_2O fertilizer applications were the major underlying factors for increasing rice yield, they were exiguously correlated positively with rice yield ($R^2 = 0.16$, $p > 0.01$, $R^2 = 0.11$, $p > 0.01$, $R^2 = 0.46$, $p > 0.01$, and $R^2 = 0.69$, $p > 0.01$ respectively) (Figure 3a,b). On the other hand, the amount of burned rice residues showed a negative correlation to rice yield ($R^2 = 0.02$, $p > 0.01$) (Figure 3c). In addition, rice yield and SOCSR was related markedly to only the amount of K_2O fertilizer application rates (Table 3).

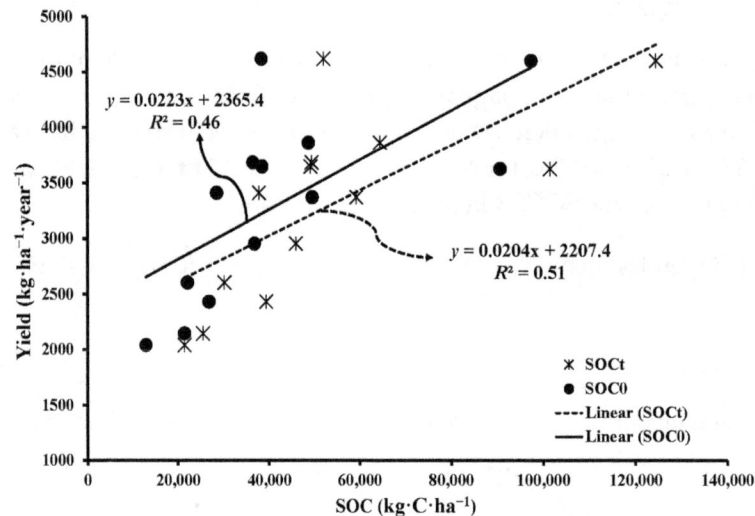

Figure 1. Relationship between rice yield and SOC_t and SOC_0 for all sites.

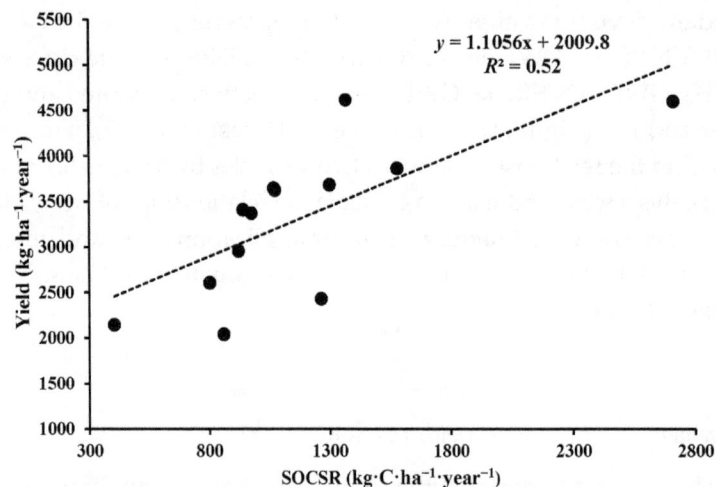

Figure 2. Relationship between rice yield and SOCSR for all sites.

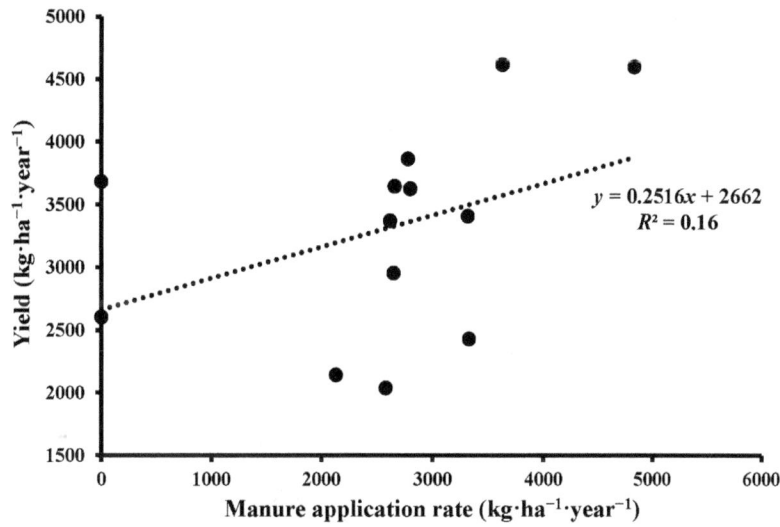

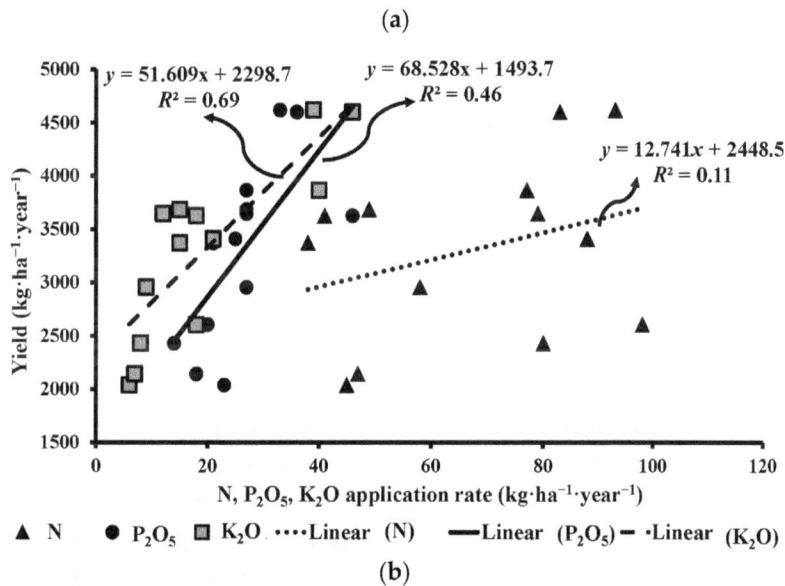

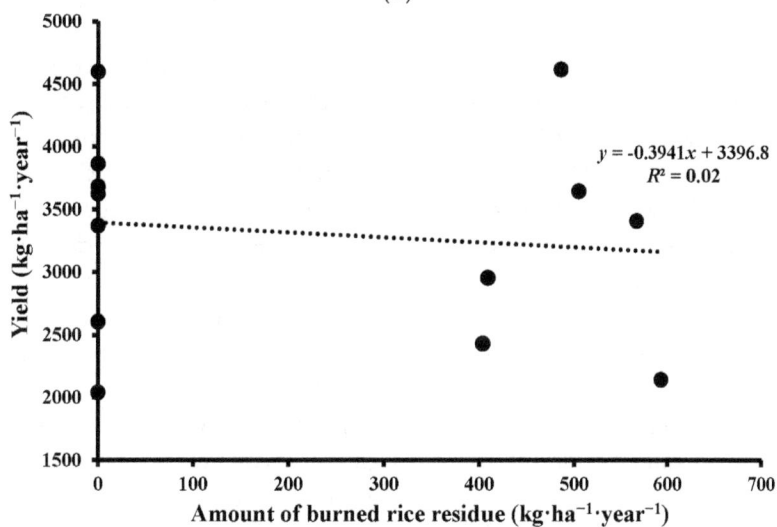

Figure 3. Relationship between pertinent practices and rice yield: (**a**) manure application; (**b**) fertilizer application; and (**c**) amount of burned rice residue.

Table 2. Pertinent management practices, rice yield, and SOCSR (mean ± standard deviation).

| Site No. | Manure Application Rate (kg·ha^{-1}·year^{-1}) | Fertilizer Application Rate (kg·ha^{-1}·year^{-1}) | | | Burned Rice Residue (kg ·ha^{-1}·year^{-1}) | Rice Yield (kg·ha^{-1} ·year^{-1}) | SOC$_t$ (kg·C·ha^{-1}) | SOC$_0$ (kg·C·ha^{-1}) | SOCSR (kg·C·ha^{-1} ·year^{-1}) |
		N	P$_2$O$_5$	K$_2$O					
I1	3320	88	25	21	566	3410	37,810	28,430	938
I2	4830	83	36	46	0	4600	124,400	97,320	2708
I3	2780	77	27	40	0	3864	64,540	48,730	1581
I4	0	49	27	15	0	3684	49,430	36,440	1299
I5	3630	93	33	39	485	4618	52,100	38,430	1367
I6	3330	80	14	8	403	2430	39,400	26,760	1264
I7	2660	79	27	12	504	3646	49,180	38,500	1068
I8	2580	45	23	6	0	2040	21,430	12,850	858
I9	2650	58	27	9	409	2954	45,960	36,770	919
Average	2864 ± 1287	72 ± 17	27 ± 6	22 ± 16	263 ± 254	3472 ± 882	53,806 ± 28,973	40,470 ± 23,542	1334 ± 568
R1	2800	41	46	18	0	3626	101,390	90,660	1073
R2	2620	38	21	15	0	3372	59,290	49,550	974
R3	2130	47	18	7	593	2144	25,460	21,430	403
R4	0	98	20	18	0	2604	30,100	22,090	801
Average	1888 ± 1290	56 ± 28	26 ± 13	15 ± 5	148 ± 297	2937 ± 684	54,060 ± 34,926	45,933 ± 32,570	813 ± 295
p-value	0.233	0.217	0.954	0.238	0.838	0.308	0.989	0.736	0.116
Overall	2564 ± 1319	67 ± 22	26 ± 8	20 ± 14	228 ± 261	3307 ± 838	53,884 ± 29,404	42,151 ± 25,330	1173 ± 547

I = Irrigated area, R = Rain-fed area, p-value indicates a significant difference of value between irrigated and rain-fed areas.

Table 3. Multiple regression equations to predict rice yield and SOCSR using manure (M), N fertilizer (N), P$_2$O$_5$ fertilizer (P), K$_2$O fertilizer (K), and burned rice residues (B).

Depended Variable	Equation
Rice Yield	Yield = 51.61 × K + 2298.72 (R^2 = 0.66, $p < 0.05$)
SOCSR	SOCSR = 31.91 × K + 549.84 (R^2 = 0.59, $p < 0.05$)

3.2. CO$_2$ Emissions

In this study, CO$_2$ emissions reflected the utilization of fossil fuels (diesel and gasoline), insecticides, and herbicides. The utilization of diesel and gasoline fuels revealed that rain-fed areas generated more CO$_2$ emissions from drainage water into the field than the irrigated areas. Moreover, at the sites where there was no burning rice residue (sites I2, I3, I4, I8, R1, R2, and R4) there was a slightly higher amount of CO$_2$ emissions from utilization of diesel fuel than burned rice residue sites (sites I1, I5, I6, I7, I9, and R3). This is because farmers need to use the machine for the incorporation of rice residues into the soil but not for burning rice residues. The CO$_2$ emissions from the utilization of diesel fuel ranged from 127 to 211 kg·CO$_2$eq·ha^{-1}·year^{-1} across sites, with averages of 152 and 188 kg·CO$_2$eq·ha^{-1}·year^{-1} for irrigated and rain-fed areas, respectively. Gasoline fuel utilization generated CO$_2$ emissions, varying from 10 to 73 kg·CO$_2$eq·ha^{-1}·year^{-1} in all sites, with averages of 30 and 51 kg·CO$_2$eq·ha^{-1}·year^{-1} for irrigated and rain-fed areas, respectively. Meanwhile, CO$_2$ emissions from the utilization of insecticides and herbicides ranged from 37 to 73 kg·CO$_2$eq·ha^{-1}·year^{-1} across sites, with averages of 52 and 43 kg·CO$_2$eq·ha^{-1}·year^{-1} for irrigated and rain-fed areas, respectively. There were significant differences for the utilization of diesel fuel ($p < 0.05$), while there were no significant differences in the utilization of gasoline fuel, or insecticides and herbicides, between irrigated and rain-fed areas ($p < 0.05$) (Table 4).

The total CO$_2$ emissions were estimated, ranging from 201 to 301 kg·CO$_2$eq·ha^{-1}·year^{-1} across sites, with averages of 233 and 281 kg·CO$_2$eq·ha^{-1}·year^{-1} for irrigated and rain-fed areas, respectively (Table 5). The land management practice of the high amount of diesel fuel utilization caused the highest total CO$_2$ emissions as seen at site R1, which was the highest amount of diesel fuel utilization at 211 kg·CO$_2$eq·ha^{-1}·year^{-1} compared with others (Table 4).

3.3. N_2O Emissions

Remarkably, N_2O emissions depended on chemical fertilizer application and the amount of burned rice residues. N_2O emissions from chemical fertilizer utilization ranged from 211 to 541 $kg \cdot CO_2eq \cdot ha^{-1} \cdot year^{-1}$ in all sites, with averages of 409 and 346 $kg \cdot CO_2eq \cdot ha^{-1} \cdot year^{-1}$ for irrigated and rain-fed areas, respectively. Irrigated areas had slightly higher N_2O emissions from chemical fertilizer utilization than the rain-fed areas, but there were no significant differences between both areas ($p < 0.05$). Meanwhile, N_2O emissions from burning rice residue were found the wide range of 0 to 11 $kg \cdot CO_2eq \cdot ha^{-1} \cdot year^{-1}$ in all sites, with averages of 5 and 3 $kg \cdot CO_2eq \cdot ha^{-1} \cdot year^{-1}$ for irrigated and rain-fed areas, respectively, and there were no significant differences between both areas ($p < 0.05$) (Table 4).

A range of total N_2O emissions values (211–541 $kg \cdot CO_2eq \cdot ha^{-1} \cdot year^{-1}$) was calculated, with averages of 414 and 349 $kg \cdot CO_2eq \cdot ha^{-1} \cdot year^{-1}$ for irrigated and rain-fed areas, respectively (Table 5). Highly positive correlations were found between N_2O emissions and N fertilizer application, with r values of 0.925 ($p < 0.01$) (Table 6). The highest total N_2O emissions were found at site R4, where the high amount of chemical was practiced. On the other hand, at site R2, where the lowest amount of chemical fertilizer was found and there was no use of burned rice residues (Table 2), the lowest total N_2O emissions were seen.

Table 4. GHG emissions within the farm gate in each activity (mean ± standard deviation).

Site No.	CO_2 Emissions ($kg \cdot CO_2eq \cdot ha^{-1} \cdot year^{-1}$)			N_2O Emissions ($kg \cdot CO_2eq \cdot ha^{-1} \cdot year^{-1}$)		CH_4 Emissions ($kg \cdot CO_2eq \cdot ha^{-1} \cdot year^{-1}$)	
	Diesel Fuel	Gasoline Fuel	Insecticide and Herbicide	Chemical Fertilizer	Burning Rice Residue	Rice Cultivation	Burning Rice Residue
I1	151	28	48	487	10	5282	43
I2	188	14	73	459	0	5418	0
I3	172	23	61	423	0	4776	0
I4	166	10	67	272	0	2404	0
I5	127	19	55	511	9	4849	37
I6	135	30	41	443	7	3616	30
I7	129	65	40	438	9	5518	38
I8	159	28	38	249	0	3805	0
I9	138	49	45	399	8	3567	31
Average	152 ± 21	30 ± 17	52 ± 13	409 ± 91	5 ± 5	4359 ± 1063	20 ± 19
R1	211	51	39	227	0	2292	0
R2	206	28	52	211	0	2153	0
R3	142	73	37	404	11	2381	45
R4	193	50	42	541	0	794	0
Average	188 ± 32	51 ± 18	43 ± 7	346 ± 157	3 ± 6	1905 ± 747	11 ± 23
p-value	0.031	0.074	0.191	0.370	0.502	0.002	0.491
Overall	163 ± 29	36 ± 20	49 ± 12	390 ± 112	4 ± 5	3604 ± 1511	17 ± 20

I = Irrigated area, R = Rain-fed area, p-value indicates a significant difference of value between irrigated and rain-fed areas.

3.4. CH_4 Emissions

Manure application, the amount of burned rice residues, and the length of rice cultivation affected the CH_4 emissions. CH_4 emissions from rice cultivation ranged from 794 to 5518 $kg \cdot CO_2eq \cdot ha^{-1} \cdot year^{-1}$ across sites, with averages of 4359 and 1905 $kg \cdot CO_2eq \cdot ha^{-1} \cdot year^{-1}$ for irrigated and rain-fed areas, respectively. There were significant differences for CH_4 emissions from rice cultivation between irrigated and rain-fed areas ($p < 0.05$). Irrigated areas had obviously higher CH_4 emissions from rice cultivation than rain-fed areas. In addition, CH_4 emissions from burning rice residue had the wide range of 0 to 45 $kg \cdot CO_2eq \cdot ha^{-1} \cdot year^{-1}$ in all sites, with averages of 20 and 11 $kg \cdot CO_2eq \cdot ha^{-1} \cdot year^{-1}$ for irrigated and rain-fed areas, respectively, but there were no significant differences between both areas ($p < 0.05$) (Table 4).

The range of total CH_4 emissions was broad, varying from 794 to 5556 kg·CO_2eq·ha^{-1}·$year^{-1}$. The average value was 3621 kg·CO_2eq·ha^{-1}·$year^{-1}$ in all sites (Table 5). Highly positive correlations were found between CH_4 emissions and manure application, GWP, net GWP, and GHGI, with r values of 0.739 ($p < 0.01$), 0.997 ($p < 0.01$), 0.932 ($p < 0.01$), and 0.604 ($p < 0.05$), respectively (Table 6). These correlations may reflect that the land management practice of applying large amounts of manure or burned rice residues, and the long rice cultivation length, would generate high CH_4 emissions. The highest and lowest CH_4 emissions were seen at site I7 and R4, respectively (Table 5).

3.5. SOCSR

The SOCSR in this study varied across sites from 403 to 2708 kg·CO_2eq·ha^{-1}·$year^{-1}$, with averages of 1334 and 813 kg·CO_2eq·ha^{-1}·$year^{-1}$ for irrigated and rain-fed areas, respectively. The average value for all sites was 1173 kg·CO_2eq·ha^{-1}·$year^{-1}$ (Table 5). The SOCSR had a highly positive correlation with rice yield ($r = 0.722$, $p < 0.01$) and K_2O fertilizer application ($r = 0.787$, $p < 0.01$), whereas a negative correlation was found with the amount of burned rice residues and GHGI, but was not statistically significant (Table 6). The results were obvious at sites I2 and R1, where the manure application was high and no burned rice residues were used. Therefore, these sites achieved high SOCSR and rice yield. However, it seems that not only can manure application and a lack of burned rice residues increase SOCSR and rice yield, but high chemical fertilizer application also can (Table 2).

3.6. Net GWP and GHGI

The evaluation of net GWP and GHGI under different land management practices is shown in Table 5. The net GWP varied across sites, ranging from 819 to 5170 kg·CO_2eq·ha^{-1}·$year^{-1}$, with an average value of 3090 kg·CO_2eq·ha^{-1}·$year^{-1}$, and GHGI ranged from 0.31 to 1.68 kg·CO_2eq·kg^{-1} yield, with an average value of 0.97 kg·CO_2eq·kg^{-1} yield. The net GWP showed a highly positive correlation with manure application, amount of burned rice residue, CH_4 emission, GWP, and GHGI ($r = 0.609$ ($p < 0.05$), 0.555 ($p < 0.05$), 0.932 ($p < 0.01$), 0.936 ($p < 0.01$), and 0.778 ($p < 0.01$), respectively). Meanwhile, GHGI had a positive correlation with the amount of burned rice residue, CH_4 emission, GWP, and net GWP, with r values of 0.656 ($p < 0.05$), 0.604 ($p < 0.05$), 0.595 ($p < 0.05$), and 0.778 ($p < 0.01$), respectively. However, this study found a negative correlation of net GWP and GHGI with CO_2 emissions ($r = -0.640$ ($p < 0.05$), and -0.662 ($p < 0.05$), respectively). This is because the sites with high net GWP and GHGI in this study generated a low amount of CO_2 emissions, but emitted a high amount of N_2O emissions from chemical fertilizer application.

Multiple regression equations to predict GHGI using manure, N, P_2O_5 and K_2O fertilizers, and burned rice residues showed that GHGI = 0.001 × burned rice residues + 0.74 ($R^2 = 0.36$, $p < 0.01$). This finding revealed that burned rice residue was the main factor determining the GHGI in this area. In addition, land management practices where the net GWP and GHGI were low involved no burned rice residues, incorporation of manure and chemical fertilizer, or application of chemical fertilizers at sites R1, R2, and R4, respectively. Meanwhile, similar land management practices had a high net GWP and GHGI as seen at sites I2, I3, I4, and I8 (Table 5), mainly due to the increased CH_4 emissions under continuous flooding.

This study revealed that 81.07% of GHG emissions came from CH_4 emissions from rice cultivation, followed by N_2O emissions from fertilizer utilization, CO_2 emissions from diesel fuel utilization, CH_4 emissions from burning rice residues, CO_2 emissions from insecticide and herbicide utilization, N_2O emissions from burning rice residues, and CO_2 emissions from gasoline fuel utilization, with sharing values of 8.57%, 3.75%, 3.72%, 1.13%, 0.93%, and 0.83%, respectively (Figure 4).

Table 5. CO_2, N_2O, and CH_4 emissions, and SOCSR, net GWP, and GHGI at all sites (mean ± standard deviation).

Site No.	Total CO_2 (kg·CO_2eq·ha⁻¹·year⁻¹)	Total N_2O (kg·CO_2eq·ha⁻¹·year⁻¹)	Total CH_4 (kg·CO_2eq·ha⁻¹·year⁻¹)	GWP (kg·CO_2eq·ha⁻¹·year⁻¹)	SOCSR (kg·CO_2eq·ha⁻¹·year⁻¹)	Net GWP (kg·CO_2eq·ha⁻¹·year⁻¹)	Rice Yield (kg·ha⁻¹·year⁻¹)	GHGI (kg·CO_2eq·kg⁻¹ Yield)
I1	227	497	5324	6048	938	5110	3410	1.50
I2	275	459	5418	6152	2708	3444	4600	0.75
I3	256	423	4776	5455	1581	3874	3864	1.00
I4	243	272	2404	2918	1299	1619	3684	0.44
I5	201	520	4886	5607	1367	4240	4618	0.92
I6	206	450	3647	4303	1264	3039	2430	1.25
I7	234	447	5556	6238	1068	5170	3646	1.42
I8	225	249	3805	4279	858	3421	2040	1.68
I9	232	407	3598	4237	919	3318	2954	1.12
Average	233 ± 23	414 ± 94	4379 ± 1070	5026 ± 1143	1334 ± 568	3693 ± 1091	3472 ± 882	1.12 ± 0.39
R1	301	227	2292	2821	1073	1748	3626	0.48
R2	286	211	2153	2650	974	1676	3372	0.50
R3	252	415	2426	3093	403	2690	2144	1.25
R4	285	541	794	1620	801	819	2604	0.31
Average	281 ± 21	349 ± 158	1916 ± 756	2546 ± 643	813 ± 295	1733 ± 765	2937 ± 684	0.64 ± 0.42
p-value	0.005	0.365	0.002	0.002	0.116	0.008	0.308	0.068
Overall	248 ± 31	394 ± 114	3621 ± 1519	4263 ± 1547	1173 ± 547	3090 ± 1351	3307 ± 838	0.97 ± 0.45

I = Irrigated area, R = Rain-fed area, p-value indicates a significant difference of value between irrigated and rain-fed areas.

Table 6. Correlation matrix of the pertinent factors and among the GHG emissions.

	Manure	N	P_2O_5	K_2O	Burning	CO_2	N_2O	CH_4	GWP	SOCSR	Net GWP	Yield	GHGI
Manure	1.00												
N	0.175	1.00											
P_2O_5	−0.335	0.447	1.00										
K_2O	0.449	0.509	0.079	1.00									
Burning	0.189	0.134	−0.224	−0.350	1.00								
CO_2	−0.216	−0.313	0.049	0.131	−0.544	1.00							
N_2O	0.216	0.925 **	0.331	0.328	0.497	−0.462	1.00						
CH_4	0.739 **	0.391	−0.088	0.436	0.322	−0.535	0.457	1.00					
GWP	0.730 **	0.452	−0.056	0.454	0.344	−0.537	0.520	0.997 **	1.00				
SOCSR	0.526	0.339	−0.126	0.787 **	−0.423	0.093	0.130	0.468	0.466	1.00			
Net GWP	0.609 *	0.372	−0.012	0.195	0.555 *	−0.640 *	0.531	0.932 **	0.936 **	0.124	1.00		
Yield	0.396	0.327	−0.14	0.832 **	−0.278	0.093	0.185	0.460	0.463	0.722 **	0.231	1.00	
GHGI	0.383	0.030	0.000	−0.323	0.656 *	−0.662 *	0.269	0.604 *	0.595 *	−0.278	0.778 **	−0.400	1.00

* = Correlation is significant at 0.05 probability level ($p < 0.05$), ** = Correlation is significant at 0.01 probability level ($p < 0.01$).

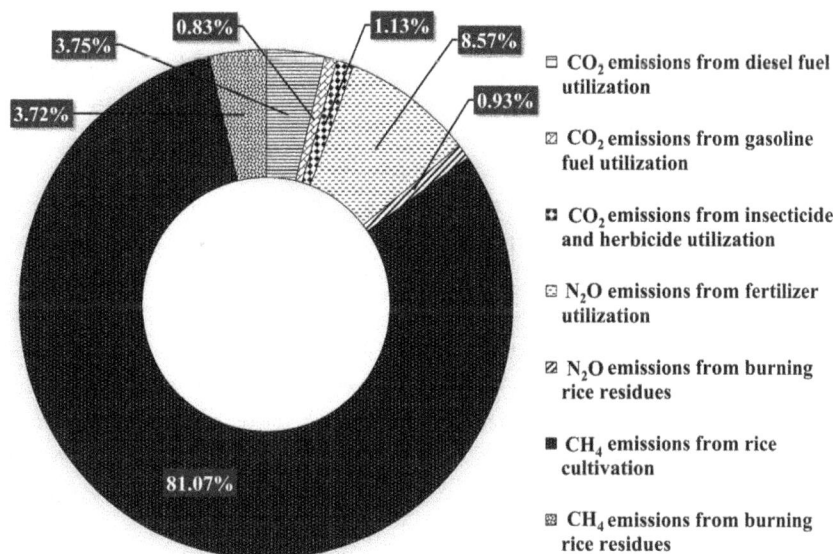

Figure 4. The contribution of GHG emission sources in each activity.

4. Discussion

4.1. Rice Yield and SOC under Different Management Practices

Proper management practices can increase SOC sequestration through increasing OM inputs to the soil. Soil organic material can be mineralized towards releasing the nutrient, which subsequently can be taken up by crops to increase crop yields. Therefore, the mineralization of soil organic matter (SOM) is a vital parameter to enhance crop yields. This is consistent with our result that rice yield had a highly positive correlation with SOC (Figure 1) and SOCSR (Figure 2). Liang et al. [41] indicated that increasing the amount of SOC could be accomplished by regular manure application with a return of more crop residues, which subsequently can lead to higher crop production. In arable land cropping systems, increasing the amount of OM in soil not only increases SOM, but also reduces net GHG emissions [42]. Moreover, studies have shown that combining both organic and chemical fertilizers can be a suitable way of enriching soil [43,44]. From this study, I2 and R1 reached the same rice yield and SOC, which was the highest overall (Table 2).

4.2. Effects of Land Management Practice on CO_2, CH_4, and N_2O Emissions

According to many previous studies, the estimation of GHG emissions from rice production varies, but they all agree that rice production is a significant contributor to overall emissions. As in flooded rice paddies generally, flooding rice fields blocks oxygen penetration into the soil, which allows bacteria capable of producing CH_4 to thrive [45]. Rice production also generates N_2O from N-fertilizer application [46]. Meanwhile, the main sources of CO_2 emissions are either cropping operations such as tillage, sowing, or harvesting, including stationary operations such as pumping water, spraying, and grain drying. The burning of rice residue is another emissions source yielding CO_2, CH_4, and N_2O [47]. This study revealed that the land management practice of highly burned rice residues generated high CH_4 and N_2O emissions, as was seen at site R3 (Table 4). In addition, more fuel consumption was found in rain-fed areas than in irrigated areas, owing to the energy needed for pumping water into rain-fed rice fields and farm operations, such as herbicide application, because the dry land would face more weeds than flooded land would. Consequently, these management practices can also produce higher amounts of CO_2eq [48].

CH_4 is produced from the decomposition of OM in anaerobic conditions by methanogens. SOM is the most common limiting factor for methanogenesis in rice fields [49]. OM arises from four main sources: animal manure, green manure, crop residues (straw, stubble, roots), and by-products of rice production (root exudates, sloughed-off root cells, and root turnover). Neue et al. [50] reported that the rice straw application of 5 $Mg \cdot ha^{-1} \cdot year^{-1}$ increased CH_4 emissions 10-fold compared to the use of urea fertilizer only. Reducing the CH_4 emissions associated with water management in rice fields in which wetting and drying cycles alternate could reduce more CH_4 emissions than continuously flooded fields [9]. These results were in complete agreement with our findings that the land management practices for site I7 (Table 5) caused the highest CH_4 emissions due to the higher manure application and amount of burned rice residues than at other sites. Additionally, irrigated areas had higher CH_4 emissions than rain-fed areas, owing to the longer period of flooding in rice fields, which was similar to previous investigations by Bhattacharyya et al. [51] and Shen et al. [52]. This result was obtained from sites I1 to I9, with an average value of 4379 $kg \cdot CO_2eq \cdot ha^{-1} \cdot year^{-1}$ for irrigated areas, and sites R1 to R4, with an average value of 1916 $kg \cdot CO_2eq \cdot ha^{-1} \cdot year^{-1}$ for rain-fed areas (Table 5). However, the addition of organic material such as rice residues and manure application leads to increasing CH_4 emissions due to anaerobic decomposition [53], but it can greatly offset the mitigation benefits of soil carbon sequestration [54]. This can obviously be found at site I2, where manure is applied but no burning rice residues (Table 2), with the high SOCSR (2708 $kg \cdot CO_2eq \cdot ha^{-1} \cdot year^{-1}$) and low GHGI (0.75 $kg \cdot CO_2eq \cdot kg^{-1}$ yield) (Table 5). Based on our study, we, therefore, support the addition of manure application and returning rice residues to the soil because these practices not only gain more C storage in the soil than is released to the atmosphere but also can enhance soil fertility through the mineralization of SOM, which in turn will increase crop productivity.

Fertilizer application is important from a climate change perspective due to energy-intensive production and the positive relationship with N_2O emissions from soils [55,56]. This was consistent with our result for the land management practices at site I5 (Table 5), which generated the highest N_2O due to the high chemical fertilizer application and amount of burned rice residues.

4.3. Effects of Land Management Practice on Net GWP and GHGI

The net GWP has been illuminated to understand agriculture's impact on radiative forcing [57]. Therefore, net GWP and GHGI need to be considered when evaluating a management strategy for mitigating GHG emissions. Our study found that more burning rice residues greatly contributed to high net GWP and GHGI. Our results were consistent with Zhang et al. [58], whose study showed that a chemical fertilizer application rate of 210 $kg \cdot N \cdot ha^{-1} \cdot year^{-1}$ was the most suitable for balancing GHG emissions and rice yield in Chongming Island, Eastern China. In this study, the average GHGI was 0.97 $kg \cdot CO_2eq \cdot kg^{-1}$ yield (Table 5). In Jiangsu province, China, the GHGI varied from 0.41 to 0.74 $kg \cdot CO_2eq \cdot kg^{-1}$ yield under annual rice–wheat rotations with integrated soil and crop system management [59]. Qin et al. [13] studied midseason drainage and organic manure incorporation in Southeast China and found that the GHGI varied from 0.24 to 0.74 $kg \cdot CO_2eq \cdot kg^{-1}$ yield, which was lower than in this study. In Thailand, the study of Yodkhum and Sampattagul [60], who applied a life cycle assessment concept and carbon footprint to determine GHG emissions of rice production in Thailand, reported that in northeast Thailand, GHG emissions of KDML 105 of NongKhai was 2.39 $kg \cdot CO_2eq \cdot kg^{-1}$ yield, which was higher than in this study. Arunrat et al. [30] estimated GHG emissions based on the concept of the life cycle assessment of the greenhouse gas emissions (LCA-GHG) of products in Phichit province of Thailand. Their results revealed that GHG emissions from rice production varied from 1.81 to 2.87 $kg \cdot CO_2eq \cdot kg^{-1}$ yield, and 1.72 to 2.70 $kg \cdot CO_2eq \cdot kg^{-1}$ yield for irrigated and rain-fed areas, respectively, which was higher than in this study. However, the report of the Office of Agricultural Economics about the GHG emissions estimation and database developments in Thailand in 2012 using the methodology of life cycle assessment of greenhouse gas emissions of products and IPCC guideline. The GHGIs ranged from 0.67 to 3.96 $kg \cdot CO_2eq \cdot kg^{-1}$ yield and averaged 2.32 $kg \cdot CO_2eq \cdot kg^{-1}$ yield [36]. Taking the country's value as the baseline, the GHGI in this study was lower than the country's average. This is because the GHG emissions that occur outside the farm gate were not included in this study such as raw materials production, the transportation of agricultural inputs from manufacturing to the farm, and the transportation of rice production from the farm to the mill and storehouse. This study emphasized the balance between GHG emissions and SOC sequestration on the effect of land management practices because the SOC content at local-scale data in the estimation of GHGI, which is usually limited due to high uncertainties in the large-scale data [61]. Reasonable land management practices are the main components for mitigating GHG emissions because CO_2, CH_4, and N_2O emissions would be negated by the benefits of SOC sequestration. It is possible that the goal of reducing the net GWP and GHGI in Thailand should focus on increasing the SOC and simultaneously decreasing burning rice residues.

5. Conclusions

This study showed that the amount of K_2O fertilizer applied is the most significant factor explaining rice yield and SOCSR in this area. The contributions of CO_2, CH_4, and N_2O to net GWP decreased in the order $CH_4 > N_2O > CO_2$ at all sites. GHGI had a positive correlation with the amount of burned rice residues, CH_4 emission, GWP, and net GWP. The land management practices that led to low GHGIs were those that returned residues to the field after harvesting and incorporated manure and chemical fertilizers. These practices are an effective way to reduce GHG emissions and contribute to sustainable rice production for food security with low GHGI and high productivity.

Acknowledgments: This study was financially supported by the Japanese government under the Japan Society for the Promotion of Science (JSPS) RONPAKU (Dissertation PhD) Program and the Thailand Research Fund (TRF): Grand No. RDG5620041. The authors' sincere gratitude is also extended to National Research Council of Thailand (NRCT) for their support of this study. The authors deeply appreciate Attaya Phinchongsakuldit, Office of Soil Resources Survey and Research, Land Development Department, Ministry of Agriculture and Cooperatives, Thailand for supporting the academic information. Furthermore, the authors would like to thank the reviewers for their helpful comments to improve the manuscript.

Author Contributions: Noppol Arunrat collected the primary data, preformed laboratory analysis, wrote and revised the manuscript. Nathsuda Pumijumnong collected the secondary data from information sources and provided advice to this study.

Conflicts of Interest: The authors declare no conflict of interest.

References

1. Zheng, X.; Han, S.; Huang, Y.; Wang, Y.; Wang, M. Re-quantifying the emission factors based on field measurements and estimating the direct N_2O emission from Chinese croplands. *Glob. Biogeochem. Cycle* **2004**, *18*, GB2018. [CrossRef]
2. Li, C.S.; Frolking, S.; Xiao, X.M.; Moore, B., III; Boles, S.; Qiu, J.; Huang, Y.; Salas, W.; Sass, R. Modeling impacts of farming management alternatives on CO_2, CH_4, and N_2O emissions: A case study for water management of rice agriculture of China. *Glob. Biogeochem. Cycle* **2005**, *19*, GB3010. [CrossRef]
3. IPCC (Intergovernmental Panel on Climate Change). *Fourth Assessment Report on Climate Change: Climate Change 2007: Impacts, Adaptation and Vulnerability*; Cambridge University Press: Cambridge, UK, 2007.
4. Jain, N.; Pathak, H.; Mitra, S.; Bhatia, A. Emission of methane from rice fields—A review. *J. Sci. Ind. Res.* **2004**, *63*, 101–115.
5. Van Hulzen, J.B.; Segers, R.; van Bodegom, P.M.; Leffelaar, P.A. Temperature effects on soil methane production: An explanation for observed variability. *Soil Biol. Biochem.* **1999**, *31*, 1919–1929. [CrossRef]
6. Kang, G.D.; Cai, Z.C.; Feng, X.Z. Importance of water regime during the non-rice growing period in winter in regional variation of CH_4 emissions from rice fields during following rice growing period in China. *Nutr. Cycl. Agroecosyst.* **2002**, *64*, 95–100. [CrossRef]
7. Mitra, S.; Wassmann, R.; Jain, M.C.; Pathak, H. Properties of rice soil affecting methane production potentials: I. Temporal patterns and diagnostic procedures. *Nutr. Cycl. Agroecosyst.* **2002**, *64*, 169–182. [CrossRef]
8. Lu, Y.; Wassmann, R.; Neue, H.U.; Huang, C.; Bueno, C.S. Methanogenic responses to exogenous substrates in anaerobic rice soils. *Soil Biol. Biochem.* **2000**, *32*, 1683–1690. [CrossRef]
9. Adhya, T.K.; Bharati, K.; Mohanty, S.R.; Ramakrishnan, B.; Rao, V.R.; Sethunathan, N.; Wassmann, R. Methane Emission from Rice Fields at Cuttack, India. *Nutr. Cycl. Agroecosyst.* **2000**, *58*, 95–105. [CrossRef]

10. Hou, A.X.; Chen, G.X.; Wang, Z.P.; Van Cleemput, O.; Patrick, W.H. Methane and Nitrous Oxide Emissions from a Rice Field in Relation to Soil Redox and Microbiological Processes. *Soil Sci. Soc. Am. J.* **2000**, *64*, 2180–2186. [CrossRef]

11. Mitra, S.; Jain, M.C.; Kumar, S.; Bandyopadhyay, S.K.; Kalra, N. Effect of rice cultivars on methane emission. *Agric. Ecosyst. Environ.* **1999**, *73*, 177–183. [CrossRef]

12. Liao, Q.; Yan, X. Statistical analysis of factors influencing N_2O emission from paddy fields in Asia. *Huan Jing Ke Xue* **2011**, *32*, 38–45. [PubMed]

13. Qin, Y.; Liu, S.; Guo, Y.; Liu, Q.; Zou, J. Methane and nitrous oxide emissions from organic and conventional rice cropping systems in Southeast China. *Biol. Fertil. Soils* **2010**, *46*, 825–834. [CrossRef]

14. Hou, H.; Peng, S.; Xu, J.; Yang, S.; Mao, Z. Seasonal variations of CH_4 and N_2O emissions in response to water management of paddy fields located in Southeast China. *Chemosphere* **2012**, *89*, 884–892. [CrossRef] [PubMed]

15. Peng, S.; Hou, H.; Xu, J.; Mao, Z.; Abudo, S.; Luo, Y. Nitrous oxide emissions from paddy fields under different water managements in southwest China. *Paddy Water Environ.* **2011**, *9*, 403–411. [CrossRef]

16. Lal, R. Soil carbon sequestration impacts on global climate change and food security. *Science* **2004**, *304*, 1623–1627. [CrossRef] [PubMed]

17. Crutzen, P.J.; Andreae, M.O. Biomass burning in the tropics: Impact on atmospheric chemistry and biogeochemical cycles. *Science* **1990**, *250*, 1669–1678. [CrossRef] [PubMed]

18. Lal, R. Soil carbon sequestration impacts to mitigate climate change. *Geoderma* **2004**, *123*, 1–22. [CrossRef]

19. Liese, B.; Isvilanonda, S.; Tri, K.N.; Ngoc, L.N.; Pananurak, P.; Pech, R.; Shwe, T.M.; Sombounkhanh, K.; Möllmann, T.; Zimmer, Y. *Economics of Southeast Asian Rice Production*; Agri Benchmark: Braunschweig, Germany, 2014.

20. Jongdee, B.; Pantuwan, G.; Fukai, S.; Fischer, K. Improving drought tolerance in rainfed lowland rice: An example from Thailand. *Agric. Water Manag.* **2006**, *80*, 225–240. [CrossRef]

21. Khush, G.S. *Terminology for Rice Growing Environments*; IRRI: Los Baños, Philippines, 1984.

22. Robertson, G.P.; Grace, P.R. Greenhouse gas fluxes in tropical and temperate agriculture: The need for a full-cost accounting of global warming potentials. *Environ. Dev. Sustain.* **2004**, *6*, 51–63. [CrossRef]

23. Mosier, A.R.; Halvorson, A.D.; Reule, C.A.; Liu, X.J. Net global warming potential and greenhouse gas intensity in irrigated cropping systems in northeastern Colorado. *J. Environ. Qual.* **2006**, *35*, 1584–1598. [CrossRef] [PubMed]

24. Robertson, G.P.; Paul, E.A.; Harwood, R.R. Greenhouse gases in intensive agriculture: Contributions of individual gases to the radiative forcing of the atmosphere. *Science* **2000**, *289*, 1922–1925. [CrossRef] [PubMed]

25. Shang, Q.Y.; Yang, X.X.; Gao, C.M.; Wu, P.P.; Liu, J.J.; Xu, Y.C.; Shen, Q.R.; Zou, J.W.; Guo, S.W. Net annual global warming potential and greenhouse gas intensity in Chinese double rice-cropping systems: A 3-year field measurement in long-term fertilizer experiments. *Glob. Chang. Biol.* **2011**, *17*, 2196–2210. [CrossRef]

26. LDD (Land Development Department). *Distribution of Salt Affected Soil in the Northeast Region 1:100,000 Map*; Land Development Department, Ministry of Agriculture and Cooperatives: Bangkok, Thailand, 1991.

27. LDD (Land Development Department). *Characterization of Established Soil Series in the Northeast Region of Thailand Reclassified According to Soil Taxonomy 2003*; Land Development Department, Ministry of Agriculture and Cooperatives: Bangkok, Thailand, 2003.

28. OAE (Office of Agricultural Economics). Agricultural Statistics of Thailand 2014. 2014. Available online: http://www.oae.go.th/download/download_journal/2558/yearbook57.pdf (accessed on 10 February 2015).

29. Walkley, A.; Black, J.A. An examination of the dichormate method for determining soil organic matter and a proposed modification of the chromic acid titration method. *Soil Sci.* **1934**, *37*, 29–38. [CrossRef]

30. Arunrat, N.; Wang, C.; Pumijumnong, N. Alternative cropping systems for greenhouse gases mitigation in rice field: A case study in Phichit province of Thailand. *J. Clean. Prod.* **2016**, *133*, 657–671. [CrossRef]

31. IPCC (Intergovernmental Panel on Climate Change). Guidelines for National Greenhouse Gas Inventories 2006. Available online: http://www.ipcc-nggip.iges.or.jp/public/2006gl/index.htm (accessed on 22 September 2014).

32. Maciel, V.G.; Zortea, R.B.; da Silva, W.M.; Cybis, L.F.A.; Einloft, S.; Seferin, M. Life Cycle Inventory for the agricultural stages of soybean production in the state of Rio Grande do Sul, Brazil. *J. Clean. Prod.* **2015**, *93*, 65–74. [CrossRef]

33. EPA (Environmental Protection Agency). Emission Factors for Greenhouse Gas Inventories. United States Environmental Protection Agency. 2014. Available online: https://www.epa.gov/sites/production/files/2015-07/documents/emission-factors_2014.pdf (accessed on 25 June 2015).

34. Lal, R. Carbon emission from farm operations. *Environ. Int.* **2004**, *30*, 981–990. [CrossRef] [PubMed]

35. Yan, X.; Ohara, T.; Akimoto, H. Development of region-specific emission factors and estimation of methane emission from rice fields in the East, Southeast and South Asian countries. *Glob. Chang. Biol.* **2003**, *9*, 237–254. [CrossRef]

36. Kanokkanjana, K.; Garivait, S. Alternative rice straw management practices to reduce field open burning in Thailand. *Int. J. Environ. Sci. Dev.* **2013**, *4*, 119–123. [CrossRef]

37. OAE (Office of Agricultural Economics). *Final Report: Project of Greenhouse Gas Emissions Database in Agriculture Sector*; Ministry of Natural Resources and Environment: Bangkok, Thailand, 2012; p. 429.

38. IPCC (Intergovernmental Panel on Climate Change). *The Physical Science Basis: Working Group I Contribution to the Fifth Assessment Report of the Intergovernmental Panel on Climate Change*; Cambridge University Press: Cambridge, UK; New York, NY, USA, 2013.

39. Liu, Y.; Zhou, Z.; Zhang, X.; Xu, X.; Chen, H.; Xiong, Z. Net global warming potential and greenhouse gas intensity from the double rice system with integrated soil-crop system management: A three-year field study. *Atmos. Environ.* **2015**, *116*, 92–101. [CrossRef]

40. Zhang, X.; Zhou, Z.; Liu, Y.; Xu, X.; Wang, J.; Zhang, H.; Xiong, Z. Net global warming potential and greenhouse gas intensity in rice agriculture driven by high yields and nitrogen use efficiency: A 5 year field study. *Biogeosciences* **2015**, *12*, 18883–18911. [CrossRef]

41. Liang, Q.; Chen, H.Q.; Gong, Y.S.; Fan, M.S.; Yang, H.F.; Lal, R.; Kuzyakov, Y. Effects of 15 year of manure and inorganic fertilizers on soil organic carbon fractions in a wheat-maize system in the North China Plain. *Nutr. Cycl. Agroecosys.* **2012**, *92*, 21–33. [CrossRef]

42. Koga, N.; Sawamoto, T.; Tsuruta, H. Life cycle inventory-based analysis of greenhouse gas emissions from arable land farming systems in Hokkaido, northern Japan. *Soil Sci. Plant Nutr.* **2006**, *52*, 564–574. [CrossRef]

43. Nie, J.; Zhou, J.M.; Wang, H.Y.; Chen, X.Q.; Du, C.W. Effect of long-term rice straw return on soil glomalin, carbon and nitrogen. *Pedosphere* **2007**, *17*, 295–302. [CrossRef]

44. Hao, X.H.; Liu, S.L.; Wu, J.S.; Hu, R.G.; Tong, C.L.; Su, Y.Y. Effect of long-term application of inorganic fertilizer and organic amendments on soil organic matter and microbial biomass in three subtropical paddy soils. *Nutr. Cycl. Agroecosyst.* **2008**, *81*, 17–24. [CrossRef]

45. Wassmann, R.; Lantin, R.S.; Neue, H.U.; Buendia, L.V.; Corton, T.M.; Lu, Y. Characterization of methane emissions from rice fields in Asia. III. Mitigation options and future research needs. *Nutr. Cycl. Agroecosyst.* **2000**, *58*, 23–36.

46. Yang, L.G.; Wang, Y.D. The impact of free-air CO_2 enrichment (FACE) and nitrogen supply on grain quality of rice. *Field Crops Res.* **2007**, *102*, 128–140. [CrossRef]

47. Duan, F.; Liu, X.; Yu, T.; Cachier, H. Identification and estimate of biomass burning contribution to the urban aerosal organic carbon concentrations in Beijing. *Atmos. Environ.* **2004**, *38*, 1275–1282. [CrossRef]

48. Snyder, C.S.; Bruulsema, T.W.; Jensen, T.L.; Fixen, P.E. Review of greenhouse gas emissions from crop production systems and fertilizer management effects. *Agric. Ecosyst. Environ.* **2009**, *133*, 247–266. [CrossRef]

49. Wang, Z.Y.; Xu, Y.C.; Li, Z.; Guo, Y.X.; Wassmann, R.; Neue, H.U.; Lantin, R.S.; Buendia, L.V.; Ding, Y.P.; Wang, Z.Z. A Four-Year Record of Methane Emissions from Irrigated Rice Fields in the Beijing Region of China. *Nutr. Cycl. Agroecosyst.* **2000**, *58*, 55–63. [CrossRef]

50. Neue, H.U.; Wassmann, R.; Lantin, R.S.; Alberto, M.C.R.; Aduna, J.B.; Javellana, A.M. Factors affecting methane emission from rice fields. *Atmos. Environ.* **1996**, *30*, 1751–1754. [CrossRef]

51. Bhattacharyya, P.; Roy, K.S.; Neogi, S.; Adhya, T.K.; Rao, K.S.; Manna, M.C. Effects of rice straw and nitrogen fertilization on greenhouse gas emissions and carbon storage in tropical flooded soil planted with rice. *Soil Tillage Res.* **2012**, *124*, 119–130. [CrossRef]

52. Shen, J.; Tang, H.; Liu, J.; Wang, C.; Li, Y.; Ge, T.; Wu, J. Contrasting effects of straw and straw-derived biochar amendments on greenhouse gas emissions within double rice cropping systems. *Agric. Ecosyst. Environ.* **2014**, *188*, 264–274. [CrossRef]

53. Chidthaisong, A.; Watanabe, I. Methane formation and emission from flooded rice soil incorporated with ^{13}C-labeled rice straw. *Soil Biol. Biochem.* **1997**, *29*, 1173–1181. [CrossRef]

54. Lu, F.; Wang, X.K.; Han, B.; Ouyang, Z.Y.; Zheng, H. Straw return to rice paddy: Soil carbon sequestration and increased methane emission. *Ying Yong Sheng Tai Xue Bao* **2010**, *21*, 99–108. [PubMed]

55. Zou, J.; Huang, Y.; Qin, Y.; Liu, S.; Shen, Q.; Pan, G.; Lu, Y.; Liu, Q. Changes in fertilizer-induced direct N_2O emissions from paddy fields during rice-growing season in China between 1950s and 1990s. *Glob. Chang. Biol.* **2008**, *15*, 229–242. [CrossRef]

56. Liu, C.; Wang, K.; Zheng, X. Responses of N_2O and CH_4 fluxes to fertilizer nitrogen addition rates in an irrigated wheat–maize cropping system in northern China. *Biogeosciences* **2012**, *9*, 839–850. [CrossRef]

57. Linquist, B.; Groenigen, K.J.; Adviento-Borbe, M.A.; Pittelkow, C.; Kessel, C. An agronomic assessment of greenhouse gas emissions from major cereal crops. *Glob. Chang. Biol.* **2012**, *18*, 194–209. [CrossRef]

58. Zhang, X.; Yin, S.; Li, Y.; Zhuang, H.; Li, C.; Liu, C. Comparison of greenhouse gas emissions from rice paddy fields under different nitrogen fertilization loads in Chongming Island, Eastern China. *Sci. Total Environ.* **2014**, *472*, 381–388. [CrossRef] [PubMed]

59. Ma, Y.C.; Kong, X.W.; Yang, B.; Zhang, X.L.; Yan, X.Y.; Yang, J.C.; Xiong, Z.Q. Net global warming potential and greenhouse gas intensity of annual rice-wheat rotations with integrated soil-crop system management. *Agric. Ecosyst. Environ.* **2013**, *164*, 209–219. [CrossRef]

60. Yodkhum, S.; Sampattagul, S. Life Cycle Greenhouse Gas Evaluation of Rice Production in Thailand. In Proceedings of the 1st Environment and Natural Resources International Conference, Bangkok, Thailand, 6–7 November 2014.

61. Wang, W.; Guo, L.P.; Li, Y.C.; Su, M.; Lin, Y.B.; de Perthuis, C.; Ju, X.T.; Lin, E.; Moran, D. Greenhouse gas intensity of three main crops and implications for low-carbon agriculture in China. *Clim. Chang.* **2015**, *128*, 57–70. [CrossRef]

Permissions

List of Contributors

Mauro Zaninelli
Department of Human Sciences and Quality of Life Promotion, Universita Telematica San Raffaele Roma, Via di Val Cannuta 247, Rome 00166, Italy

Hans Thodsen, Gitte Blicher-Mathiesen, Ruth Grant, Hans Estrup Andersen and Dennis Trolle
Department of BioScience, Aarhus University, Vejlsøvej 25, 8600 Silkeborg, Denmark

Csilla Farkas and Alexander Engebretsen
Bioforsk, Division for Soil, Water and Environment, Frederik A. Dahlsvei 20, 1430 Ås, Norway

Jaroslaw Chormanski and Ignacy Kardel
Department of Hydraulic Engineering, Faculty of Civil and Environmental Engineering, Warsaw University of Life Sciences-SGGW, ul. Nowoursynowska 166, 02-787 Warszawa, Poland

Laura Vincent-Caboud and Joséphine Peigné
Department of Agroecology and Environment, ISARA-Lyon (member of the University of Lyon), 23 rue Jean Baldassini, F-69364 Lyon Cedex 07, France

Marion Casagrande
Department of Agroecology and Environment, ISARA-Lyon (member of the University of Lyon), 23 rue Jean Baldassini, F-69364 Lyon Cedex 07, France
ITAB, Quartier Marcellas, F-26800 Etoile sur Rhône, France

Erin M. Silva
Department of Agronomy, University of Wisconsin-Madison, Madison, 53706 WI, USA

Erin M. Silva
Department of Plant Pathology, University of Wisconsin-Madison, 1630 Linden Dr., Madison, WI 53706, USA

Kathleen Delate
Departments of Agronomy and Horticulture, Iowa State University, 106 Horticulture Hall, Ames, IA 50011, USA

Emilio Chiodo and Giuseppe Martino
Faculty of Bioscience and Agro-Food and Environmental Technology, University of Teramo, 64100 Teramo, Italy

Maria Angela Perito
Faculty of Bioscience and Agro-Food and Environmental Technology, University of Teramo, 64100 Teramo, Italy
ALISS, UR1303, INRA, F-94205 Ivry-sur-Seine, France

Marcello De Rosa and Luca Bartoli
Department of Economics and Law, University of Cassino and Southern Lazio, 03043 Cassino, Italy

Fumiaki Takakai, Takashi Sato and Yoshihiro Kaneta
Faculty of Bioresource Sciences, Akita Prefectural University, 241-438 Aza Kaidobata-Nishi, Shimoshinjo Nakano, Akita 010-0195, Japan

Shinpei Nakagawa Kensuke Sato Kazuhiro Kon
Akita Prefectural Agricultural Experiment Station, 34-1, Aza Genpachizawa, Yuwa Aikawa, Akita 010-1231, Japan

Atena Shadmehr, Hossein Ramshini and Ali Izadi Darbandi
Department of Agronomy and Plant Breeding Sciences, Agricultural College of Aburaihan,
University of Tehran, Emam reza Blvd, Pakdasht, Tehran 3391653755, Iran

Mehrshad Zeinalabedini and Mohammad Reza Ghaffari
Systems Biology Department, Agricultural Biotechnology Research Institute of Iran, Agricultural Research, Education and Extension Organization (AREEO), 31359-33151 Karaj, Iran

Masoud Parvizi Almani and Mahmoud Fooladvand
Department of Biotechnology, Cane Development and Sidelong Industrial Research and Education Institute, Golestan Blvd, Khuzestan 1465834581, Iran

Ilaria Zambon, Lavinia Delfanti, Alvaro Marucci, Roberto Bedini, Walter Bessone, Massimo Cecchini and Danilo Monarca
Department of Agricultural and Forestry Sciences, DAFNE Tuscia University, Via San Camillo de Lellis snc, 01100 Viterbo, Italy

Yohann Fare, Marc Dufumier and Myriam Loloum
Unité d'Enseignement et de Recherche Agriculture Comparée et Développement Agricole, AgroParisTech. 16, rue Claude Bernard, F-75231 Paris CEDEX 05, France

Fanny Miss
École Nationale du Génie Rural, des Eaux et des Forêts, AgroParisTech, 19 Avenue du Maine, 75732 Paris CEDEX 15, France

Alassane Pouye and Ahmat Khastalani
Ecole Nationale Supérieure d´Agriculture (ENSA) de Thiès, B.P A 296-Thiès, Sénégal

Adama Fall
SOS SAHEL International, 21001 Thiès, Senegal

Teresa del Rosario Ayora-Talavera, Cristina A. Ramos-Chan, Ana G. Covarrubias-Cárdenas, Angeles Sánchez-Contreras and Neith A. Pacheco L.
Centro de Investigación y Asistencia en Tecnología y Diseño del Estado de Jalisco CIATEJ Unidad Sureste. Parque Científico Tecnológico de Yucatán, Km 5.5 Carretera Sierra Papacal-Chuburná puerto, Mérida-CP 97302, México

Ulises García-Cruz
Centro de Investigación y Estudios Avanzados del Instituto Politécnico Nacional. Unidad Mérida Km. 6 Antigua carretera a Progreso Apdo. Postal 73, Cordemex, Mérida Yuc. 97310, México

Guruswamy Prabhakaran, Subhash Janardhan Bhore and Manikam Ravichandran
Department of Biotechnology, Faculty of Applied Sciences, AIMST University, Bedong-Semeling Road, Semeling, 08100 Bedong, Kedah, Malaysia

Ana Paula Matei
Secretaria de Desenvolvimento Tecnológico (SEDETEC/UFRGS), Porto Alegre 90040-020, Brazil

Paul Swagemakers
Department of Applied Economics, University of Vigo, Ourense 32004, Spain

Maria Dolores Dominguez Garcia
Department of Applied Economics, Complutense University of Madrid, Pozuelo de Alarcón 28223, Spain

Leonardo Xavier da Silva
Departamento de Economia e Relações Internacionais, Universidade Federal Do Rio Grande Do Su, Porto Alegre 90040-000, Brazil

Flaminia Ventura and Pierluigi Milone
Dipartimento di Ingegneria Civile ed Ambientale (DICA, Department of Civil and Environmental Engineering), University of Perugia, Perugia 06125, Italy

Noppol Arunrat
Laboratory of Soil Science, Graduate School of Agriculture, Hokkaido University, Kita 9 Nishi 9, Kita-ku, Sapporo 060-8589, Japan
Faculty of Environment and Resource Studies, Mahidol University, Salaya, Phutthamonthon, Nakhon Pathom 73170, Thailand

Nathsuda Pumijumnong
Faculty of Environment and Resource Studies, Mahidol University, Salaya, Phutthamonthon, Nakhon Pathom 73170, Thailand

Index

www.ingramcontent.com/pod-product-compliance
Lightning Source LLC
Chambersburg PA
CBHW050442200326
41458CB00014B/5033